Controller Design for Industrial Applications

Scrivener Publishing
100 Cummings Center, Suite 541J
Beverly, MA 01915-6106

Publishers at Scrivener
Martin Scrivener (martin@scrivenerpublishing.com)
Phillip Carmical (pcarmical@scrivenerpublishing.com)

Controller Design for Industrial Applications

Edited by

Arindam Mondal

and

Souvik Ganguli

Scrivener Publishing

WILEY

This edition first published 2025 by John Wiley & Sons, Inc., 111 River Street, Hoboken, NJ 07030, USA and Scrivener Publishing LLC, 100 Cummings Center, Suite 541J, Beverly, MA 01915, USA

For more information about Scrivener publications please visit www.scrivenerpublishing.com.

Wiley Global Headquarters
111 River Street, Hoboken, NJ 07030, USA

For details of our global editorial offices, customer services, and more information about Wiley products visit us at www.wiley.com.

Library of Congress Cataloging-in-Publication Data

ISBN 9781394287079

Front cover images supplied by Adobe Firefly
Cover design by Russell Richardson

Set in size of 11pt and Minion Pro by Manila Typesetting Company, Makati, Philippines

Printed in the USA

10 9 8 7 6 5 4 3 2 1

Contents

Preface

Industrial processes are often complex and dynamic, making it challenging to design controllers that can maintain stable and optimal operation. Traditional controllers, such as PID controllers, have been widely used in industrial applications but have limitations in handling non-linear and uncertain systems. Intelligent controllers offer an alternative solution that can adapt to changing system dynamics and disturbances.

Intelligent controllers utilize advanced control theory and techniques, such as fuzzy logic control, neural network control, and model predictive control, to achieve optimal control performance. They are capable of learning from data and experience, making them suitable for handling non-linear and uncertain systems. Furthermore, they can improve the robustness and flexibility of the control system and enhance the overall performance.

The use of intelligent controllers in industrial applications has gained increasing attention in recent years, with numerous successful implementations in various fields, such as process control, robotics control, HVAC control, power systems control, and autonomous vehicles control. However, the design and implementation of intelligent controllers require careful consideration of hardware and software requirements, as well as simulation and testing procedures to ensure reliable and safe operation.

In recent years, there has been a growing interest in the development and implementation of intelligent controllers for industrial applications. Intelligent controllers are capable of adapting to changing system dynamics and disturbances, resulting in improved performance and robustness. In the rapidly evolving industrial landscape, it is essential to develop advanced control techniques to enhance productivity, minimize costs, and ensure safety. Traditional control methods often struggle to handle complex systems and unpredictable environments.

However, with the emergence of intelligent control techniques, there is a great opportunity to improve industrial automation and control systems. This book aims to provide a comprehensive understanding of intelligent controller design for industrial applications, from theoretical concepts to practical implementation. It will cover the fundamental concepts of intelligent control theory and techniques, their application in various industrial fields, and the practical implementation and design considerations. It is suitable for researchers, engineers, and students in the field of control engineering and industrial automation.

Chapter 1 deals with the practical applications of Fuzzy Logic Control (FLC) in various industries, highlighting its ability to manage imprecise and uncertain data effectively. It underscores the adaptability of FLC to dynamic and complex systems, illustrating its utilization in automotive, consumer electronics, robotics, and other sectors.

Chapter 2 discusses the application of Artificial Neural Networks (ANNs) in various industrial contexts, emphasizing their capacity to handle complex data and improve decision-making processes. It also explains the basic structure of ANNs, including different types like multilayer perceptron and recurrent networks, and how they are used in sectors such as manufacturing, energy management, and process control. Moreover, the chapter highlights the role of ANN in enhancing productivity and cost-efficiency through examples like predictive maintenance and quality control, demonstrating the technology's broad applicability and effectiveness across different industries.

Chapter 3 deliberates an innovative approach that combines an Artificial Neural Network (ANN) based observer with a Sliding Mode Controller (SMC) to enhance control over non-linear systems. This integration aims to utilize the capability of ANN to accurately estimate unknown system states and disturbances, improving the robustness and performance of the SMC. The effectiveness of this combined strategy is demonstrated through simulations, particularly using a single-link robot dynamical model, showing significant improvement in handling system uncertainties and disturbances.

Chapter 4 presents a comprehensive examination of the Finite Control Set Model Predictive Control (FCSMPC) approach for Permanent Magnet Synchronous Motor (PMSM) drives, focusing on the application in electric vehicles (EVs) and its integration with renewable energy sources. It covers the evolution from traditional motor control strategies like Field-Oriented Control and Direct Torque Control to the advanced FCSMPC, detailing its advantages in handling complex, dynamic performance requirements. The chapter also elaborates on the mathematical models and the real-time implementation challenges of FCSMPC, underscoring its effectiveness in reducing computational complexity and improving system responsiveness and efficiency.

Chapter 5 explores the kinematic and dynamic modeling of walking robots, focusing on creating accurate simulations using MATLAB to better understand the complexities of robotic locomotion. It also emphasizes the importance of integrating mechanical, electrical, and control systems to develop robots capable of handling real-world tasks. The chapter further highlights the potential applications of walking robots in various fields such as healthcare, manufacturing, and service industries, underscoring the ongoing innovations and challenges in the field of robotics.

Chapter 6 discusses the design and implementation of a hybrid FUZZY-(1+PD)-FOPID controller for a two-area power system, integrating thermal, nuclear, and non-conventional energy sources. It highlights the use of the Tree-Seed Algorithm (TSA) for optimizing controller parameters to enhance system stability and response characteristics, such as settling time and overshoot. The effectiveness of this controller is demonstrated through MATLAB simulations, showing superior performance over traditional PID controllers in managing frequency and tie-bar power deviations within the power system.

Chapter 7 explores the implementation of the Tree Seed Algorithm (TSA) for tuning a Model Predictive Control (MPC) system aimed at enhancing the performance of a two-area interconnected hybrid power system. This hybrid system incorporates both conventional (thermal and nuclear) and non-conventional (ocean thermal and solar) energy sources. The TSA optimizes the MPC parameters to minimize power system oscillations, effectively improving stability and response times in the face of load changes and system uncertainties, as demonstrated through MATLAB simulations.

Chapter 8 outlines the development and implementation of Wide Area Monitoring, Protection, Automation, and Control (WAMPAC) systems in response to the integration of renewable energy sources and the risks of power outages. It further emphasizes the role of Phasor Measurement Units (PMUs) and Intelligent Electronic Devices (IEDs) in enhancing grid visibility and control by providing real-time data, which aids in maintaining system stability and efficiency. The chapter also discusses various phasor estimation techniques and the critical use of communication protocols like DNP3, IEC61850, and others to ensure seamless data flow and reliable grid operation.

Chapter 9 discusses the design and implementation of a smart prepaid interface for power distribution in industrial settings, focusing on the integration of microgrid technology with prepaid systems. This approach enhances consumer empowerment by allowing them to purchase electricity efficiently from the closest available power station, thereby optimizing the electricity flow and improving cost-effectiveness. The system also utilizes a dynamic fusion of centralized and decentralized features to streamline user experiences and tailor electricity distribution based on location and priority, ultimately promising a more resilient and consumer-centric future in energy distribution.

Chapter 10 presents a study on the implementation of the Grey Wolf Optimization (GWO) algorithm for maximum power point tracking (MPPT) in photovoltaic (PV) systems under partial shading conditions. It contrasts the GWO method with the traditional Perturb and Observe (P&O) method, highlighting the effectiveness of GWO in avoiding local maxima and efficiently tracking the global maximum power point, even with variable irradiance. The analysis includes simulation results validating the enhanced performance and reliability of the GWO algorithm, which optimizes the output of the PV system by adapting dynamically to changes in environmental conditions.

Chapter 11 introduces an efficient optimization approach for solving the Relay Coordination Problem (RCP) in power distribution systems, emphasizing the need for precise configuration of over-current relays (OCRs) to handle real-time fault mitigation. It discusses the optimization of relay settings to achieve minimal fault clearance times while maintaining system integrity, using Particle Swarm Optimization (PSO) among other techniques. The chapter further details various strategies and algorithms used historically and currently, providing a comprehensive look at advancements in relay coordination to enhance the reliability and efficiency of power systems.

Chapter 12 focuses on the advanced control strategies for energy-efficient HVAC systems, particularly the integration of intelligent control techniques using machine learning and artificial intelligence. It discusses the complexities of modeling HVAC systems due to dynamic, interconnected components and external influences like weather changes and occupancy variations. The chapter also emphasizes the enhancement of HVAC control through predictive models and adaptive algorithms that improve energy efficiency, occupant comfort, and system responsiveness. Case studies are presented to demonstrate the significant impacts of these intelligent control systems in reducing energy consumption and maintaining optimal indoor environmental quality in various settings.

Chapter 13 introduces the Closest Obstacle Avoidance and A* (COAA*) Algorithm, a robust and efficient obstacle avoidance and navigation solution tested on the Heavy Lift Experimental (HLX) unmanned aerial vehicles, effectively combining the strengths of both offline and online methods while ensuring optimal performance in both simulations and real-world applications.

Chapter 14 develops a novel control scheme called Steady-State Integral Proportional Integral Controller (SIPIC) for racing-grade multi-rotors, designed to optimize system actuation with minimal overshoot and oscillations, and evaluates its performance using a nonlinear dynamic model simulator. Additionally, it employs a Robust and Perfect Tracking (RPT) controller for accurate tracking of position, velocity, and acceleration setpoints, demonstrating remarkable robustness and performance with a straightforward tuning process.

Chapter 15 introduces a novel cascaded fuzzy logic-based approach for path planning and collision avoidance in autonomous vehicles, utilizing multiple fuzzy inference systems (FIS) to enhance navigation and safety. The approach employs LiDAR sensor data to dynamically adjust vehicle movement, aiming to effectively manage real-time navigation challenges due to sensor inaccuracies and environmental unpredictability. The techniques are further validated through MATLAB simulations, emphasizing their potential in improving autonomous vehicle control systems.

Chapter 16 delves with the integration of artificial intelligence (AI) in optimizing electric vehicle (EV) charging systems to address the increased demand on power grids and improve charging infrastructure. It explores how AI can dynamically manage charging schedules, balance demand, and provide predictive maintenance to enhance efficiency and reliability. Additionally, the chapter also considers the implications of AI-driven systems on data privacy, interoperability, and regulatory challenges, aiming to provide a robust framework for sustainable electric vehicle integration.

Chapter 17 discusses the various wireless power transfer (WPT) methods for electric vehicles (EVs), focusing on dynamic wireless charging systems that operate through electromagnetic induction. It reviews three types of WPT—inductive, capacitive, and resonant inductive—detailing their operational principles, efficiencies, and system designs. The chapter emphasizes the advantages of dynamic, contactless EV charging stations that integrate renewable energy sources, aiming to enhance charging infrastructure and address electric vehicles' range and energy storage challenges.

1

Fuzzy Logic Control for Industrial Applications

Srabanti Maji[1] and Souvik Ganguli[2]*

[1]School of Computing, DIT University, Dehradun, Uttarakhand, India
[2]Department of Electrical Engineering, Thapar Institute of Engineering & Technology, Patiala, Punjab, India

Abstract

Fuzzy control has emerged as the most important and practical field of study in fuzzy set theory, particularly for industrial developments that do not use conventional control techniques due to the lack of readily available, precise data on input-output relationships. The foundation of the fuzzy control system is fuzzy logic. Basically, this logical system is considerably more analogous to how people think and speak. A linguistic control approach built on expert knowledge can be transformed into an automatic control strategy using the fuzzy logic–based controller, which is based on the fuzzy logic concept. The understanding of membership functions, in particular, is well-known and plays a significant role in a variety of applications such as industry, traffic, and medical science. While probability theory deals with randomness, the main objective of fuzzy set theory is to provide a technique for the analysis of uncertainty originating in human subjectivity. The fuzzy logic system focuses primarily on decision-making and fuzzy inference systems. It includes fuzzification and defuzzification strategies, fuzzy control rules, fuzzy implication, and an analysis of fuzzy reasoning appliances. There are multiple applications of fuzzy logic controls observed in the industrial process. Without the operator having any prior knowledge of the system to be controlled, the industrial controller's hardware aids in fine-tuning a PID controller. Fuzzy logic supervisory control is utilized in software in the process industry to enhance the functioning of the sintering oven through an advanced integration of priority management and deviation-controlled timing. It can also be utilized in intelligent control modeling, and programmable logic controller (PLC), where programmed to function in a cost-effective manner. Additionally, it can be used

**Corresponding author*: souvik.ganguli@thapar.edu

Arindam Mondal and Souvik Ganguli (eds.) Controller Design for Industrial Applications, (1–20)

in a fuzzy image processing scheme, medical engineering vessel segmentation, the development of fuzzy controllers for maintaining an inter-vehicle safety headway, and knowledge-based gear-position decision, among other applications. All these industrial applications are discussed in this chapter.

Keywords: Fuzzy set theory, fuzzy logic control, programmable logic controller (PLC), fuzzy image processing

1.1 Introduction

In the field of modern industrial automation, the search for improved control techniques capable of managing intricate and unpredictable systems has resulted in the development of novel methodologies. Fuzzy logic control (FLC) is an effective and adaptable technique that has actually attracted considerable passion. Unclear reasoning stems from human reasoning together with assumption efficiently dealing with vague and unclear information, making it a very ideal choice for a large range of useful applications.

Proportional-integral-derivative (PID) controllers have actually been typically utilized in commercial setups for control objectives. Although these methods have actually shown efficiency in numerous circumstances, they regularly come across problems when faced with nonlinear, uncertain, and vibration-based systems. The fundamental restrictions of conventional controllers mandated the growth of advanced and also smart approaches, leading to the appearance of FLC.

Lotfi A. Zadeh created fuzzy logic in the 1960s as a mathematical device rooted in the concept of unclear collections [1]. Unclear collections vary from traditional collection concepts because they permit partial subscription, whereas, in traditional collection concepts, a component either completely comes from a collection or does not belong in any way. Unclear reasoning makes it possible for the depiction and control of obscurity as well as vagueness in a way that carefully mirrors human cognitive procedures. FLC might efficiently imitate human decision-making by using etymological variables as well as unclear guidelines to replicate fancy communications between inputs as well as outcomes.

FLC features by utilizing a collection of language guidelines that express the professional understanding relating to the system under control. FLC is thus an efficient knowledge-based control technique that is created from the experience of the specialists and also domain-specific understanding. The control technique uses certain if-then regulations in which inputs as well as outcomes are related to fuzzy linguistic principles like "low," "medium," and "high" [2]. The FLC (fuzzy logic control) procedure includes three important actions: fuzzification, rule assessment, and defuzzification. Throughout the fuzzification stage, exact input worths are changed right into unclear collections and connected to appropriate language principles. After this, the rule assessment phase gauges the degree of fulfillment for each and every regulation by taking into consideration the fuzzy input values. Lastly, the defuzzification treatment incorporates the fuzzy outcomes to acquire exact control activities.

Meanwhile, FLC has proven its worth through its ability to handle uncertainty and disturbances effectively. The traditional control approaches typically require an exact mathematical model of the system which in turn makes these approaches susceptible to errors when implemented in real life. In contrast, FLC possesses the ability to handle imprecision adequately and as such is of great importance in cases when the system's performance or unfavorable behavior is unknown or cannot be predicted accurately. These modifications make FLC adaptable in such a way that it can respond to dynamic situations while ensuring optimal performance online and so on and so forth. This characteristic has given rise to the use of PID as a satisfactory controller in complex industrial systems especially where they experience variable operating conditions [3].

Control techniques available for linear systems face insurmountable complexities when it comes to nonlinearities. The underlying nonlinear nature of FLC makes it an effective tool in solving complicated control problems. It can represent complex interactions between inputs and outputs which are hardly accounted for by the usual linear control methodologies. It has the advantage of being intuitive, enabling its prompt application and understanding of control rules. By including language variables and rules, the control logic becomes more easily understandable for specialists in the subject, thereby allowing them to actively participate in designing and refining processes. The transparency of FLC promotes a clear understanding and examination of the decision-making process while enhancing confidence and facilitating problem-solving within industrial applications.

FLC has widely been applied in many industrial sectors including process control, robotics, energy management, and intelligent transport systems. FLC is a very successful method for controlling various parameters in industrial processes such as temperature, pressure, humidity, and flow rate; it can handle uncertainties and nonlinearities that make it suitable for chemical plants, oil refineries, and power generation facilities. It plays a fundamental role in robotics and automation fields where accuracy of control and adaptability are vital. It is used to control robot motion, identify objects, plan paths or trajectories, and avoid collisions so that robots can operate safely and efficiently in changing environments. FLC has the capability to enhance energy efficiency in industrial facilities through the regulation of equipment operation and load management. This makes it easier to distribute energy effectively thus promoting the economical utilization of energy in different applications including smart grids, renewable energy systems; HVAC (heating ventilation air-conditioning) control. In traffic control systems, FLC has proven itself by effectively regulating traffic flow as well as timing of signals. To mitigate congestion and enhance collective traffic efficiency, the system can change signal timings quickly depending on the traffic state. FLC is a tough control strategy whose applications in various industries are very promising. It is an attractive alternative for guiding complex systems as it can handle unpredictability, nonlinearity, and adaptiveness. The intuitive nature of FLC, its transparency, and robustness make it suitable for professionals and easily understood by them for practical use [4].

1.2 The Evaluation of Fuzzy Logic Control: From Theory to Industrial Applications

1.2.1 The Origin of Fuzzy Logic: A Paradigm Shift in Control Theory

Fuzzy logic dates back to 1965 when Lotfi A. Zadeh developed the concept of fuzzy sets [1]. The scholar argued that the binary definition of set membership was not suitable for real-life cases and should be substituted with partial membership based on the degree of truthfulness. The consideration above revolutionized control theory by introducing a new paradigm that could accurately work with vague and uncertain information.

1.2.2 Mamdani's Fuzzy Logic Control: The First Practical Implementation

In the early 1970s, Mamdani and Assilian implemented the first practical application of fuzzy logic control based on fuzzy logic controllers, known as the Mamdani model [5]. They demonstrated the successful employment of FLC for the control of a steam engine, illustrating its potential for solving complicated control problems in the real world. The Mamdani model relied on linguistic variables, fuzzy rules, and membership functions to create a relationship between the inputs and outputs, creating a system that is interpretable.

1.2.3 Takagi-Sugeno-Kang (TSK) Fuzzy Models: Advancing the Practicality of FLC

In the late 1980s, Takagi, Sugeno, and Kang proposed the TSK fuzzy model, which enhanced the applicability and computational efficiency of FLC [6]. The TSK model is distinguished from the Mamdani model because it uses a weighted linear summation of input variables to generate output. This introduced the opportunity for a reduced rule base with less complicated computation and eased the practical implementation of real-time control systems.

1.2.4 Hybrid and Adaptive Fuzzy Logic Control: Addressing Complexity and Uncertainty

With the rising popularity of FLC, researchers have started investigating hybrid and adaptive methods to cope with the difficulties put forward by complex and unpredictable systems.

1.2.4.1 Hybrid Fuzzy Logic Control

Hybrid FLC combines fuzzy logic with other control methods to allow them to cooperate with their best abilities [7]. The best example is integrating the fuzzy logic controller with neural networks, allowing neuro-fuzzy systems to appear. Such systems combine the ability to learn presented by neural networks with the ability to interpret and illuminate operations of fuzzy logic. It enables enhanced control and increased flexibility.

1.2.4.2 Adaptive Fuzzy Logic Control

Adaptive FLC implies the ability to independently change the fuzzy rules and membership functions concerning the development of the system's dynamics or environmental conditions. The FLC system changes and refines the control system through adaptive procedures, including fuzzy rule interpolation, fuzzy clustering, and genetic algorithms, among others. In scenarios where the system parameters change or are unknown, this system is more effective [8].

1.3 Basics of Fuzzy Logic Control: A Comprehensive Overview

FLC is a versatile and resilient control method that has attracted attention for its ability to handle unpredictable systems effectively. Fuzzy logic, which mirrors cognition and understanding, is also employed to manage data that lacks precision or clarity. This summary will explore the core components of FLC, such as its principles, ideas, and the sequential steps involved in the control procedure.

1.3.1 Fuzzy Sets and Linguistic Variables

The core concept of FLC is centered on the idea of sets. Unlike set theory, fuzzy sets permit membership rather than strict binary classification of elements as fully belonging or not belonging to a set. Fuzzy sets excel at handling uncertainty and vagueness, making them ideal for modeling real-world systems. In a set, membership is determined by a boundary, whereas, in a fuzzy set, membership can exist to a certain degree.

Fuzzy logic uses linguistic variables to denote the input and output variables in a control system. Linguistic variables allow the use of human-like linguistic expressions to describe the states of the system. For example, in controlling the speed of a car, language such as "low," "medium," and "high" can be used to describe the speed values.

1.3.2 Membership Functions

The membership functions do play a critical role in the FLC by demarking the degree of membership of an element in a fuzzy set. A membership function maps the crisp input/output values to their corresponding degrees of membership in a fuzzy set. A fuzzy set usually allows a member to have a partial degree of membership, which can be mapped into a universe of membership values. Let us assume that we have a fuzzy set B, and if an element x is a member of this fuzzy set, the mapping can be represented as

$$\mu_B(x) \in \left[0,\ 1\right] (\mathrm{B} = (\mathrm{x}, \mu_B(x)|\ x \ \in \ \mathrm{X}) \tag{1.1}$$

A fuzzy subset B with an element x has a membership function of $\mu_B(x)$. In the situation when the universe of discourse X is discrete and finite, this mapping can be denoted by Equation 1.2.

$$B = \frac{\mu_B(x_1)}{x_1} + \frac{\mu_B(x_2)}{x_2} + \ldots\ldots\ldots = \sum_i \frac{\mu_B(x_i)}{x_i} \tag{1.2}$$

In the situation when universe X is continuous and infinite, at that situation, fuzzy set B can be written as

$$B = \int \frac{\mu_B(x)}{x} \tag{1.3}$$

There are several shapes of membership functions that can be used, such as triangular, trapezoidal, Gaussian, or sigmoidal, depending on the nature of the variables and the desired representation.

1.3.3 Fuzzy Rule Base

The fuzzy rule base represents the knowledge and expertise of human operators or domain experts. It is composed of several instances of fuzzy if-then rules that connect the fuzzy input and fuzzy output variables. Each rule consists of an antecedent (if-part) and a consequent (then-part). The antecedent uses fuzzy logic operators (e.g., AND, OR) to evaluate the degree of satisfaction of the rule based on the input variables' membership values. The outcome depicts the fuzzy output variables and their corresponding membership functions.

For example, a fuzzy rule that describes a situation when it is necessary to control the speed of a vehicle is: IF the distance to an obstacle is large AND the speed of the vehicle is slow THEN increase the speed. Here, the antecedent checks the level of satisfaction using the membership degree of "far" and "low." The consequent, in turn, defines the action to be taken, which is increasing the speed for this particular case.

1.3.4 Fuzzification

This stage involves transforming the crisp input values into fuzzy sets by applying the membership functions of each of them. Translation to linguistic phrases is conducted based on the membership degrees. The process of fuzzification is characterized by the need to determine the values of membership of each linguistic term dependent on the input values [10].

1.3.5 Rule Evaluation

Fuzzy paradigm thinking relies on fundamental information derived from fuzzy IF-THEN rules. After the input variables have been fuzzified, the rule evaluation step calculates the level of satisfaction for each rule by considering the fuzzy input values. The process involves the utilization of fuzzy logic operators (AND, OR) to merge the membership degrees linked to the antecedent linguistic phrases of each rule. The outcome is a level of activation for each rule, denoting the intensity or significance of the rule in the specific circumstance [11].

1.3.6 Rule Aggregation

The rule aggregation procedure combines the activated rules to determine the overall fuzzy output, after evaluating the degree of satisfaction for each rule. Multiple aggregation techniques, such as maximum, average, or weighted average, can be employed based on the particular application and control criteria.

1.3.7 Defuzzification

Defuzzification is the last phase in FLC when the fuzzy outcome received from rule aggregation is changed right into an accurate control activity. Defuzzification is the procedure of transforming the subscription levels of the fuzzy outcome back right into the crisp outcome room. Defuzzification can be achieved by making use of various methods consisting of centroid defuzzification. This method involves calculating the center of gravity of the fuzzy output and using it as the precise output value. Below are the enumerated defuzzification procedures.

- Center of sums method (COS)
- Center of gravity (COG)/centroid of area (COA) method
- Center of area/bisector of area method (BOA)
- Weighted average method
- Maxima methods
    - First of maxima method (FOM)
    - Last of maxima method (LOM)
    - Mean of maxima method (MOM)

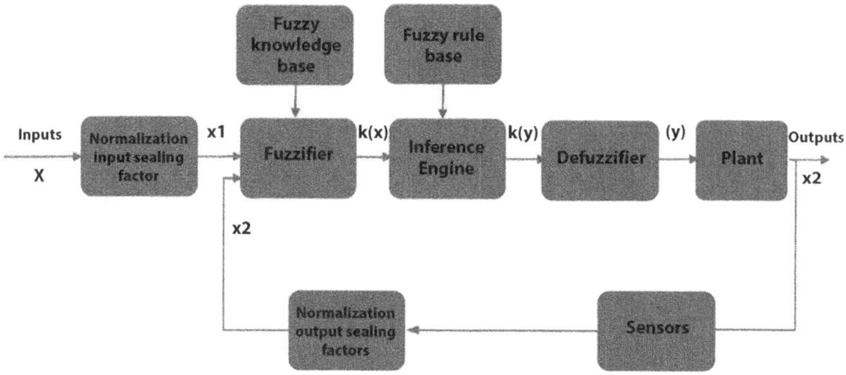

Figure 1.1 Basic architecture of a fuzzy logic controller.

1.3.8 Feedback and Iteration

FLC can be enhanced by incorporating feedback mechanisms. Feedback loops enable the system to adjust its control actions based on the observed output and desired set points. By continuously monitoring and comparing the system's performance, FLC can adapt and refine its control strategy over time.

1.3.9 Architecture of Fuzzy Logic Controller

The required parameters that are employed in the fuzzy logic controller are normally defined with the help of membership functions. A simple architecture of a fuzzy logic controller is thus shown in Figure 1.1.

A fuzzifier, a fuzzy rule base, a fuzzy knowledge base, an inference engine, and a defuzzifier are the core elements of an FLC system. It also has normalization-related parameters. When a plant control action is not the output of the defuzzifier, then the system is a fuzzy logic decision system. The fuzzifier in place transforms sharp values into fuzzy ones. Knowledge of the functioning of the domain expertise process is stored in the fuzzy rule base. The information regarding all fuzzy input-output relationships is kept in the fuzzy knowledge base.

1.4 Merits of Fuzzy Logic Control

FLC [9] offers several advantages that make it well-suited for industrial applications. Its ability to handle imprecise and uncertain information,

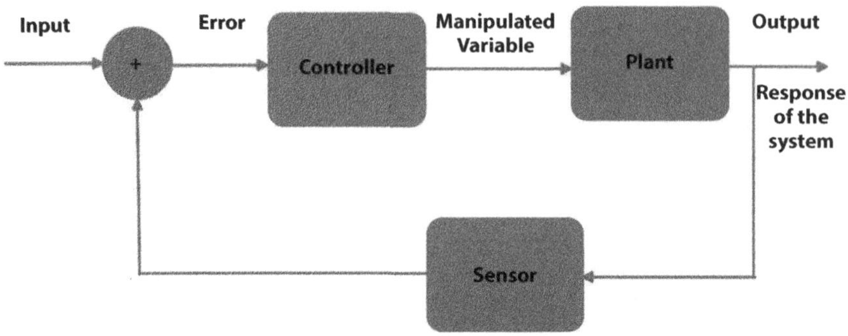

Figure 1.2 Block diagram of the closed-loop control system.

adapt to changing conditions, and provide robust control makes it a powerful tool in various domains. In numerous industrial applications, there is an approximation of information regarding the objectives and procedures. The type of knowledge utilized is FLC, which is employed to create a user-friendly interface [12]. Figure 1.2 illustrates the closed-loop control system.

In the next section, some of the key advantages of FLC in industrial applications are explored.

1.4.1 Ability to Handle Imprecise and Uncertain Information

An inherent benefit of FLC is its capacity to effectively manage imprecise and unclear data. Creating precise mathematical versions in various commercial procedures may be tested as a result of detailed characteristics or an insufficient understanding of system actions. Fuzzy reasoning control (FLC) assists in the representation and handling of etymological variables coupled with unclear collections, thus promoting the combination of skilled expertise with unsteady measurements. The adaptability of FLC allows it to successfully attend to the details and unpredictability of real-world circumstances, making certain strong as well as dependable control.

1.4.2 Linguistic Representation and Human-Like Decision-Making

FLC makes use of linguistic variables as well as fuzzy collections to precisely reveal and adjust information. This language depiction allows the expression of control concepts in a kind that is both legible as well as reasonable to human beings. FLC can duplicate human decision-making procedures by utilizing linguistic expressions such as "low," "medium," and "high." This attribute is particularly useful when involving human drivers, as it makes it possible for clear along with open interaction and also minimizes the trouble in discovering what is normally connected with standard control systems.

1.4.3 Adaptive and Self-Learning Capabilities

The FLC displays flexible and also independent knowing abilities permitting it to get used to dynamic situations as well as improve its effectiveness considerably. FLC controllers can integrate approaches that allow them to get as well as change their policy base or criteria by making use of real-time responses. The adaptability of FLC allows it to properly reply to modifications in system characteristics, outside situations, or running factors. FLC can boost control activities together with boosting system efficiency in commercial applications via recurring understanding as well as change.

1.4.4 Nonlinear and Complex System Control

Industrial procedures often present nonlinear and complex actions posturing obstacles for control utilizing traditional approaches. FLC shows remarkable efficiency in handling nonlinearity and sophisticated system characteristics as a result of its essential versatility and the ability to approximate nonlinear links. FLC controllers can properly catch and also stand for the complicated behavior of complex systems. This makes them appropriate for a selection of commercial procedures that show nonlinear as well as uncertain functions.

1.4.5 Robustness and Fault Tolerance

FLC has innate strength as well as fault tolerance, making it appropriate for applications defined by regular system failings or unpredictability. The FLC system has the capacity to take care of fuzzy or missing information, noisy measurements, and specification modifications without an obvious reduction in efficiency. The strong resilience of FLC controllers allows them to work dependably in commercial setups, where unpredictability as well as disruptions prevail. Additionally, the capacity of FLC to handle partial system failings or sensing unit mistakes ensures that the control system can continue operating even when mistakes exist, therefore boosting system dependability coupled with safety and security.

1.4.6 Intuitive Rule-Based Design

FLC controllers are made utilizing a rule-based method which makes them understandable together with usage for control system developers. The control regulations depend on experienced understanding together with can be easily revealed utilizing linguistic principles along with fuzzy collection procedures. The usage of a rule-based style enables control designers to incorporate their specialized understanding right into the layout of the control system, causing the development of controllers that remain in line with their cognitive structures as well as specifications. The openness and interpretability of the control regulations additionally improve the capability to examine, debug, and also enhance the system.

1.4.7 Reduced Development and Implementation Time

FLC supplies a minimized duration for both growth and application as contrasted to traditional control methods. Due to the absence of dependence on detailed mathematical designs and the lack of a requirement for precise criterion change, the layout procedure for FLC can be structured. The rule-based layout and user-friendly nature of FLC make it possible for control system designers to quickly model prototypes and implement control solutions. This advantage is particularly substantial in high-speed commercial setups where the rate of bringing an item to market and releasing a system is crucial.

1.4.8 Integration with Conventional Control Methods

FLC can be efficiently integrated right into conventional control techniques making it possible for the production of crossbreed control systems that profit from the benefits of both the techniques. The combination of FLC with traditional control approaches such as PID (proportional-integral-derivative) control flexible control or design anticipating control can cause boosted total system efficiency. This combination enables FLC to successfully take care of specific control requirements while gaining from the security together with the strength of conventional control strategies.

FLC gives various advantages in commercial applications. The effective and flexible control technique is defined by its ability to handle approximate as coupled with unpredictable details, language depiction, flexible abilities strength together with compatibility with nonlinear, and also intricate systems. The fundamental rule-based layout as well as lowered advancement time in addition improved its appearance. As a result, FLC is extensively utilized in different industries supplying reliable and reputable control remedies.

FLC can be smoothly incorporated into traditional control methods, enabling the creation of hybrid control systems that capitalize on the advantages of both approaches. The integration of FLC with conventional control methods such as PID (proportional-integral-derivative) control, adaptive control, or model predictive control can result in improved overall system performance. This integration allows FLC to effectively handle particular control needs while benefiting from the stability and resilience of traditional control techniques.

1.5 Industrial Applications of Fuzzy Logic Control

FLC is an adaptable and durable control system technique that has actually been extensively made use of in varied sectors [13]. FLC provides a mathematical structure to deal with approximate and fuzzy input, making it well fit for areas where standard control techniques counting on specific reasoning are poor. This section will certainly take a look at a number of uses of FLC in various sectors.

1.5.1 Automotive Industry

FLC has contributed highly to the field of the automobile industry. Fuzzy logic–based engine control units (ECUs) can dynamically alter the air-fuel ratio, ignition timing, and fuel injection timing in response to real-time inputs such as engine speed, temperature, and exhaust gas composition. The use of FLC results in precise control of engine operation, which, in turn, causes higher fuel economy, lower emissions, and better drivability [14].

1.5.2 Consumer Electronics

FLC is implemented in several consumer electronic devices that can help to enhance their working and user experience. The air-conditioning system is a good example of that. Fuzzy logic controllers process inputs such as room temperature and humidity, as well as user inputs, to alter the temperature and fan speed in order to make it a better place to live, as well as to save energy under fluctuating ambient conditions [15].

1.5.3 Robotics

FLC plays a pivotal role in robotics because it is essential for precise control and intelligent decision-making. Fuzzy logic controllers are applied to control the motion of robots, regulate grasping forces, and follow the objects. Robots with FLC can cover irregular terrain, handle delicate items, and respond to changing environments. This makes them more versatile and resilient [16].

1.5.4 Industrial Process Control

FLC has been adopted widely in a variety of industrial process control schemes. It has been successfully employed in various manufacturing systems, where many related production parameters such as temperature, pressure, and flow rates are tailored to ensure maximum performance and quality. Fuzzy logic controllers are known to be used in systems with complex dynamics and nonlinearity. FLC is able to cope with uncertainties and imprecise measurements accurately. Hence, this form of control is robust as well as stable [17].

1.5.5 Power Systems

Recently, FLC has been established for the purpose of power systems control and optimization. FLCs can help the grid maintain stability and reliability through power generation, power transport, and power delivery. FLCs are also applied in the context of fault detection and diagnosis in order to outflank the possibility of triggering cascading failures and blackouts. FLC-based optimization algorithms are implemented for load prediction and demand-side management to increase energy use [18].

1.5.6 Traffic Control Systems

FLC has greatly influenced traffic control systems, especially in urban areas. Fuzzy logic controllers can change traffic signal timings according to current traffic conditions, which ultimately facilitates smoother traffic and less congestion. It can analyze factors such as traffic volume, vehicle density, and road conditions to make well-informed decisions and adapt to changing traffic patterns [19].

1.5.7 Medical Systems

In healthcare, FLC has found applications across numerous medical systems to improve diagnostic and therapeutic procedures [19]. One such application is that of anesthesia delivery systems, using fuzzy logic–based controllers for the accurate and consistent delivery of medication. However, some studies have experimented with FLC in medical imaging for quality control, especially in reducing noise and artifacts in images [20]. Another clinical application is decision support systems for disease diagnosis and planning treatment using the clinician's expertise and the patient's case specifics.

1.5.8 Agriculture

Agriculture: In the area of agriculture, FLC has been used in promoting better irrigation systems to enable higher crop yields [21]. FLCs can control water flow rates by taking into account factors such as soil moisture, temperature, and other weather predictions. Another application also includes designing intelligent crop disease identification and pest control systems using fuzzy logic, which could help in taking timely actions when there are any signs of a disease or an insect attack and prevent heavy crop losses [22].

1.5.9 Aerospace and Defense

FLC has found application in the aerospace and defense industries. Controllers made using fuzzy logic examples are extensively used in the control systems of aircraft autopilots to manipulate the dynamics of flights, maintain dynamic stability, and adapt to uncertain and changing situations. FLC is also applied in target tracking and missile guidance systems to track mobile targets accurately and robustly [23].

1.5.10 Financial Systems

In financial systems, FLC has been used to assess risk in financial system decision-making. Fuzzy logic models may evaluate financial performance indicators and market statistics to identify investment opportunities and risks in a portfolio based on them. The use of FLC in this sphere allows for deeper risk identification as it considers fuzzy data and supports well-informed financial decisions [24]. The application areas of FLC are much broader, even though these are merely a few examples. The unique ability of fuzzy logic to handle fuzzy, unpredictable input data, adapt to changing circumstances, and provide robust control makes it an essential tool in many industries [25]. With the advancement of technology and our deepening comprehension of complex systems, it is foreseeable that FLC will have a broader range of uses in the future. This will lead to improved automation, efficiency, and decision-making across many industries on a global scale.

1.6 Discussions and Future Scope of Work

FLC is a robust and flexible control approach that shows great promise for use in industrial settings [26]. The capacity to manage unpredictability, nonlinearity, and adaptability has rendered it an appealing option for regulating intricate systems across many industries. FLC offers a control method that is easily understood and implemented by both experts and operators, thanks to the utilization of linguistic variables, membership functions, fuzzy rule bases, and defuzzification approaches. The development of FLC has witnessed progress in integrating with other intelligent techniques, such as neural networks and metaheuristic optimization algorithms, as well as the creation of adaptive mechanisms to enhance control performance over time. The progress in technology has significantly enhanced the functionality and usefulness of FLC, making it possible to effectively apply it in many industrial systems.

The use of fuzzy logic controllers (FLCs) in industrial environments has produced remarkable outcomes. It has been utilized in various fields including process control, robotics and automation, energy management systems, and intelligent transportation systems, leading to enhanced efficiency, safety, and cost-effectiveness. Within the domain of process management, fuzzy logic control (FLC) has demonstrated its worth in effectively managing intricate and nonlinear systems. It has effectively been utilized to regulate variables like as temperature, pressure, flow rate, and pH levels in chemical plants, power generation facilities, and manufacturing operations. FLC's capacity to manage uncertainty and nonlinearity enables robust control and enhances system performance in such applications.

FLC has proven to be highly effective in achieving precise control and adaptability in the field of robotics and automation. It has been utilized in the control of robot movement, identification of objects, determination of paths, and prevention of collisions, allowing robots to carry out intricate jobs with effectiveness and security. FLC's capacity to manage uncertain and dynamic settings makes it highly suitable for robotics applications that require real-time decision-making. FLC has also made substantial contributions to energy management systems. FLC plays a crucial role in enhancing energy efficiency and cutting costs by optimizing equipment operation, load control, and energy distribution. It has been utilized in smart grids, renewable energy systems, and HVAC control, enabling intelligent and adaptable energy management tactics.

FLC has significantly enhanced the efficiency of intelligent transportation systems. By implementing fuzzy logic control (FLC) techniques in traffic management systems, it is possible to efficiently regulate traffic flow and alleviate congestion. Flexible lane control (FLC) allows for immediate modification of signal timings according to traffic conditions, resulting in greater traffic efficiency, decreased travel time, and improved safety on road networks. The future potential of utilizing FLC in industrial applications is quite promising. The fusion of artificial intelligence and machine learning techniques offers the potential to improve control performance and also facilitate decision-making autonomously in intricate systems. By integrating FLC's interpretability with the learning and adaptation capabilities of these approaches, control systems can enhance their intelligence and adaptability. Researching the management of systems on a big scale is another area of focus for the future. It is essential to create FLC algorithms and techniques that can scale and perform computations efficiently in order to regulate intricate industrial processes that have multiple inputs, outputs, and intricate interconnections.

FLC continues to have difficulties in managing uncertainty and missing information. Advancements in fuzzy rule interpolation, fuzzy clustering, and fuzzy inference can improve the ability of fuzzy logic controllers (FLCs) to handle circumstances with inaccurate or sparse information. Real-time execution and refinement of FLC algorithms are crucial for industrial control systems. Research can prioritize the development of algorithms and hardware designs that facilitate the real-time execution of FLC. This would enable prompt control actions in dynamic contexts. Enhancing the resilience and ability to handle errors of FLC systems is of utmost importance in industrial settings where system failures or disruptions may arise. Integrating fault detection and recovery procedures into the FLC architecture can manipulatively work to enhance the dependability and efficacy of the system. Moreover, investigating the capacity for human-machine cooperation within FLC systems can result in efficient decision-making procedures and enhanced operator-automation engagement in industrial environments. By integrating human skills and preferences into the control system, FLC can facilitate operators to collaborate effectively with automated control systems, resulting in enhanced operational efficiency and reliability. Therefore, FLC has demonstrated its worth as a valuable control technology for industrial applications. The capacity to manage uncertainty, nonlinearity, and adaptability, together with its transparency and interpretability, renders it an appealing option for governing intricate systems. Through ongoing research and development, FLC is expected to make significant progress in areas such as integration with advanced techniques, scalability, handling uncertainty, real-time implementation, fault tolerance, and human-machine collaboration. These advancements will further contribute to the optimization and automation of industrial processes in the future.

1.7 Conclusions

FLC has demonstrated its efficacy as a beneficial control technology in industrial applications. This progress will be achieved by integrating advanced techniques, improving scalability, effectively managing uncertainty, implementing real-time systems, ensuring fault tolerance, and enhancing collaboration between humans and machines. It excels in several industries such as aerospace and defense, finance, agriculture, medical imaging, traffic control, industrial process control, power systems, robotics, human-machine cooperation, consumer electronics, and more. Therefore, FLC has demonstrated its worth as a significant control mechanism in industrial applications and other disciplines, offering enhanced performance and precision.

References

1. Zadeh, L.A., Fuzzy logic. *Computer*, 21, 83–93, 1988.
2. Driankov, D., Hellendoorn, H., Reinfrank, M., *An introduction to fuzzy control*, Springer Science & Business Media, Berlin, Germany, 2013.
3. Yager, R.R. and Zadeh, L.A., *An introduction to fuzzy logic applications in intelligent systems*, Springer Science & Business Media, New York, USA, 2012.
4. Chen, G., Pham, T.T., Boustany, N., Introduction to fuzzy sets, fuzzy logic, and fuzzy control systems. *Appl. Mech. Rev.*, 54, B102–B103, 2001.
5. Iancu, I., A Mamdani type fuzzy logic controller, in: *Fuzzy logic-controls, concepts, theories and applications*, vol. 15, pp. 325–350, 2012.

6. Rezaee, B. and Zarandi, M.F., Data-driven fuzzy modeling for Takagi–Sugeno–Kang fuzzy system. *Inf. Sci.*, 180, 241–255, 2010.
7. Erginer, B. and Altuğ, E., Design and implementation of a hybrid fuzzy logic controller for a quadrotor VTOL vehicle. *Int. J. Control Autom. Syst.*, 10, 61–70, 2012.
8. Elmas, C., Deperlioglu, O., Sayan, H.H., Adaptive fuzzy logic controller for DC–DC converters. *Expert Syst. Appl.*, 36, 1540–1548, 2009.
9. Bai, Y. and Wang, D., Fundamentals of fuzzy logic control—fuzzy sets, fuzzy rules and defuzzifications, in: *Advanced fuzzy logic technologies in industrial applications*, pp. 17–36, 2006.
10. Dzitac, I., The fuzzification of classical structures: A general view. *Int. J. Comput. Commun. Control*, 10, 12–28, 2015.
11. Tikk, D., Johanyák, Z.C., Kovács, S., Wong, K.W., Fuzzy rule interpolation and extrapolation techniques: Criteria and evaluation guidelines. *J. Adv. Comput. Intell. Intell. Inform.*, 15, 254–263, 2011.
12. Albertos, P., Sala, A., Olivares, M., Fuzzy logic controllers. Advantages and drawbacks, in: *VIII international congress of automatic control*, International Federation of Automatic Control Laxenburg, Austria, pp. 833–844, 1998.
13. Larsen, P.M., Industrial applications of fuzzy logic control. *Int. J. Man Mach. Stud.*, 12, 3–10, 1980.
14. Von Altrock, C., Fuzzy logic technologies in automotive engineering, in: *Proceedings of WESCON'94*, IEEE, pp. 110–117, 1994.
15. Kumar, S., A review of smart volume controllers for consumer electronics. *IEEE Trans. Consum. Electron.*, 51, 600–605, 2005.
16. Qureshi, M.S., Swarnkar, P., Gupta, S., A supervisory on-line tuned fuzzy logic based sliding mode control for robotics: An application to surgical robots. *Rob. Auton. Syst.*, 109, 68–85, 2018.
17. Murthy, B.V., Kumar, Y.P., Kumari, U.R., Fuzzy logic intelligent controlling concepts in industrial furnace temperature process control, in: *2012 IEEE International Conference on Advanced Communication Control and Computing Technologies (ICACCCT)*, IEEE, pp. 353–358, 2012.
18. Kumar, N.K. and Gandhi, V.I., Implementation of fuzzy logic controller in power system applications. *J. Intell. Fuzzy Syst.*, 36, 4115–4126, 2019.
19. Mohanaselvi, S. and Shanpriya, B., Application of fuzzy logic to control traffic signals, in: *AIP Conference Proceedings*, AIP Publishing, 2019.
20. Vlamou, E. and Papadopoulos, B., Fuzzy logic systems and medical applications. *AIMS Neurosci.*, 6, 266, 2019.
21. Ali, R.B., Aridhi, E., Abbes, M., Mami, A., Fuzzy logic controller of temperature and humidity inside an agricultural greenhouse, in: *2016 7th International Renewable Energy Congress (IREC)*, IEEE, pp. 1–6, 2016.
22. Septiarini, F., Dewi, T., Rusdianasari, Design of a solar-powered mobile manipulator using fuzzy logic controller of agriculture application. *Int. J. Comput. Vision Rob.*, 12, 506–531, 2022.

23. Coskun, M. and Tasdemir, S., Fuzzy Logic Based Threat Assessment Application In Air Defense Systems. *IEEE Trans. Aerosp. Electron. Syst.*, 59, 3, 2245–2251 2022.
24. Hachicha, N., Jarboui, B., Siarry, P., A fuzzy logic control using a differential evolution algorithm aimed at modelling the financial market dynamics. *Inf. Sci.*, 181, 79–91, 2011.
25. Bojadziev, G. and Bojadziev, M., *Fuzzy logic for business, finance, and management*, World Scientific, London, UK, 2007.
26. Srivastava, U.K. and Prakash, N., A systematic review on real time video compression and enhancing quality using fuzzy logic. *JCSE*, 6, 653–665, 2018.

2

Artificial Neural Network for Industrial Applications

Biswajit Saha and Gour Sundar Mitra Thakur*

Department of CSE (AIML), Dr. B. C. Roy Engineering College, Durgapur, West Bengal, India

Abstract

Artificial neural networks (ANNs), inspired by the structure of the human brain and the way in which the brain functions, are central to understanding the domain of artificial intelligence, machine learning, and pattern recognition. Artificial neurons, also known as perceptrons, form the foundation of ANNs. They process input signals and apply weights and activation functions according to the requirements. These neurons are organized into the following layers: input, hidden, and output. Data are received by the input layers. The processing and output are done by the hidden layers and the output layers, respectively. Commercial implementations of ANNs find excellent use in quality control, predictive maintenance, and anomaly detection. Vast datasets are analyzed to identify patterns. This, in turn, helps to make real-time decisions so that efficiency is increased and costs arc reduced. The complexity of data in various industrial applications is efficiently managed by ANNs. For example, manufacturing defects can be identified early. This helps to reduce product flaws as well as predict equipment failures. Consequently, preemptive maintenance can be carried out to avoid costly downtimes. Their ability to adapt, learn, and handle complex data generated from industrial processes boosts productivity and profitability.

***Keywords*:** Artificial neural network, ANN in industry, supervised learning, unsupervised learning, reinforcement learning

**Corresponding author*: cse.gsmt@gmail.com

Arindam Mondal and Souvik Ganguli (eds.) Controller Design for Industrial Applications, (21–40)

2.1 Introduction

The advent of artificial neural networks (ANNs) has ushered in a new era in industrial technology, marking a significant shift in how data are analyzed and decisions are made across various sectors. At the heart of this transformation is the deep neural network (DNN), a sophisticated variant of ANNs characterized by multiple hidden layers that enable complex data processing and pattern recognition. DNNs have been demonstrated to be exceptionally useful in extracting high-level features from raw data, which is crucial in tackling complex industrial challenges such as predictive maintenance, automated quality control, and supply chain optimization. Figure 2.1 illustrates the architecture of a typical DNN, showcasing its layered structure and the interconnected neurons within each layer. This chapter delves into the principles of ANNs, their industrial applications, and the profound impact they have had on revolutionizing traditional processes and systems.

ANNs can be of various types, each with its own characteristics and training methods. One popular type is the multilayer perceptron (MLP) (Rosenblatt, 1958), which is typically trained using the backpropagation algorithm (Linnainmaa, 1970). Other types include learning vector quantization (Kohonen and Kohonen, 1995), radial basis function (RBF) (NCTIONS, 1988), Hopfield (Hopfield, 1982), and Kohonen networks (Kohonen, 1982). ANN can be categorized as either feedforward (Ivakhnenko, Lapa *et al.*, 1967) or recurrent (Amari, 1972), depending on the way in which data are passed through the network and processed, with recurrent networks incorporating feedback loops. Another classification criterion for ANN is their learning method, where some employ

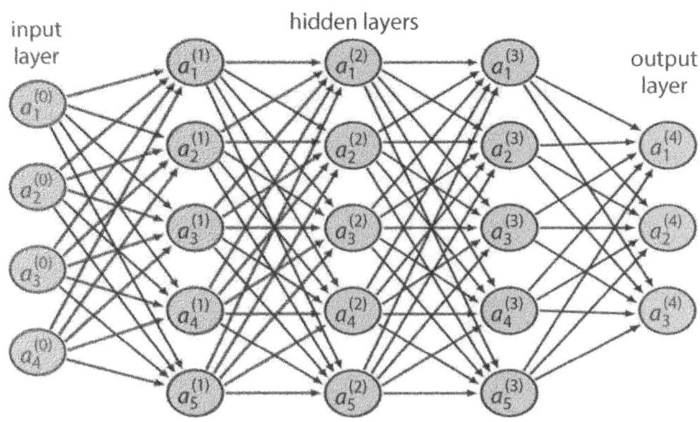

Figure 2.1 Architecture of a typical deep neural network.

supervised training, while others are known as semi-supervised, unsupervised, or self-organizing networks (Meireles *et al.*, 2003).

Industrial applications can vary, encompassing manufacturing automation, process control, and management of energy. Process control finds utility in diverse industries like food, chemical, petrochemical, and nuclear. Automation of the manufacturing industry involves numerically controlling machinery along with extensive systems that process data. Control applications are applicable to various sectors like distribution control, highways, railways, shipping, gas and oil transportation, and also airways. Despite relatively low rates of information required for these applications in comparison to fiber capacity, control systems based on fiber-optic offer numerous benefits above electronic wire systems. Advantages include immunity to electromagnetic interference, increased quality of signal, ample scope for future upgrades, and safety in risky environments, for instance, oil or gas industries. Additionally, remote control can be effortlessly achieved, along with effective maintenance of fiber-optic systems using secondary sources of power during power failures (Tariyal and Cherin, 2003).

According to *Precedence Research*, the market size of deep learning (DL) globally was estimated to be at US$52.13 billion in 2022, and the future projection is that it is expected to touch around US$978.88 billion by 2032 (Figure 2.2). This accounts for an expansion at a CAGR of 34.08% during the period of forecast from 2023 to 2032.

The organization of the remainder of this book chapter is as follows: Section 2.2 discussed the application of different types of ANN models

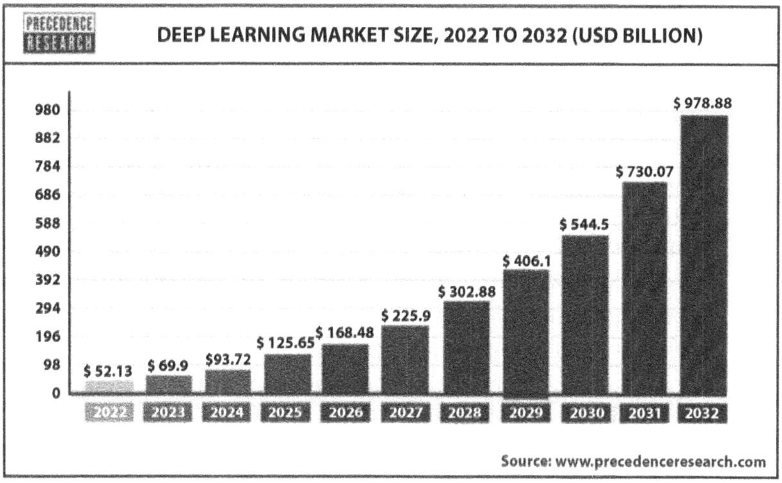

Figure 2.2 Deep learning market size.

used in different industrial applications. In section 2.3, different challenges are faced in implementing various ANN models in different industries. Industry use cases and a sample case study are discussed in section 2.4. Section 2.5 gives an idea about future directions and merging trends in this domain, and finally, section 2.6 gives the conclusion.

2.2 Neural Network Models for Industrial Applications

Over the years, a number of neural network models have been developed or proposed for industrial applications based on the following types of machine learning methods:

- **Supervised Learning:** It is characterized by the occurrence of training data that are labeled. The term implies the presence of a 'supervisor' that guides the learning system in assigning labels to training instances. These labels usually represent class categories in classification tasks. Algorithms for supervised learning derive models from this annotated training data, enabling them to classify new, unlabeled data points (Cunningham *et al.*, 2008).
- **Semi-Supervised Learning:** Unsupervised learning techniques lack input guidance during training, leading to decreased labeling expenses. Nevertheless, their performance typically exhibits lower accuracy. In contrast, semi-supervised learning approaches offer a way to capitalize on the precision attained by supervised learning while mitigating costs through reduced labeling efforts. These learning approaches can be broadly categorized into two main groups: methods based on data augmentation and methods based on semi-supervised mechanisms (Chen *et al.*, 2023).
- **Unsupervised Learning:** Unsupervised learning focuses on the process of training systems to understand and represent patterns in input data based on the underlying statistical patterns present in the entire dataset. Unlike supervised learning or reinforcement learning (RL), unsupervised learning does not require predefined desired outputs or external evaluations for each input. Instead, the unsupervised learner relies on biases that are inherent in

determining the features of the input's structure to be captured in the generated output (Dayan *et al.*, 1999).

- **Reinforcement Learning:** It operates within the space between supervised and unsupervised learning. It focuses on mastering sequential decision-making tasks where feedback is sparse (Wiering and Van Otterlo, 2012).

Supervised Learning in Industrial Applications

In the research carried out by Kordkheili and Rahimpour (2023), the focus was on enhancing temperature prediction in an industrial-scale gasoil HDT (hydrodesulfurization) reactor. Two methods were investigated for their predictive capabilities: neural network and semi-empirical regression models. The feed-forward backpropagation algorithm coupled with the learning function was employed to build up the ANN models. To determine the optimal configuration, there was a variation in the number of neurons in the hidden layer within the range of 7 to 13, and the most reliable ANN models were selected based on an experimental approach. In order to evaluate the performance of these models, root mean square error (RMSE) and average absolute deviation (AAD) were used as metrics. Upon comparison with the regression models, it was found that the ANN model outperformed other methods in predicting the reactor bed temperature profiles. Therefore, the neural network model was identified as the superior approach for achieving improved prediction accuracy in the temperature variation of the gas oil HDT reactor.

Zamaniyan *et al.* (2013) applied ANN using backpropagation and three layers to model an industrial hydrogen plant. Operating data are gathered by simulating the industrial plant and varying key parameters such as carbon dioxide to methane ratio, reformer pressure, feed temperature, and steam to carbon ratio to train the ANN. The tangent sigmoid transfer function is employed by the hidden and output layers, and the training is conducted using a gradient descent algorithm. The mean square error (MSE) is minimized in order to identify the optimal number of neurons in the hidden layer. By altering these key parameters, the model can predict the product's mole fraction of hydrogen and carbon monoxide, pressure, and temperature. These predictions provide valuable insights and allow for the optimization of hydrogen production plants, enhancing overall efficiency and performance.

The growing demand for process quality, efficiency, and higher products is motivating organizations to seek innovative ways to optimize their production. Even minor variations in the product state during production can lead to costly rework or scrappage in high-tech manufacturing.

In order to satisfy quality requirements and remain competitive, it is essential to describe the product state of an individual throughout the entire manufacturing process, incorporating all related information, including adjustments of process parameters in-process. It is possible to analyze this information directly, and if critical trends or events are identified, appropriate actions, such as triggering alarms, can be taken. Conventional methods based on causal sequence modeling face challenges due to the increase in complexity and high dimensionality of contemporary manufacturing practices. New approaches are needed to handle this complexity and high dimensionality while producing practical results efficiently. Wuest *et al.* (2014) proposed the possibility of creating a system by combining supervised machine learning and cluster analysis using data on product state along the manufacturing process.

Semi-Supervised Learning in Industrial Applications

Matrix-assisted laser desorption-ionization (MALDI) in combination with a time-of-flight (TOF) detector forms a technique known as MALDI-TOF mass spectrometry which holds promise for efficiently identifying data patterns and building a robust methodology for rapid and accurate identification of microorganism. The complexity of data poses challenges, often leading to difficulties in correct labeling. To tackle this issue, González *et al.* (2023) proposed using a semi-supervised paradigm as a pioneering approach to classify MALDI-TOF mass spectrometry data. Their experiment involved analyzing mass spectrometry data from healthy and infected salmon with the *Piscirickettsia salmonis* pathogen. By utilizing both labeled and unlabeled data, the projected algorithm based on self-training demonstrated superior results as compared to conventional machine learning methods such as naive Bayes (NB), random forest (RF), and support vector machine (SVM). The semi-supervised learning approach achieved balanced performance across all evaluation metrics even with only a small percentage of labeled instances (25%). Results of experiments revealed that self-training with an RF classifier achieved an accuracy of 0.9, sensitivity of 0.75, and specificity of 1. Additionally, the identification of 15 potential biomarkers was allowed by feature selection that accurately distinguished between healthy and infected salmon profiles. Overall, these findings highlight the possibility of the projected semi-supervised learning methodology as a useful key for classifying MALDI-TOF mass spectrometry data and addressing the hurdles posed by data complexity and labeling issues.

Automatic surface inspection of ferrite beads is crucial for enhancing quality and ensuring proper functionality. While DL-based defect detection methods show promise (Dechter, 1986), they often demand substantial costly annotated data for effective training, limiting their practicality. Addressing this concern, Li *et al.* (2023) introduced a feeble supervised learning algorithm for detecting defects. It achieved high accuracy in identifying and localizing defects, relying solely on image-level labels. A spatial association module (SAM) that employs superficial features to enhance the performance of the network was introduced to harness texture-based defect location data. A training enhancement approach was proposed that leverages guide crop and object ignore algorithms. These algorithms extract key defects and background areas from images, aiding in generating the most favorable decisions. Additionally, an optimal inference method that enhances localization comprehensiveness without compromising accuracy was presented, offering more reasonable visual explanations for defect recognition.

Unsupervised Learning in Industrial Applications

In various industries, the process of assuring visual quality was from being manual, labor-intensive, and prone to errors. It has evolved to become a highly automated and accurate evaluation of industrial quality. The transformation has been facilitated by advancements in machine learning, particularly supervised learning. However, a key limitation of most supervised learning methods is their ability to only recognize predefined categories, such as specific types of defects in manufactured items. They struggle to identify new, previously unseen types of defects. To address this issue, Zipfel *et al.* (2023) delved into the domain of unsupervised models built on DNNs. Unlike their supervised counterparts, these models are not restricted to predefined categories and can holistically assess the overall quality of objects. To exemplify this approach, the study focuses on a quality inspection scenario involving a European car manufacturer. The research evaluates the performance of three unsupervised models (Skip-GANomaly, PaDiM, and PatchCore) in detecting defects. Through a comprehensive assessment, the study demonstrates that fully unsupervised methods can yield dependable results, and, in fact, they can even compete with the performance of supervised models. In summary, the shift toward automated visual quality assurance owes much to machine learning, and while supervised learning has limitations, this work introduces unsupervised DNN models as a promising solution. The evaluation conducted on a real-world quality inspection case underscores the efficacy of these unsupervised approaches in achieving reliable results, comparable to those obtained through supervised methods.

Industrial control systems rely heavily on the communication and capabilities of data exchange facilitated by industrial control protocols, which have the possibility of being intricate and even incorporating encryption. The act of flip engineering these protocols has emerged as a vital subject in the realm of industrial security research. Ning *et al.* (2023) introduced PREIUD, a tool designed for the reverse engineering of industrial control protocols, leveraging DNN techniques and unsupervised learning. The reverse engineering process is compartmentalized into distinct phases. Initially, the bootstrap voting expert algorithm is employed to deduce the keyword segment boundaries within the following practices while taking into consideration the properties of symmetry. Subsequently, a conditional random field which is a bidirectional long short-term memory is integrated with a mechanism of attention to not only categorize the protocols but also extract their structural and semantic characteristics. For the purpose of training and testing the model, data sample sets that were manually constructed were considered including six generally utilized industrial protocols. The model's performance was then evaluated against two complex protocol reverse engineering tools, namely, Discoverer and MSERA. The outcomes indicated that PREIUD achieved an average precision enhancement of 7.4% as compared to MSERA, along with a notable 15.4% improvement compared to Discoverer. Additionally, the proposed approach managed to maintain a balance between computational efficiency and conciseness. In essence, the proposed methodology constituted a substantial stride forward in the domain of reverse engineering industrial control protocols. Their approach bears practical implications for enhancing the security of industrial control systems.

To differentiate between various aerogel samples, "Principal Components Analysis" (PCA) was conducted by Younes *et al.* (2023). This analysis considered both the chemical/physical properties of the aerogels and their adsorption characteristics. Multiple approaches and data pre-treatments were employed to mitigate the occurrences of any statistical biases. The outcome revealed that the aerogel specimens were situated in the middle of the biplot, surrounded by diverse physical/chemical and adsorption properties. This suggests a comparable efficacy among investigated aerogels—whether derived from nanocellulose, chitosan, or graphene—for ion removal. In summary, the PCA method demonstrated uniform efficiency in ion removal across all examined aerogels. Notably, this technique's strength lies in its ability to identify similarities and differences among numerous factors, eliminating the limitations of laborious and time-intensive two-dimensional data visualization techniques.

In order to facilitate intricate industrial processes in modern manufacturing, the adoption of cyber-physical systems (CPSs) is on the rise. These CPSs are responsible for executing repetitive and recurring steps within processes like production, packaging, and transportation. As these processes become more complex, the significance of each individual step in achieving the desired outcome becomes increasingly critical. Any failure occurring within one step can lead to a completely flawed process, causing disruptions in the functioning of the CPS or even the entire manufacturing process. Hence, ensuring the effectiveness and safety of modern CPS requires constant monitoring of each individual process step, allowing for the prompt detection of potential anomalies. The identification of anomalies can help prevent failures from occurring or expedite the resolution of any encountered issues. Goetz and Humm (2023) introduced a novel approach for autonomously detecting process anomalies in CPS while adhering to industry-specific limitations. The primary focus was on recurring procedure tasks with fixed time durations considering various industry constraints such as real-time demands, prediction accuracy, data-driven limitations, processing capabilities, customizable design, and communication interface limitations. The proposed idea was put to the test using an industrial application that imitates a transportation process. The results of the experiment validated the effectiveness of the introduced concept in successfully identifying anomalies within the analyzed process. Consequently, the presented concept offered a promising avenue for detecting anomalies in repetitive process tasks within CPS. This approach holds the potential for enhancing the reliability and performance of CPSs operating under industry-specific constraints.

Reinforcement Learning in Industrial Applications

The flavor industry places a significant emphasis on taste, aligning with societal trends and behaviors. Developing new flavoring agents and molecules is a crucial aspect of this sector. However, the emphasis on natural flavors is particularly crucial in today's world. To address this, Queiroz *et al.* (2023) introduced an innovative framework that leverages scientific machine learning to tackle a growing issue in flavor engineering and the industry. The proposed framework integrates generative and RL models to create a comprehensive system. This research introduces a unique approach to designing novel flavor molecules. The molecules are assessed based on factors such as synthetic feasibility, atom count, and resemblance to pseudo-natural or natural substances. Notably, the work contributes by implementing a web scraping code to gather data from a database of flavors and by integrating two sophisticated scientific machine learning techniques within a composite framework. Compared to using only a generative model, the utilization of this complex system led to a 10% increase in molecules that met optimal criteria. The molecules generated by the RL model were further evaluated for their presence in the market and their usage within the flavor industry. This study validates the framework's potential in identifying molecules suitable for the development of flavor-based products.

RL has achieved major accomplishments in guiding sequential choice-making across various domains such as marketing, robotics, autonomous vehicles, and the gaming industry. This achievement has spurred interest in employing RL for controlling energy systems of buildings, which are growing more intricate due to the requirement to optimize multiple, potentially disagreed objectives like energy consumption, occupant comfort, and grid interaction. When applied to real-world scenarios, RL comes with numerous limitations, including the requirement for substantial training time and data, as well as control behavior which is unstable during initial exploration phases. These drawbacks can render RL impractical for direct implementation in building control tasks. An approach rooted in imitation learning is employed by Dey *et al.* (2023) to tackle these challenges. In this method, RL agents begin by inheriting policies from established rule-based and heuristic strategies. This strategy proves effective in reducing training duration, curbing unstable early exploration behaviors, and enhancing performance compared to accepted rule-based policies. These outcomes collectively render RL an additional viable control strategy for applications of the real world in the realm of control building.

Esteso *et al.* (2023) explored how RL techniques are utilized in the field of production planning and control (PPC) in their study. The study focuses on various aspects of PPC, including capacity planning, purchase and supply management, facility resource planning, inventory management, and production scheduling. The analysis delves into key attributes of RL, such as the methods employed, contextual aspects, states, actions, rewards, and notable features. The investigation also considers factors like the number of agents involved, applications, and specific RL software tools. This includes aspects like platforms, programming languages, RL frameworks, and application programming interfaces. Through a review of 181 articles, the findings revealed that RL finds its primary application in production scheduling challenges, followed by purchase and supply management concerns. The most frequently studied RL algorithms are those falling under the model-free and single-agent categories. These algorithms are often applied to simplified PPC scenarios. Despite this simplification, their outcomes demonstrate promise in comparison to traditional mathematical programming and heuristic/metaheuristic solutions. This is especially noticeable when dealing with uncertainty or nonlinear characteristics in the problem. Among the RL approaches, value-based methods are the most popular. Notably, Q-learning and its variations are extensively employed. Additionally, in recent times, actor-critic approaches have gained significant traction. These include techniques such as proximal policy optimization, advantage actor-critic, trust region policy optimization, and deep deterministic policy gradient.

Guo *et al.* (2023) observed that a significant portion of current research predominantly relies on simulating specific real-world conditions within virtual environments. This approach seldom involves real sensor systems or actual disassembly operations, limiting the broader applicability of RL in disassembly lines across various commercial contexts. Conversely, the disassembly process places substantial demands on factors like overall efficiency, recovery costs, and dismantlement profitability. Consequently, RL often remains confined to the exploratory and research stages due to the scarcity of real-world cases. To facilitate the integration of RL into the realm of disassembly lines, several critical obstacles must be addressed. Firstly, continual enhancements to the theoretical foundations of RL algorithms are necessary to enhance their robustness and global convergence capabilities. Secondly, it is imperative to establish mathematical models for disassembly challenges, broaden the range of applicable scenarios, and establish interfaces that link multiple problems to ensure comprehensive coordination. Moreover, there is a pressing need for collaborative research and support pertaining to policy issues such as carbon emissions and remanufacturing. These elements are essential for advancing the utilization of RL within the disassembly line domain.

Table 2.1 below shows some more recent and important applications of ANN in different industries.

2.3 Challenges and Limitations

Lennox *et al.* (2001) presented the findings of a comprehensive 2-year industrial investigation focused on the implementation of ANN for monitoring and control of processes. This research was a joint effort involving the University of Newcastle-upon-Tyne, EDS, and 24 UK-based international companies. The study included various significant activities, such as utilizing neural networks for detecting faults in a vitrification process and,

Table 2.1 Applications of artificial neural networks in various industries.

Industry	Application	Reference
Public Health	Forecasting COVID-19 cases and death rates	Tamang *et al.*, 2020
Economics	Analysis of economic impact during COVID-19	Jena *et al.*, 2021
Economic Forecasting	Inflation rate prediction	Talakua *et al.*, 2022
Healthcare	Predicting adverse drug reactions	Imai *et al.*, 2020
Machine Learning	Image recognition with optical neural networks	Xu *et al.*, 2021
Pharmaceutical	Disease diagnosis, drug design, studying drug interactions	Teja *et al.*, 2022
Cancer Research	Diagnosis and treatment of lung cancer	Prisciandaro *et al.*, 2023
Finance	Predicting financial distress in companies	Dube *et al.*, 2023

in the case of a gasoline engine, employing model-based predictive control. Throughout the program, the researchers encountered practical challenges while applying neural networks in industrial settings. Nevertheless, they gained valuable insights and lessons from completing the project. The main takeaway was that neural networks have the potential to enhance industrial process monitoring and control systems. The extent of improvement achievable greatly depends on the specific problem being addressed, and in certain cases, the use of neural networks may not be acceptable.

According to Sharifani and Amini (2023), as these advancements progress, it becomes imperative to confront the ethical considerations linked to their implementation, encompassing concerns about prejudice, inequity, and confidentiality. Scholars and professionals within this domain must collaborate to construct machine learning and DL models that are characterized by enhanced clarity, comprehensibility, and responsibility. Furthermore, it is essential to guarantee that both individuals and entities are equipped with the essential expertise to proficiently engage with and comprehend these technologies. Achieving this goal can be realized through educational initiatives and training schemes, coupled with fostering partnerships that bridge academia, industry, and governmental entities.

2.4 Industry Use Cases and a Sample Case Study

A case study was carried out at a manufacturing plant in Sweden by Leberruyer *et al.* (2023) to assess the effectiveness of a detection strategy of defects for products that were susceptible to misclassification. The study also dealt with an imbalanced dataset containing only a limited number of defects. The main objective of the research was to implement artificial intelligence (AI) in an industrial setting to gain valuable insights and determine the essential requirements to achieve zero defect manufacturing (ZDM). The combination of ZDM and AI presented promising prospects for enhancing quality management and problem-solving, offering novel opportunities for advancement in the manufacturing industry. Despite the potential benefits of this combination, there is a lack of research in real-world settings, hindering widespread deployment of the concept. The researchers employed a semi-supervised learning approach, utilizing vibration properties to distinguish confirmed defects from products that were approved. This method allowed them to calculate a defect similarity ratio, which predicted how closely newly manufactured items resembled defective products. This study helped to identify four critical prerequisites and four valuable insights essential for developing an AI solution to support ZDM. The main finding emphasized the complementary nature of conventional and pioneering quality methods. The results highlighted the significance of initiating data science projects promptly to guarantee data quality and enable the ZDM detection strategy to accumulate knowledge, facilitating the growth of proactive measures like defect prediction and prevention.

In a new case study, Bezrucav *et al.* (2023) established the integration of AI task planning with computer vision systems, leading to a process of manufacturing that is completely autonomous. Additionally, the research showcased the adaptation of low-cost robots to handle various system requirements efficiently. The researchers introduced a completely integrated automation solution for a complete yet straightforward manufacturing process. Instead of focusing on developing new environment cognition or task planning methodologies, they implemented state-of-the-art algorithms in the projected system. The novelty of their approach lies in integrating these modules into a high-level, easily configurable control system, enabling an independent and highly flexible application. In this particular application, collaborative mobile manipulators were utilized to select screws from logistics stations and place them onto a rotary table while handling challenges such as noisy sensors and unpredictable human behaviors throughout the process. Subsequently, a serial manipulator, directly mounted to the workstation, was employed to fasten the screws.

In yet another case study, to address the challenges posed by potential errors in the manufacturing process of memory chips, Micron Technology, a leading semiconductor company, recognized the need for advanced technologies and solutions to improve efficiency, reduce downtime, and enhance product quality. Relying solely on human watchfulness to identify quality issues and mechanical problems is time-consuming, error-prone, and costly. In order to mitigate these challenges, Micron employed various strategies and technologies as stated below (tradecloud, 2023):

- **Machine Vision**: Micron Technology integrated machine vision technology into its photo lithographic cameras during the circuitry etching process on wafers. This technology is designed to identify common defects by scanning the surfaces. When a flaw is identified, engineers are notified. The time it takes for an alert to be issued ranges from 15 seconds to 15 minutes, depending on the specific defect. To address the challenge of manual defect classification, Micron developed the auto-defect classification (ADC) system. This system utilizes DL techniques to automatically classify and group millions of detected flaws into different categories.
- **Thermal Imaging**: In order to enhance the efficiency and precision of AI technology, Micron Technology integrated thermal imaging into its manufacturing process monitoring. They created "heat maps" that depict the regular operating conditions of the factory environment. These heat maps are superimposed onto a "digital twin," which is basically a virtual replica of the actual plant setting. This digital twin serves as a reference point for comparing real-time infrared images of the plant. Should the system detect any inconsistencies, such as unusual temperature variations compared to the digital twin's baseline, an alarm is triggered to alert the relevant personnel.

- **Acoustic Listening**: Perhaps the most unexpected of the three innovations, Micron Technology has developed an AI-driven solution to detect abnormal sounds during the manufacturing process. Analogous to how peculiar sounds in a car can signify issues, unusual noises from machinery often indicate problems. Micron's AI system has been taught to identify anomalies in sound frequencies by translating sound into visual data points. To capture the sounds of specific equipment in noisy surroundings, auditory sensors are positioned near machines or pumps. Engineers then take charge of categorizing the sounds and determining potential underlying causes.
- **Benefits**: The implementation of AI solutions at Micron Technology has led to a noticeable enhancement in manufacturing efficiency and precision. The safety of workers has also improved. The utilization of AI solutions has provided their engineers with more available time to concentrate their efforts on other tasks. The integration of AI solutions into the process of manufacturing has extended to various other aspects of the business, including tasks like predicting product demand. This expansion has resulted in a notable improvement of 10% to 20% in forecasting accuracy.

2.5 Future Directions and Emerging Trends

Despite encountering certain challenges, current RL remains a pivotal component of AI technology. It has demonstrated successful implementation across diverse scenarios and application domains, progressively garnering greater attention. While theoretical and technical hurdles persist, there is an unwavering confidence in its future advancement. Regarding the future trajectory of RL development, several key aspects merit consideration. Firstly, RL should adhere to established standards when defining fundamental elements to enhance its abstraction capabilities, thus enabling model generalization. Secondly, the neural network architecture of deep RL (DRL) should be synergized with advancements in DL and alternative methodologies to enhance learning prowess. This encompasses aspects like training speed and accuracy. Ultimately, the practical deployment of RL involves forging stronger links between virtual environments and the real world, thereby bridging the gap between research and industrial technology. This necessitates establishing interfaces that facilitate seamless integration. Consequently, a greater emphasis must be placed on the application of RL to drive progress in this direction (Guo *et al.*, 2023).

2.6 Conclusion

In summary, the integration of ANNs into industrial settings marks a significant technological leap, transforming operations across various sectors. These models excel in handling complex tasks, improving decision-making, and enhancing efficiency. As industries pivot toward automation and data-centric strategies, ANNs emerge as key drivers of innovation, optimizing processes, reducing costs, and elevating quality control. A multidisciplinary approach is necessary for their implementation which combines expertise in data science, engineering, and domain-specific knowledge. Addressing ongoing challenges like model interpretability, robustness, and ethical issues is critical as the technology evolves. The transition of ANNs from research labs to industrial applications has been challenging, yet the progress made demonstrates their transformative potential. Future advancements depend on collaborative efforts between academia and industry, focusing on refining and customizing these models for specific needs. Looking ahead, ANNs are poised to revolutionize various aspects of the industry, from predictive maintenance to product innovation. Their adaptability and learning capabilities make them integral to the fourth industrial revolution. Ultimately, ANNs are more than just technological tools; they are pivotal in shaping a future where industries operate with greater precision, agility, and sustainability, paving the way for innovative and competitive business practices.

References

Rosenblatt, F., The perceptron: A probabilistic model for information storage and organization in the brain. *Psychol. Rev.*, *65*, 6, 386, 1958.

Linnainmaa, S., The representation of the cumulative rounding error of an algorithm as a taylor expansion of the local rounding errors, [Doctoral dissertation, Master's Thesis (in Finnish), Univ. Helsinki], 1970.

Kohonen, T., Learning vector quantization, in: *Self-organizing maps*, pp. 175–189, 1995.

NCTIONS, R. B. F., Roval signals & radar, 1988.

Hopfield, J.J., Neural networks and physical systems with emergent collective computational abilities. *Proc. Natl. Acad. Sci.*, *79*, 8, 2554–2558, 1982.

Kohonen, T., Self-organized formation of topologically correct feature maps. *Biol. Cybern.*, *43*, 1, 59–69, 1982.

Ivakhnenko, A.G., Lapa, V.G., *et al.*, *Cybernetics and forecasting techniques*, Naukova Dumka Press, 1967.

Amari, S.-I., Learning patterns and pattern sequences by self-organizing nets of threshold elements. *IEEE Trans. Comput.*, *100*, 11, 1197–1206, 1972.

Meireles, M.R., Almeida, P.E., Simões, M.G., A comprehensive review for industrial applicability of artificial neural networks. *IEEE Trans. Ind. Electron.*, *50*, 3, 585–601, 2003.

Tariyal, B. and Cherin, A., Optical fiber communications, in: *Encyclopedia of physical science and technology*, third edition, R.A. Meyers (Ed.), pp. 271–294, Academic Press, 2003, https://doi.org/https://doi.org/10.1016/B0-12-227410-5/00521-4.

Cord, M. and Cunningham, P. (Eds.)., Machine learning techniques for multimedia: case studies on organization and retrieval. Springer Science & Business Media, 2008.

Chen, T., Sampath, V., May, M.C., Shan, S., Jorg, O.J., Aguilar Martín, J.J., Stamer, F., Fantoni, G., Tosello, G., Calaon, M., Machine learning in manufacturing towards industry 4.0: From 'for now'to 'four-know'. *Appl. Sci.*, *13*, 3, 1903, 2023.

Dayan, P., Sahani, M., Deback, G., Unsupervised learning, in: *The MIT encyclopedia of the cognitive sciences*, pp. 857–859, 1999.

Wiering, M.A. and Van Otterlo, M., Reinforcement learning, in: *Adaptation, learning, and optimization*, vol. 12(3), p. 729, 2012.

Kordkheili, M.S. and Rahimpour, F., Artificial neural network and semi-empirical modeling of industrial-scale gasoil hydrodesulfurization reactor temperature profile. *Math. Comput. Simul.*, *206*, 198–215, 2023.

Zamaniyan, A., Joda, F., Behroozsarand, A., Ebrahimi, H., Application of artificial neural networks (ann) for modeling of industrial hydrogen plant. *Int. J. Hydrogen Energy*, *38*, 15, 6289–6297, 2013.

Wuest, T., Irgens, C., Thoben, K.-D., An approach to monitoring quality in manufacturing using supervised machine learning on product state data. *J. Intell. Manuf.*, *25*, 1167–1180, 2014.

González, C., Astudillo, C.A., Lóopez-Cortés, X.A., Maldonado, S., Semi-supervised learning for maldi–tof mass spectrometry data classification: An application in the salmon industry. *Neural Comput. Appl.*, *35*, 13, 9381–9391, 2023.

Dechter, R., Learning while searching in constraint-satisfaction problems, 1986.

Li, Y., Wu, X., Li, P., Liu, Y., Ferrite beads surface defect detection based on spatial attention under weakly supervised learning. *IEEE Trans. Instrum. Meas.*, *72*, 1–12, 2023.

Zipfel, J., Verworner, F., Fischer, M., Wieland, U., Kraus, M., Zschech, P., Anomaly detection for industrial quality assurance: A comparative evaluation of unsupervised deep learning models. *Comput. Ind. Eng.*, *177*, 109045, 2023.

Ning, B., Zong, X., He, K., Lian, L., Preiud: An industrial control protocols reverse engineering tool based on unsupervised learning and deep neural network methods. *Symmetry*, *15*, 3, 706, 2023.

Younes, K., Kharboutly, Y., Antar, M., Chaouk, H., Obeid, E., Mouhtady, O., Abu-Samha, M., Halwani, J., Murshid, N., Application of unsupervised machine learning for the evaluation of aerogels' efficiency towards ion removal—a principal component analysis (pca) approach. *Gels*, *9*, 4, 304, 2023.

Goetz, C. and Humm, B.G., Unsupervised process anomaly detection under industry constraints in cyber-physical systems using convolutional autoencoder, in: *Computational intelligence for engineering and management applications: Select proceedings of ciema 2022*, pp. 659–674, Springer, 2023.

Queiroz, L.P., Rebello, C.M., Costa, E.A., Santana, V.V., Rodrigues, B.C., Rodrigues, A.E., Ribeiro, A.M., Nogueira, I.B., A reinforcement learning framework to discover natural flavor molecules. *Foods*, *12*, 6, 1147, 2023.

Dey, S., Marzullo, T., Zhang, X., Henze, G., Reinforcement learning building control approach harnessing imitation learning. *Energy AI*, *14*, 100255, 2023.

Esteso, A., Peidro, D., Mula, J., Díaz-Madroñero, M., Reinforcement learning applied to production planning and control. *Int. J. Prod. Res.*, *61*, 16, 5772–5789, 2023.

Guo, X., Bi, Z., Wang, J., Qin, S., Liu, S., Qi, L., Reinforcement learning for disassembly system optimization problems: A survey. *Int. J. Netw. Dyn. Intell.*, 2, 1, 1–14, 2023.

Tamang, S., Singh, P., Datta, B., Forecasting of covid-19 cases based on prediction using artificial neural network curve fitting technique. *Global J. Environ. Sci. Manage.*, *6*, Special Issue, Special Issue (Covid-19), 53–64, 2020.

Jena, P.R., Majhi, R., Kalli, R., Managi, S., Majhi, B., Impact of covid-19 on gdp of major economies: Application of the artificial neural network forecaster. *Econ. Anal. Policy*, *69*, 324–339, 2021.

Talakua, M.W., Ilwaru, V.Y.I., Tomasouw, B.P., Limba, S.Z., Inflation forecasts in ambon using neural network applications backpropagation. *BAREKENG: J. Ilmu Mat. Terapan*, *16*, 2, 483–496, 2022.

Imai, S., Takekuma, Y., Kashiwagi, H., Miyai, T., Kobayashi, M., Iseki, K., Sugawara, M., Validation of the usefulness of artificial neural networks for risk prediction of adverse drug reactions used for individual patients in clinical practice. *PLoS One*, *15*, 7, e0236789, 2020.

Xu, X., Tan, M., Corcoran, B., Wu, J., Boes, A., Nguyen, T.G., Chu, S.T., Little, B.E., Hicks, D.G., Morandotti, R., *et al.*, 11 tops photonic convolutional accelerator for optical neural networks. *Nature*, *589*, 7840, 44–51, 2021.

Teja, T.B., Sekar, M., Pallavi, T., Mettu, S., Murthy, T., Mat Rani, N.N.I., Bramhachari, P.V., Bonam, S.R., Role of artificial neural networks in pharmaceutical sciences. *J. Young Pharm.*, *14*, 1, 6-14, 2022.

Prisciandaro, E., Sedda, G., Cara, A., Diotti, C., Spaggiari, L., Bertolaccini, L., Artificial neural networks in lung cancer research: A narrative review. *J. Clin. Med.*, *12*, 3, 880, 2023.

Dube, F., Nzimande, N., Muzindutsi, P.-F., Application of artificial neural networks in predicting financial distress in the jse financial services and manufacturing companies. *J. Sustain. Finance Invest.*, *13*, 1, 723–743, 2023.

Lennox, B., Montague, G.A., Frith, A.M., Gent, C., Bevan, V., Industrial application of neural networks—an investigation. *J. Process Control*, *11*, 5, 497–507, 2001.

Sharifani, K. and Amini, M., Machine learning and deep learning: A review of methods and applications. *World Inf. Technol. Eng. J.*, *10*, 07, 3897–3904, 2023.

Leberruyer, N., Bruch, J., Ahlskog, M., Afshar, S., Toward zero defect manufacturing with the support of artificial intelligence—insights from an industrial application. *Comput. Ind.*, *147*, 103877, 2023.

Bezrucav, S.-O., Mandischer, N., Corves, B., Artificial intelligence task planning of cooperating low-cost mobile manipulators: A case study on a fully autonomous manufacturing application. *Procedia Comput. Sci.*, *217*, 306–315, 2023.

Tradecloud, Ml in the manufacturing process [https://https://www. tradecloud1.com/en/ai-case-study-4-machine-learning-in-the-manufacturing-process/ [Accessed: (13/08/2023)]], 2023.

3

Artificial Neural Network–Based Sliding Mode Controller for a Class of Nonlinear System

Sheetla Prasad[1], Rammurti Meena[2] and Vipin Chandra Pal[2*]

[1]*Department of Electrical, Electronics and Communication Engineering, Galgotias University, Greater Noida, Uttar Pradesh, India*
[2]*Department of Instrument and Control Engineering, NIT Silchar, Silchar, Assam, India*

Abstract

An artificial neural network (ANN) can easily detect and adapt variations in complex and nonlinear systems. This ability enhances the robustness of the estimation model. On the other hand, the sliding mode controller (SMC) ensures that the system stays on a predefined sliding surface, making it less sensitive to uncertainties, disturbances, and nonlinear changes. Thus, an ANN observer–based SMC is proposed and analyzed for a category of nonlinear systems. The ANN observer is designed to estimate unknown states and unknown disturbances. The observer is developed using the structure of a neural network comprising three layers. The ANN observer precisely estimated the states, and the estimation error converges to 0. The SMC controller is developed based on an ANN observer, and convergence of the law is also demonstrated through the Lyapunov theorem. The presented law improves the system's performance. The effectiveness of the controller is demonstrated through simulations on a single-link robot dynamical model.

Keywords: Artificial neural network (ANN) observer, sliding mode controller (SMC)

[*]*Corresponding author:* vipin.vchandra@gmail.com

Arindam Mondal and Souvik Ganguli (eds.) Controller Design for Industrial Applications, (41–54)

3.1 Introduction

In a nonlinear system, the system's response is not directly proportional to the input it receives. A controller monitors the controlled process and minimizes the error signal that arrives through the feedback path. Controller design is essential in order to predict the system output for a nonlinear system. The design is very challenging and difficult, particularly when the process is dynamic in nature. SMC is a method that is used in modern electrical and automated systems to predict future value. SMC is a nonlinear approach that changes the behavior of the system by applying a perturbation that compels the system to react in a linear fashion and, hence, can be predicted correctly. It has been proved that the SMC method is robust as it is insensitive to system uncertainty. Due to its effectiveness, SMC has been widely used in many applications. SMC can be realized using an artificial neural network (ANN).

A networked control system (NCS) is a setup that communicates and controls remotely with the plant through independent paths. Introducing a communication channel in a network brings many challenges, particularly the network delay, which makes the system unstable. Network delay may be analyzed using a fixed delay, a random delay, or a delay with a known probability [1]. In [2], an aerial work platform is controlled using a sliding mode controller that employs neural networks (NNs). In [3], a robust SMC is presented for a nonlinear discrete-time delayed system, employing NNs with a delay that is fixed. The authors claim that the SMC technique ensures the trajectory of the system state aligns with the designed sliding surface. An SMC method based on NN for a hysteresis nonlinear system with a backlash-like model is discussed in [4]. The NN-based SMC controller does not require tracking error. Adaptive control [5] can be used instead of SMC, but the adaptive controller offers more accuracy than the latter to compensate for the error and disturbance that arise due to external reasons. Observer-based sliding mode control for discrete delayed nonlinear systems has been reported in [6]. Observer-based SMC for state-delayed nonlinear systems has been proposed in [7]. The approach diminishes the need for accessible state variables for control purposes.

In [8], a method employing fuzzy multiple hidden layer neural SMC is presented. The SMC reduces the reference tracking error and eventually goes to zero, and the fuzzy NN eliminates the chattering problem. The multiple feedback loop is introduced with a NN to estimate the unknown nonlinear terms in the dynamics. In [9], the robust decoupled SMC and adaptive feedback linearization are combined with optimum fuzzy logic to eliminate a nonlinear underactuated system's external disturbance and uncertainty. The robust nonlinear super twisting algorithm SMC is proposed to eliminate the effects of parametric uncertainties.

External disturbances affecting a single-phase voltage source inverter interfaced with decentralized energy resources are discussed in [10]. In [11], the SMC is combined with model-free adaptive control using a data-driven control algorithm. The author introduces a novel sliding variable for robust stability and vanishes the error in tracking despite the existence of external influences of a discrete-time nonlinear system.

The SMC is capable of eliminating the negative impact of uncertainties and nonlinear change in states [12]. However, SMC does not require an accurate model of linear and nonlinear systems and is less sensitive to nonlinearities compared to some other control techniques. As a result, SMC is also utilized effectively, as demonstrated in [13], even in the presence of bad data vulnerability in the system due to advanced measuring devices such as sensors. The chattering issues are suppressed via proper optimization of the limiting terms through any suitable optimization techniques, such as linear matrix inequality (LMI) and genetic algorithms [14]. Hence, chattering suppression improves the practical implementation of the SMC strategy.

The NN has basic application in the field of prediction and mitigation strategy. The NN-based predictor can excel in estimating any type of linear and nonlinear systems. In [15], the authors converted an infinite-dimensional nonlinear system into a finite linear system via the application of an NN observer. The uncertain nonlinear trajectories are predicted and regulated by the application of NN observer and LMI, as given in [16]. The layers in the NN are used to balance between accuracy and convergence speed. To predict real-time trajectories, a multilayer NN observer is developed and used to regulate the system performance, as demonstrated in [17]. An iterative-based learning algorithm is developed to predict nonlinear system trajectories with a faster convergence rate, as claimed in [18]. Furthermore, an interval-based NN adaptive observer is developed to predict a higher degree of variations in a nonlinear system and given a guaranteed convergence, as mentioned in [19]. Thus, NN observers can be seamlessly integrated with robust as well as optimal control strategies in the literature.

By integrating ANN-based estimation with SMC control, the overall control system can benefit from the improved accuracy of the estimated states, leading to enhanced control performance. As a result, it enables an adaptive nature law to allow the system to adapt to changing conditions and uncertainties, improving overall system performance and stability.

The adaptive nature of both ANN-based estimation and SMC control is effectively utilized in the available studies [20–23]. Here, the authors used an NN-based observer to predict sensitive variables, and after these, estimated variables are utilized in SMC law in such a way that system trajectories are free from any type of system vulnerabilities. To mitigate the demerits and enhance the merits of SMC and NN observers, respectively, both are integrated together to propose a robust law for robot manipulator–based nonlinear systems and achieve better performance [24]. Thus, the combined approach is versatile and applied to an extensive range of systems, including those with complex and unknown dynamics.

As discussed in the above paragraphs, the SMC is designed using estimated states of the system through NN. In this work, the NN observer has been designed for nonlinear networked systems, and SMC has been designed for tracking control in the presence of time-varying disturbance. The accuracy of intelligent observer and SMC performance are shown in the simulation result, and the result has been compared with the existing one.

The paper is structured as follows: Section 3.2 outlines the problem formulation; section 3.3 details the design of the ANN structure, followed by the ANN observer–based sliding mode control law approach discussed in section 3.4. In section 3.5, the simulation result is discussed, followed by the conclusion in section 3.6.

3.2 Problem Formulation

Nonlinear systems often have complex and intricate dynamics that can be challenging to model accurately. The nonlinear system has several limitations, such as multiple solutions, limited range of stability, large computation, generalization issues, and difficulty to control. However, consider a nonlinear networked system as follows:

$$\begin{cases} \dot{p}_1 = p_2(t) \\ \dot{p}_2 = p_3(t) \\ \vdots \\ \dot{p}_n = f(p(t)) + g(p(t))u(t) + q(t) \\ o = p_1(t) \end{cases} \tag{3.1}$$

where $p(t) = \begin{bmatrix} p_1(t) & p_2(t) & \cdots p_n(t) \end{bmatrix} \in R^n, u(t) \in R, o(t) \in R$ are the state, the control action, and the response of the plant, respectively; unknown disturbance $q(t)$ with upper bound q_d;$f(p)$ and $g(p)$ are unknown smooth functions.

The above equation is expressed as

$$\begin{aligned} \dot{p} &= M_s p + N[f_0(p) + g(p)u(t) + q \\ o &= v^T p \end{aligned} \tag{3.2}$$

where

$$M_s = \begin{bmatrix} 0 & 1 & & & \\ & \ddots & \ddots & & \\ & & \ddots & \ddots & \\ & & & 0 & 1 \\ -m_n & \cdots & \cdots & -m_2 & -m_1 \end{bmatrix}, N = \begin{bmatrix} 0 \\ \vdots \\ \vdots \\ 0 \\ 1 \end{bmatrix}, v = \begin{bmatrix} 1 \\ 0 \\ \vdots \\ \vdots \\ 0 \end{bmatrix}$$

Moreover, $f_o(p) = f(p) + \sum_{i=1}^{n} m_{n-i+1} p_i$ for M_s is stable with the consideration of parameters m_i. For stabilization of the above system, the following Lyapunov condition must be satisfied $M_s^T I_s + I_s M_s = -G$ for positive definite matrices I_s.

The main contributions of this work are outlined as follows:

(i) To develop an ANN-based observer having the efficiency of estimating the actual state with the proper control input u(t) in the presence of bounded time-varying disturbances.
(ii) The application of the Lyapunov stability theorem has been employed to verify the closed-loop stability of the proposed system.
(iii) The simulation results demonstrate the effectiveness of the proposed ANN observer–based sliding mode controller.

The desired trajectory vector is given as

$$p_d(t) = \begin{bmatrix} o_d(t) & o_d(t) & \ldots & o_d^{(n-1)}(t) \end{bmatrix} \tag{3.3}$$

In the accomplishment of the desired objective, the following assumptions have been considered as (A_1) the continuous presence of $p_d(t)$ for any time of measurement with its bounded characteristics $p_d(t)$ and $g(p)>0$ [2].

3.3 Artificial Neural Network Structure

The most effective strategy for identifying any unknown parameter in unknown dynamics is the use of ANN. As observed, the NN observer can be seamlessly integrated with a robust SMC control strategy. In a three-layer ANN, each input is linked to a node within the hidden layer through a weight, and similarly, the hidden node is connected to the output through another weight. A three-layer feed-forward ANN output is represented as

$$\hat{c}_i = \sum_{q=1}^{l_2} \left[u_{qr}\sigma \left[\sum_{r=1}^{l_1} h_{sr} i_s + \varepsilon_{vs} \right] \right]_{s=1,2,3,\ldots\ldots\ldots l_3} \tag{3.4}$$

where s is the input, and r and q are the nodes of the hidden and output layers, respectively. l_1, l_2, l_3 indicates the count of neurons in the output, hidden, and input layers. The dimension of the input and output layers of the ANN structure is $b = \left[i_1 i_2 \ldots\ldots i_{l_3} \right]^T, \hat{c} = \left[\hat{c}_1 \hat{c}_2 \ldots\ldots \hat{c}_{l_1} \right]$. h_{sr} and u_{qr} represents the weight of the node, and weight matrices are $h^T = \left[h_{sr} \right], u^T = \left[u_{qr} \right]$.

The threshold offset express as ε_{vi} and its dimension is $\varepsilon = \left[\varepsilon_1 \varepsilon_2 \ldots\ldots \varepsilon_{N_2} \right]$. The activation function $\sigma(\bullet)$ is represented as a sigmoidal function and express as

$$\sigma(b_r) = \left[\frac{1}{1+e^{-2a_r}} - 1 \right] \tag{3.5}$$

If $\varepsilon_v = 1$ is introduced in the input vector, then the ANN structure can be written as

$$c(i,t) = u^T \sigma(h^T b) \tag{3.6}$$

The modified expression of Equation 3.2 can be expressed as

$$\begin{aligned} \dot{\hat{P}} &= M_s \hat{p}(t) + N[f_0(p) + \hat{c}(p)u(t) + q] + R(o(t) - \hat{o}(t)) \\ \hat{o} &= v^T \hat{p}(t) \end{aligned} \tag{3.7}$$

where $\hat{p}$ indicates as state variable of an observer and $\hat{o}$ is output variable.

If gain of the observer is selected such that $L \in \Re^{n \times m}$ and observer matrix $(M - RO)$ becomes Hurwitz matrix, then Equation 3.6 can be expressed as

$$\hat{c}(\hat{i}, t) = \hat{u}^T \sigma(\hat{h}^T \hat{b}) \tag{3.8}$$

where $(\hat{c})$ represents the estimated variable.

3.4 Neural Network Observer–Based SMC

To overcome the demerits with the integration of the merits of SMC and NN observer respectively, both are integrated together to propose a robust law for a nonlinear system for better performance and accuracy. The first NN observer is used to predict states, and then these predicted states are utilized in the design of SMC law. The NN observer has the ability to diminish nonlinearities and system disturbances accurately. Then, let us consider NN observer to estimate the nonlinear disturbances as

$$\dot{\hat{p}} = A\hat{p} + g(p)\begin{bmatrix} \hat{f}(\hat{p}) + \hat{g}(\hat{p})u(t) \\ +qsign(y(t) - C\hat{p}) \end{bmatrix} + L(y(t) - C\hat{p}) \tag{3.9}$$

where variables $\hat{f}(\hat{p}), \hat{g}(\hat{p})$ and $\hat{p}$ are estimated nonlinear functions and system states, respectively. The term L and $d(t)$ is known as observer gain and disturbance uncertainties, respectively. The observer gain L is obtained at the polynomial characteristic equation of $(A - LC)$ to be a Hurwitz. Then, observation error dynamics $e = p - \hat{p}$ is written as

$$\dot{e}(t) = (A - LC)e(t) + g(p)\begin{bmatrix}(f(p) - \hat{f}(\hat{p})) + q(1 - sign(Ce)) \\ +(g(p) - \hat{g}(\hat{p}))u(t)\end{bmatrix}$$

$$\dot{e}(t) = (A - LC)e(t) + g(p)\begin{bmatrix}\tilde{f}(\tilde{p}) + q(1 - sign(Ce(t))) \\ +\tilde{g}(\tilde{p}))u(t)\end{bmatrix} \tag{3.10}$$

where $\tilde{f}(\tilde{p}) = f(p) - \hat{f}(\hat{p}),\ \tilde{g}(\tilde{p}) = g(p) - \hat{g}(\hat{p})$.

The matrix (A, C) is assumed as an observable for a stable system matrix A. Hence, observable gain L is obtained using the Lyapunov equation for a positive matrix P_1 as

$$(A - LC)P_1 + P_1(A - LC)^T = -Q_2 \tag{3.11}$$

where matrix $Q_2 = Q_2^T > 0$.

The sliding mode controller provides guaranteed robust control through the switching capability of control effort on a predefined surface against the system uncertainties. The system uncertainties could be classified as internal and external uncertainties. The switching condition is divided into the reaching and sliding phases. However, sliding mode controller performance is dependent on how to select the switching surface [25]. Hence, choosing the switching surface as

$$\rho = \kappa\left(p - \hat{p}\right) + \left(\dot{p} - \dot{\hat{p}}\right) \tag{3.12}$$

where the term κ is a positive constant parameter.

Apply sliding mode controller reaching and sliding phase conditions in Equation 3.12 above, and after simplification, it gives

$$u(t) = \left(\frac{1}{g(p)}\right)\begin{Bmatrix}-\kappa(y(t) + \mu(p - \hat{p}) - \dot{\hat{p}}) - \hat{f}(\hat{p}) \\ +q(1 - sign(C(p - \hat{p}))) - \eta sign(\rho) - \lambda(p - \hat{p})\end{Bmatrix} \tag{3.13}$$

where η, μ, λ are positive constant tuning parameters.

Proof: Consider a Lyapunov function as

$$\varepsilon = 0.5e^T Pe \tag{3.14}$$

The derivative with respect to time:

$$\dot{\varepsilon} = 0.5e^T P\dot{e} + 0.5\dot{e}^T Pe \tag{3.15}$$

Using Equations 3.10, 3.11, and 3.13, it gives

$$\dot{\varepsilon} = e^T \tilde{g}(\tilde{p})\begin{pmatrix} -Q - \kappa(y(t) + \mu(p - \hat{p}) - \dot{\hat{p}}) \\ -\hat{f}(\hat{p}) - \eta sign(\rho) - \lambda(p - \hat{p}) \end{pmatrix}$$

Hence, $\dot{\varepsilon} < 0$ and converges asymptotically.

3.5 Simulation and Demonstrations

The nonlinear system has several limitations, such as multiple solutions, limited range of stability, large computation, generalization issues, and difficulty to control. However, in this simulation section, a single link robot nonlinear system is taken to verify proposed law efficacy. The plant observer model is effectively verified in this section using a single-link robot dynamical model. The model is stated using thc following equation:

$$\begin{cases} M\ddot{\theta} + \frac{1}{2}mgl\sin\theta = u + q \\ o = \theta \end{cases} \tag{3.16}$$

where m, l, θ, M, g represents the mass, length, angle, and moment of inertia of the link, respectively. The input torque and external disturbance are indicated by u, q.

The state-space representation of the model equation is

$$\begin{cases} \dot{p} = \begin{bmatrix} 0 & 1 \\ -1 & -2 \end{bmatrix} p + \begin{bmatrix} 0 \\ 1 \end{bmatrix} [f_0(p) + g(p)u + q] \\ o = p_1 \end{cases} \tag{3.17}$$

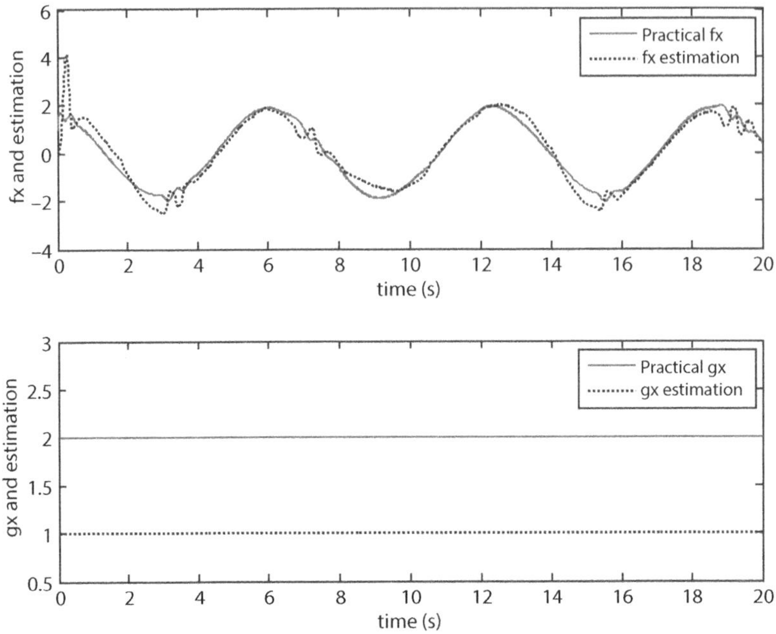

Figure 3.1 Accuracy representation of proposed law.

where $p_1 = \theta, p_2 = \dot{\theta}, q = \frac{1}{2}\sin(0.5\pi t)$ and $f_0(p) = -\frac{1}{2}mgl\sin(p_1/(M + p_1 + 2p_2)), g(p) = 1/M$. The following parameters of the model are considered for the simulation as $M = 0.5, l = 0.4, m = 0.4, g = 9.8$.

First, to test the efficacy and its error-tracking capability, the practical and predicted function $f_0(p)$ and $g(p)$ is observed using the proposed control law, as given in Figure 3.1.

It is evident that the robot function value (first subplot of Figure 3.1) is predicted accurately even in a 50% lower value of $g(p)$, as represented in the second subfigure of Figure 3.1.

However, the proposed NN observer–based SMC performs satisfactorily even in the presence of change in system constant terms. Now, the robot state trajectory x_1(position) and x_2(velocity) is given in Figure 3.2 against lower mass change. It is detected that the observer state $x_1$and x_2 almost coincides with the estimated state. Figure 3.2 indicates that the observer also accurately estimates the $f_0(p)$ and $g(p)$. Position and velocity signals coincide in Figure 3.2 and they track the desired state almost entirely as compared to existing adaptive controllers [5]. Figure 3.3 contains the required control signal. It is observed that the control signal has

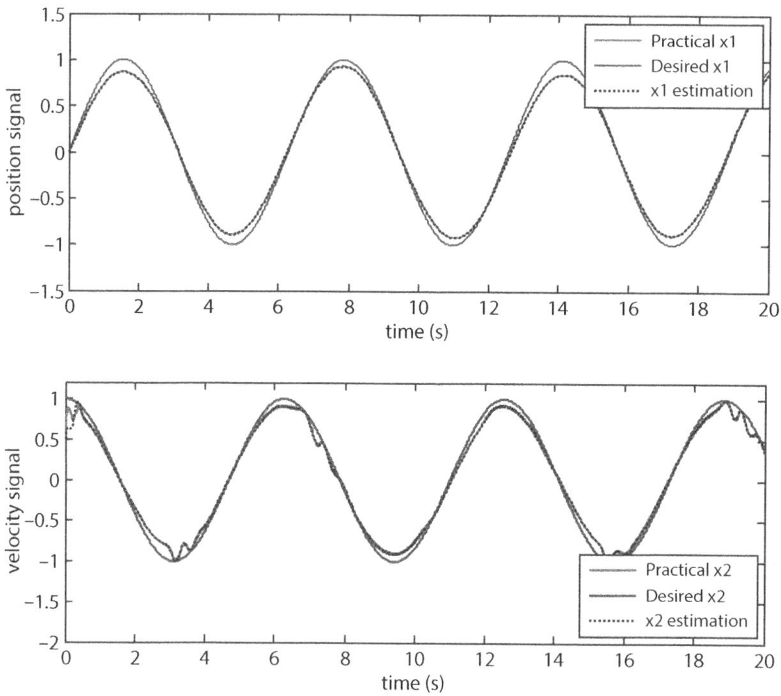

Figure 3.2 State trajectories with proposed law.

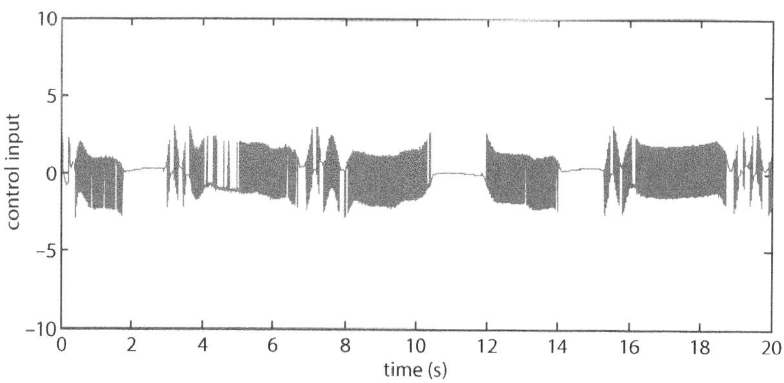

Figure 3.3 Control signal.

chattering due to a change in dynamics. Thus, the proposed law, shown in Figure 3.3, has the capability in terms of adaptive and efficacy even in the presence of robot nonlinearities.

Based on the above simulation results, the error indices, such as ISE, IAE, and ITAE, are obtained within the simulated time window. These error performances in position are 7.9079, 11.3147, and 112.5308, respectively. These terms are relatively larger due to large variations in terms of sinusoidal way.

3.6 Conclusion

In this paper, a robust SMC is designed based on an ANN observer. The three-layer feed-forward ANN observer is realized to assess the nonlinear system's unmatched states and unmatched interferences. The proposed SMC controller is implemented on a single-link robot dynamical model in the presence of disturbances. The efficacy and performance of the controller are assessed and confirmed by analyzing the outcomes of simulation experiments. It exhibits that tracking performance is significantly improved using this proposed controller.

References

1. Liu, F.C. and Yao, Y., Modeling and analysis of networked control systems using hidden Markov models. *International Conference on Machine Learning and Cybernetics*, vol. 2, pp. 928–931, 2005.
2. Hu, H., Cai, N., Cui, L., Ren, Y., Yu, W., A neural network–based sliding mode controller of folding-boom aerial work platform. *Adv. Mech. Eng.*, 9, 10, 168781401772087, 2017.
3. Goyal, V., Deolia, V., Sharma, T., Robust Sliding Mode Control for Nonlinear Discrete-Time Delayed Systems Based on Neural Network. *Intell. Control Autom.*, 06, 01, 75–83, 2015.
4. Liu, R. and Gao, X., Neural Network Identification and Sliding Mode Control for Hysteresis Nonlinear System with Backlash-Like Model. *Complexity*, 2019, 1–10, 2019.
5. Gao, X. and Liu, R., Multiscale Chebyshev Neural Network Identification and Adaptive Control for Backlash-Like Hysteresis System. *Complexity*, 2018, 1–9, 2018.
6. Guan, X., Hu, J., Cui, Y., Xu, L., Observer-based sliding mode control for discrete nonlinear systems with packet losses: an event-triggered method. *Syst. Sci. Control Eng.*, 8, 1, 175–188, 2020.
7. Niu, Y., Lam, J., Wang, X., Ho, D., Observer-based sliding mode control for nonlinear state-delayed systems. *Int. J. Syst. Sci.*, 35, 2, 139–150, 2004.

8. Zhuo, J., An, C., Fei, J., Fuzzy Multiple Hidden Layer Neural Sliding Mode Control of Active Power Filter With Multiple Feedback Loop. *IEEE Access*, 9, 114294–114307, 2021.
9. Mahmoodabadi, M. and Soleymani, T., Optimum fuzzy combination of robust decoupled sliding mode and adaptive feedback linearization controllers for uncertain under-actuated nonlinear systems. *Chin. J. Phys.*, 64, 241–250, 2020.
10. Barzegar-Kalashani, M., Tousi, B., Mahmud, M., Farhadi-Kangarlu, M., Robust nonlinear sliding mode controllers for single-phase inverter interfaced distributed energy resources based on super twisting algorithms. *ISA Trans.*, 123, 61–75, 2021.
11. Corradini, M., A Robust Sliding-Mode Based Data-Driven Model-Free Adaptive Controller. *IEEE Control Syst. Lett.*, 6, 421–427, 2021.
12. Prasad, S., Purwar, S., Kishor, N., Non-linear sliding mode load frequency control in multi-area power system. *Control Eng. Pract.*, 61, 81–92, 2017.
13. Chugh, M. and Prasad, S., Mitigation of False Data Injection on Frequency Regulation using Non-linear Sliding Mode Control. *2019 International Conference on Power Electronics, Control and Automation (ICPECA)*, pp. 1–6, New Delhi, India, 2019.
14. Zhang, M., Song, G., Mao, J., Wang, F., Zhou, J., Song, A., Chattering suppression and hydrodynamic disturbance estimation of underwater manipulators using adaptive fuzzy sliding mode control. *Trans. Inst. Meas. Control*, 46, 1, 155–166, 2024.
15. Meng, Y., Huang, J., Qiu, Y., Koopman operator learning using invertible neural networks. *J. Comput. Phys.*, 501, 112795, 2024.
16. Chen, S., Cai, S., Chen, T., Xu, C., Chu, J., Neural Observer with Lyapunov Stability Guarantee for Uncertain Nonlinear Systems. *IEEE Trans. Neural Netw. Learn. Syst.*, 35, 8, 1–15, August 2024, doi: 10.1109/TNNLS.2023.3262820.
17. Yin, Q., Zhang, H., Mu, Q., Yang, J., Ma, Q., Multilayer-neural-network observer with compensator and command-filter-based adaptive backstepping tracking control of switched nonlinear systems. *J. Franklin Inst.*, 360, 4, 2976–3000, 2023.
18. Lv, Y., Ren, X., Tian, J., Zhao, X., Inverse-model-based iterative learning control for unknown MIMO nonlinear system with neural network. *Neurocomputing*, 519, 187–193, 2023.
19. Yi, Z., Xie, W., Liu, L., Xu, B., A neural network adaptive interval observer design for nonlinear systems. *IET Control Theory Appl.*, 16, 6, 615–624, 2022.
20. Kumar, M., Prasad, S., Ansari, M.R., Mohapatra, B., Resonance attacks detection and mitigation control scheme on frequency regulation in multi-area smart grid. *Int. J. Control*, 96, 9, 2212–2229, 2023.
21. Prasad, S. and Ansari, M.R., Frequency regulation using neural network observer based controller in power system. *Control Eng. Pract.*, 102, 104571, 2020.

22. Prasad, S., Chapter 9 - Robust frequency regulation against cyberattack uncertainties in modern power system grids, in: *Advanced Frequency Regulation Strategies in Renewable-Dominated Power Systems*, S. Dhundhara, Y. Arya, R.C. Bansal (Eds.), pp. 181–214, Academic Press, United States, 2024.
23. Liu, X., Hu, P., Chen, Y., Adaptive Hierarchical Sliding Mode Control Based on Extended State Observer for Underactuated Robotic System. *Int. J. Control Autom. Syst.*, 22, 3, 1036–1049, 2024, https://doi.org/10.1007/s12555-022-0357-4.
24. Truong, T.N., Vo, A.T., Kang, H.-J., Neural network-based sliding mode controllers applied to robot manipulators: A review. *Neurocomputing*, 562, 126896, 2023.
25. Prasad, S., Purwar, S., Kishor, N., Load frequency regulation using observer based non-linear sliding mode control. *Int. J. Electr. Power Energy Syst.*, 104, 78–93, 2019.

4

Finite Control Set Model Predictive Control for Permanent Magnet Synchronous Motor Drives

Ravi Eswar Kodumur Meesala*, Phani Teja Bankupalli and Chinta Praveen Kumar

Department of Electrical and Electronics Engineering, SRM Institute of Science and Technology, Kattankulathur, Chennai, Tamil Nadu, India

Abstract

For many industry drive utilizations, the permanent magnet synchronous motor (PMSM) offers an attractive choice owing to its key factors: high power density and efficiency. In situations that require high dynamic motor performance, commonly used approaches include "field-oriented control (FOC)" or "direct torque control (DTC)." With the recent emergence of powerful and fast processors, computationally efficient model predictive control (MPC) approaches have gained popularity in motor drives. Due to the evident advantages, such as strong dynamic control, simple incorporation of nonlinearities and constraints, and multivariable operation, MPC is a viable method in electric drives. In this chapter, the optimal control concepts of finite control set MPC (FCSMPC) for the PMSM drive are discussed with the support of discrete mathematical models. The constraints in FCSMPC for motor drives are presented, and considering this, the currently available research on modified FCSMPC is discussed. Finally, the real-time implementation of FCSMPC for motor drives is deliberated.

***Keywords*:** PMSM, DTC, FOC, MPC, FCSMPC

**Corresponding author*: ravieswm@srmist.edu.in

Arindam Mondal and Souvik Ganguli (eds.) Controller Design for Industrial Applications, (55–88)

4.1 Introduction

The conventional form of traditional internal combustion engines (ICEs) to electric vehicles (EVs) not only reduces the emissions but also supports renewable energy sources (RESs), which contribute to a more supportable energy ecosystem as the electricity grid uses RESs like solar and wind more frequently. Therefore, aligning RES with EV adoption enhances the overall environmental impact and reduces global warming; in addition, it creates a complete method in the direction of reducing the carbon footmark of the transport sector. This transition consequently not only addresses the environmental challenges posed by traditional vehicles but also fosters a more resilient and sustainable energy future. The motor drive system is the heart of an EV, and it mainly consists of an electric motor, power electronic converters, and a controller. The present state of the art in EV drive system is provided below.

4.1.1 Literature Survey on EV Motors

In the early stages of the 1880s, there was a frequent use of DC (direct current) motors for EVs. However, there are challenges in terms of size, weight, and maintenance requirements. Therefore, during the period from 2010 to 2020, the EV industry has transformed from the initial use of DC motors to the predominant adoption of AC (alternating current) motors and also addresses the challenges of high-performance and cost-effective propulsion systems [1, 2]. The utilization of AC motors for EVs marked a significant advancement for the industry by recognizing the need for efficiency and performance at high speeds and enabling regenerative braking systems. Therefore, currently, the EV industry uses permanent magnet synchronous motors (PMSMs), induction motors (IMs), and switched reluctance motors (SRMs) for the upliftment of the EV Industry [1, 3]. Furthermore, we will discuss in detail PMSM, IM, and SRM drives.

The PMSM is classified into two distinctive configurations as indicated in Figure 4.1: interior PMSM (IPMSM) and surface PMSM (SPMSM). Each motor type has its own characteristics, influencing the overall performance and suitability for various applications especially within the context of the EVs industry. In applications where dynamic performance is important, such as in EVs requiring swift acceleration and deceleration, the configuration of IPMSM is preferred where the permanent magnets are fixed within the rotor core. The main advantage of this design is the reduction in rotor inertia; as a result, it optimizes the motor's efficiency, responsiveness at dynamic performance, and high torque density. Therefore, IPMSM configuration emerges as an advantageous choice in EV [4]. The SPMSM structure is attained by placing the permanent magnets on the surface of the rotor directly facing the stator. The SPMSM configuration is also prominent for impressive efficiency, high power density, and distinguished regenerative braking capabilities. This future is important in the EV used for optimizing energy efficiency [1, 5, 6]. The SRM drive is economical for implementation in EVs. However, it offers poor control performance and low torque density, which is the main concern [7].

The IMs are primarily categorized into squirrel cage (SC) and wound rotor (WR). Renowned for their robust and simple construction, SCIMs are highly reliable and require low maintenance. They find extensive use in EVs, particularly in applications prioritizing cost-effectiveness and reliability. However, SCIMs exhibit overall low efficiency. On the other hand, WRIMs have access to rotor windings which provide enhanced control over starting torque and speed. WRIMs have a more complex construction and require additional maintenance. As a result, WRIMs are less common in EVs and are often more suited for industrial applications [8].

The PMSM, in particular, stands out for its economic attractiveness to automakers, requiring less electrical energy from the battery for equivalent mechanical output. Thus, this will enable a reduction in battery costs without compromising vehicle range. This ongoing evolution underscores the industry's commitment to advancing technology for more efficient and sustainable EVs. Both types of PMSM mentioned in Figure 4.1 rely on costly permanent magnets, contributing to the overall manufacturing expenses in EVs. In light of this, the EV industry is actively seeking methods to reduce permanent magnet costs without compromising performance.

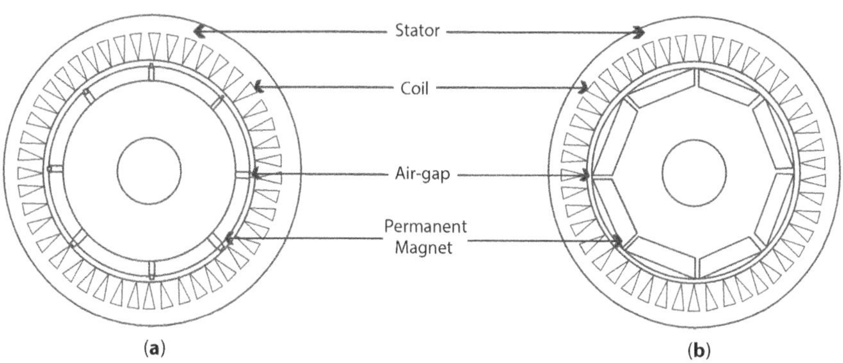

Figure 4.1 PMSM models (a) surface magnets and (b) interior magnets.

4.1.2 Literature Survey on Motor Control

Specifically focusing on PMSMs, a variety of control strategies have emerged to efficiently manage electrical parameters, thereby optimizing overall motor performance. Field-oriented control (FOC), also known as vector control, is widely employed to decouple torque and flux components, enabling precise control in the d-q frame. Direct torque control (DTC) stands out for its ability to control torque and flux without coordinate transformations, ensuring rapid dynamic response. Furthermore, the recent trending advanced control strategy named model predictive control (MPC) offers predicting the future system behavior and considering control actions accordingly for optimal performance. A brief view of all these control techniques is given below.

4.1.2.1 FOC Operation

The concept of FOC was introduced in the early 1970s by K. Hasse and F. Blaschke. The FOC garnered more recognition and adoption in the 1980s and 1990s. It has proven to be a flexible and effective control strategy for PMSM in enhancing performance. In the FOC of the PMSM drive shown in Figure 4.2, the dynamic behavior is enhanced by employing specific control actions as follows: *i.* shifting stator currents to a synchronous rotating frame and *ii.* alignment of flux-linkage space vector with a d-axis reference frame [9–12].

Through the implementation of these control measures, the motor torque is expressed as a function of independent flux- and torque-producing currents. Thus, this control achieves high dynamic performance in PMSM drive which is similar to a DC motor drive by maintaining constant flux

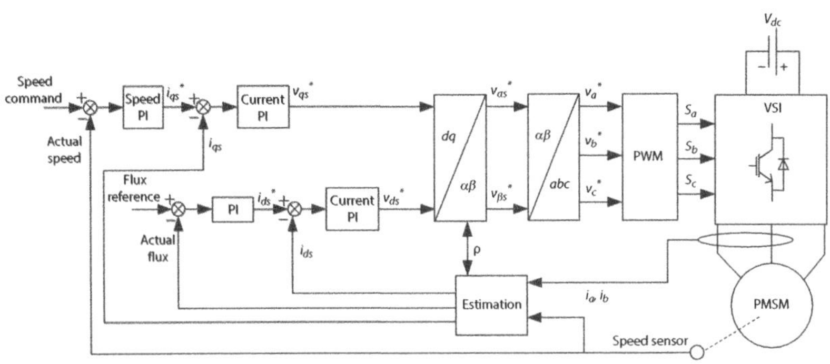

Figure 4.2 Block diagram of PMSM drives using FOC.

component of current while varying the torque component of current. The control actions involved in FOC include the utilization of proportional-integral (PI) controllers to regulate the torque and flux components. However, it is important to acknowledge certain limitations: *i.* dependent on the accurate determination of the flux linkage angle (ρ), and *ii.* need for additional current PI regulators and coordinate frame transformations which creates cumbersome for motor drive operation.

4.1.2.2 DTC Operation

The DTC was introduced in the 1980s as a control strategy for electric motors. The performance evaluation in DTC primarily focuses on torque and flux control. DTC operates without the need for coordinate transformations and current PI regulators, simplifying the control structure. DTC is often considered less sensitive to parameter variations. Figure 4.3 illustrates a block diagram for the DTC of the PMSM drive.

The DTC incorporates a predetermined look-up table drawing upon essential data such as stator flux (dλ) and torque (dT_m) errors. This process is complemented by the identification of the stator flux sector (S_n). The meticulously crafted look-up table serves as a guiding reference for selecting an optimal voltage vector (VV) throughout the control process. A notable feature is the direct control over torque and flux, enabled by applying appropriate VVs (V_s) to induce changes in stator flux and torque [13–15].

However, DTC does come with certain limitations in its implementation. Firstly, the use of hysteresis controllers in DTC introduces a potential constraint as it may require a high sampling frequency in a digital platform.

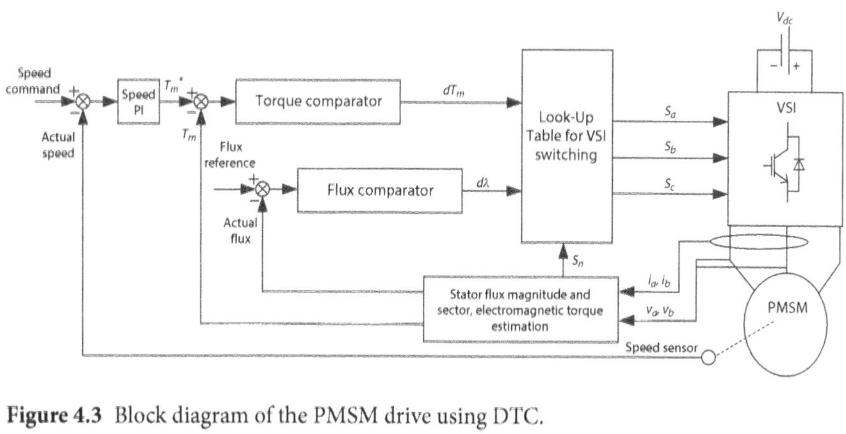

Figure 4.3 Block diagram of the PMSM drive using DTC.

Another limitation associated with DTC is variable switching frequency during operation. DTC may also exhibit large ripples in the flux, current, and torque responses during operation. These fluctuations can impact the smoothness and precision of control, potentially leading to undesirable effects such as increased audible noise and vibrations in the motor system. Poor flux regulation at lower speeds can result in performance issues, particularly in applications where low-speed operation is crucial. Recognizing these limitations, active research is currently underway in the field of DTC to address and overcome these challenges [16, 17].

4.1.2.3 MPC Operation

MPC emerged in the 1970s. This approach follows the use of dynamic models to predict the system's future behavior and optimize control actions. This control scheme found applications across various industries, including process control, automotive systems, robotics, and power electronics. In power electronics, MPC has been particularly influential in the control of converters and inverters since the 1980s.

Despite its computational complexity, MPC offers distinct advantages, such as handling multivariable systems, explicitly addressing constraints, and optimizing control inputs over a specified prediction horizon. In the context of power electronics, MPC made its debut in converter control. The evolution of MPC in this domain gained momentum with advancements in computational capabilities, notably with the widespread use of powerful microprocessors. These microprocessors play a pivotal role in enabling the real-time implementation of MPC algorithms in power electronic systems. This breakthrough marked a significant development, allowing MPC to become a practical and efficient control strategy for power electronics [18].

The various categorizations in MPC are [19]: "Continuous control set (CCS)" and "finite control set (FCS)." In the CCS scheme, the reference modulating signal is produced during its operation and undergoes pulse width modulating (PWM) generation for applying control pulses to inverter switches. Thus, the CCS scheme requires a modulating stage that facilitates fixed switching frequency operation. On the other side, the FCS scheme does not require any PWM stage and the control operation directly outputs discrete switching states of the inverter which is established from cost function optimization. The elementary model of the FCS-operated PMSM drive is presented in Figure 4.4. The prediction model is framed using discrete PMSM equations. From this, the prediction outputs are determined with the information of future inputs, present outputs from the system, and past inputs. The cost function comprises a number

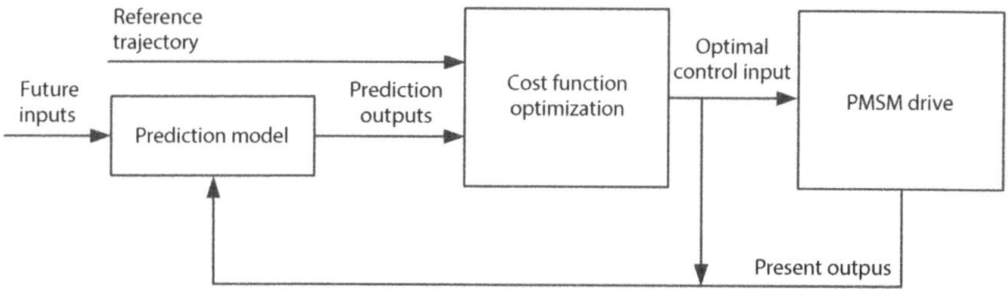

Figure 4.4 Basic block diagram of FCSMPC for the PMSM drive.

of constraints with suitable weights. The constraints are a function of their references and prediction outputs. The future input is considered for which the minimum cost function is pertained and applied to the PMSM drive.

4.1.2.3.1 Brief Discussion on Recent Advancements in FCSMPC Operation

In the FCSPTC method [20], the objectives are associated with torque and flux. Therefore, to improve the control performance and to achieve the weight factor-less operation, each objective is considered as a separate cost function. Thus, two cost functions for torque (G_T) and flux (G_F) are defined. These are computed for seven VVs of two-level VSI. Based on the error values of cost functions, the top three VVs representing errors in ascending order form a set. In both sets, the common VV is chosen as the optimal. Therefore, this VV is applied to the PMSM drive to reduce torque and flux error during the subsequent sample period. Hence, the optimization approach does not depend on the weight factor as the conventional FCSPTC method. If a VV set is considered with the top two VVs, then there is a chance of not achieving the common vector for both torque and flux and this impacts the performance of both torque and flux. To overcome this, the research [20] considers more than two VVs and it will consider only the top three VVs to enhance computational efficiency.

The new control strategy called "modified parallel structure P-FCSPTC" was adopted [21] to overcome the challenge of complex tuning which is associated with the weight factor in the conventional FCSPTC method. The existing P-PTC has certain problems as it requires additional parameters for tuning and lacks sufficient theoretical support. Due to these limitations with existing P-PTC, a modified P-PTC has been developed. This version eliminates the need for tuning extra parameters and it is specifically designed for the PMSM drive. The designed P-PTC procedure first begins with the construction of two parallel cost functions named torque and flux. These designed functions can independently identify three VV sets, each one of them having an optimized impact on torque and flux. After that, the operation involves three cases based on the intersections between the torque and flux control sets to select the final optimal VV.

The general FCSPTC method [22] is adapted to operate a PMSM drive with multi-level VSI. To address the limitation of the weight factor, two separate cost functions are evaluated sequentially. This research firstly assesses the torque cost function with 19 VVs for identifying the VV with the lowest torque error and is also used to determine the sector number. Using the information of the sector, the five specific VVs are identified to evaluate the flux cost function. Among these five VVs, the VV with optimized flux cost function is finalized for the operation of the PMSM drive. Therefore, this research contributes to reducing ripples in both torque and flux, improving the steady-state response of the control system.

In the aspect of motor control, reducing ripples in torque and flux is crucial. The research [23] presents FCSPTC without a weight factor and the operation is parameter-robust. The torque and flux control are attained without the need for a hysteresis controller by predicting the reference VV in a stationary frame. The procedure of creating reference VV is simple and, significantly, it does not compromise the robustness of the control system. The optimal VV generation and switching actions are determined by switching tables. The switching table is a function of reference VV position in a sector. As a result, the control algorithm efficiently adapts to achieve the desired motor performance by utilizing the nearby discrete VV to predicted reference VV. To curtail the switching frequency of two-level VSI fed by the SPMSM drive and be able to improve switching efficiency, the redundant switching combinations of null VV are applied effectively. Therefore, the control scheme enhances the steady-state response of the SPMSM drive along with switching loss reduction.

The traditional FCSMPC has a limitation of applying a single VV for every control period. Therefore, this approach led to major torque variation, known as torque ripple. This affects the performance of the control system. Therefore, to enhance the steady-state performance, the research studies introduced the utilization of two VVs for a control period. However, there are challenges associated with a two-vector-based control scheme, such as the involvement of complexity and computational burden in selecting the optimal VV pair. Therefore, the research [24] investigates introducing three-vector-based FCSMPC for achieving a low complexity and precisely determines active VVs based on the predicted torque error and modified switching table. This method is significant in decreasing the computational complexity and simplifying the control algorithm. Also, the application of the duty cycle approach enhances steady-state performance and attains deadbeat torque and flux control.

The research [25] introduces a modified "Model Predictive Current Control (MPCC)" for the operation of a five-phase PMSM drive. The basic MPCC uses all possible VV control sets of a two-level, five-phase VSI to evaluate the cost function. Thus, the operation has the disadvantage of longer processing time due to extensive computation of prediction and evolving of the cost function. This demands a sample time increment for the control operation. As a result, it affects well the overall efficiency of the control system. Therefore, this research aims to minimize the number of calculations, enhancing the response of torque and flux. This method considers the variation of the stator current vector position and its magnitude for selecting an optimal VV. This research is implemented with novel criteria for preselecting the VVs. Therefore, the method of MPCC emphasizes the positive impact on motor drive performance with less computational complexity and effectively reduces ripple in torque and flux.

In this chapter, the organization is as follows: In section 4.1, a brief introduction and literature about present EV motors and control technologies are provided. Furthermore, the key recent advancements in FCSMPC operation are discussed in detail. In section 4.2, the dynamic mathematical models of the machine, i.e., PMSM drive are elucidated, and then discretized equations are framed for the analysis of MPC operation. In section 4.3, the basic optimal control concepts of FCSMPC for motor drives are discussed. The basic constraints in the FCSMPC operation of the motor drive are presented in section 4.4. Finally, in section 4.5, the real-time implementation of FCSMPC for motor drive operation is deliberated.

4.2 Mathematical Model of PMSM Drive

4.2.1 Dynamic Mathematical Model of PMSM

The three-phase supply system is extensively employed in various electric motors, including IMs and PMSMs. Dynamic modeling for PMSM is vital in the realm of vector control. This model is essential for understanding the dynamic behavior of PMSM across various operating conditions. Engineers rely on these models to design and fine-tune complicated control algorithms effectively. Moreover, these mathematic models play a vital role in optimizing system performance. By utilizing mathematical models, engineers can ensure the seamless integration of PMSM systems into larger applications. Figure 4.5(a) illustrates the distribution of electrical voltages

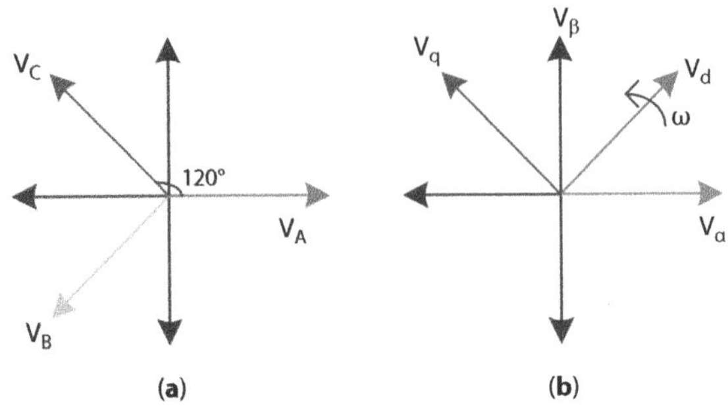

Figure 4.5 Three-phase system to two-phase system (Clark and Park transformations).

through the three-phase system where they are electrically spaced 120° apart within a 360° plane.

In dynamic modeling, the interdependence of three phases often complicates system modeling. In order to address this challenge and streamline the modeling process, strategic analysis is conducted in a two-phase coordinate system, as shown in Figure 4.5(b). Consequently, the three-phase PMSM electrical quantities, such as flux, current, and voltage, are shifted to a two-phase stationary reference frame, commonly denoted as the (α-β/d-q) reference frame. This transformation not only simplifies the modeling of the system but also enhances the understanding and analysis of its behavior in a more manageable and comprehensible manner.

The conversion of three-phase quantities into a two-coordinate system was introduced early to make analysis and control more manageable. Equations derived from Clark and Park transformations from the ABC frame facilitate the modeling of AC machines in a stationary (αβ)/rotating (dq) reference frame [26]. With the support of these coordinate transformations, the dynamic modeling of interior permanent magnet (IPM) machines within the rotor reference frame is expressed as follows (see Equations 4.1 to 4.4):

$$\begin{bmatrix} V_d \\ V_q \end{bmatrix} = R_s \begin{bmatrix} i_d \\ i_q \end{bmatrix} + \frac{d}{dt} \begin{bmatrix} \Psi_d \\ \Psi_q \end{bmatrix} + \omega_e \begin{bmatrix} -\Psi_q \\ \Psi_d \end{bmatrix} \tag{4.1}$$

$$\begin{bmatrix} \Psi_d \\ \Psi_q \end{bmatrix} = \begin{bmatrix} L_d & 0 \\ 0 & L_q \end{bmatrix} \begin{bmatrix} I_d \\ I_q \end{bmatrix} + \begin{bmatrix} \Psi_m \\ 0 \end{bmatrix} \tag{4.2}$$

$$T_e = \frac{3p}{2}\left(\Psi_m I_q - I_d I_q \left(L_q - L_d\right)\right) \tag{4.3}$$

$$T_e = T_l + J\frac{d\omega}{dt} \tag{4.4}$$

where the rotor flux, stator flux linkages, stator currents, and voltages are expressed as Ψ_m, Ψ_{dq}, i_{dq}, and V_{dq}, respectively. p is the number of pole pairs, R_s is the phase resistance, ω_e is the electrical angular speed (rad/s), T_e is the electromagnetic torque (N-m), and L_{dq} are the dq-axis inductances (H).

The desire to streamline the control process to reduce the intricacies associated with rotating coordinate frame transformations is gained by adopting a stationary reference frame model. This approach allows to operate without trigonometric functions and simplifies the system model analysis. Consequently, this contributes to a more straightforward and manageable control architecture [20].

The stationary frame model for an IPMSM in the αβ coordinate system can be described as follows (see Equations 4.5 to 4.8):

$$\vec{V}\alpha,\beta = \begin{bmatrix} V_\alpha \\ V_\beta \end{bmatrix} = R_s\begin{bmatrix} i_\alpha \\ i_\beta \end{bmatrix} + \frac{d}{dt}\left(\begin{bmatrix} \Psi_\alpha \\ \Psi_\beta \end{bmatrix}\right) \tag{4.5}$$

$$\begin{bmatrix} \Psi_\alpha \\ \Psi_\beta \end{bmatrix} = L_{\alpha\beta}\begin{bmatrix} i_\alpha \\ i_\beta \end{bmatrix} + \Psi_m\begin{bmatrix} \cos\theta_e \\ \sin\theta_e \end{bmatrix} \tag{4.6}$$

$$L_{\alpha\beta} = \begin{bmatrix} L_s - L_v\cos 2\theta_e & -L_v\sin 2\theta_e \\ -L_v\sin 2\theta_e & L_s + L_v\cos 2\theta_e \end{bmatrix} \tag{4.7}$$

$$T_e = \frac{3p}{2}\left(\Psi_\alpha i_\beta - \Psi_\beta i_\alpha\right) \tag{4.8}$$

where the rotor inductance (L_s) and inductance difference (L_v) are defined as $(L_d+L_q)/2$ and $(L_q-L_d)/2$, respectively. θ_e is the electrical rotor angle.

The SPMSM with a non-salient rotor configuration is characterized by the equality of the direct and quadrature axis inductances, denoted as $L_d = L_q = L_s$. In this specific configuration, the dynamic modeling of the PMSM can be expressed in the stationary reference frame. The stationary frame model for an SPMSM in the αβ coordinate system can be described as follows (see Equations 4.9 to 4.12) [20]:

$$\vec{V}\alpha,\beta = \begin{bmatrix} V_\alpha \\ V_\beta \end{bmatrix} = R_s \begin{bmatrix} i_\alpha \\ i_\beta \end{bmatrix} + \frac{d}{dt}\left(\begin{bmatrix} \Psi_\alpha \\ \Psi_\beta \end{bmatrix}\right) \tag{4.9}$$

$$\begin{bmatrix} \Psi_\alpha \\ \Psi_\beta \end{bmatrix} = L_{\alpha\beta}\begin{bmatrix} i_\alpha \\ i_\beta \end{bmatrix} + \Psi_m \begin{bmatrix} \cos\theta_e \\ \sin\theta_e \end{bmatrix} \tag{4.10}$$

$$L_{\alpha\beta} = \begin{bmatrix} L_s & 0 \\ 0 & L_s \end{bmatrix} \tag{4.11}$$

$$T_e = \frac{3p}{2}\left(\Psi_\alpha i_\beta - \Psi_\beta i_\alpha\right) \tag{4.12}$$

Considering the above analysis, the derivative form of state variables such as stator current and flux can be represented using Equations 4.13 and 4.14, respectively. This representation is beneficial when converting the continuous time model into a form suitable for discrete-time analysis. Here, the terms V_s and i_s represent the stator's voltage and current space vectors.

$$\frac{di_s}{dt} = \frac{V_s - R_s i_s - j\Psi_m \omega_e e^{j\theta_e}}{L_s} \tag{4.13}$$

$$\frac{d\Psi_s}{dt} = V_s - R_s i_s \tag{4.14}$$

With the use of Euler's method, the discretization of Equations 4.13 and 4.14 can be achieved. Euler's formula (4.15) is a generalized representation, denoted by the variable 'z', which describes the state of the system at both the previous instant 'k-1' and the present instant 'k' samples. In this method, 'T_s' represents the sample time, which is the time interval between consecutive samples engaged in the system.

$$\frac{dz}{dt} \cong \frac{z(k) - z(k-1)}{T_s} \tag{4.15}$$

Equation 4.15 is a way of estimating state values using backward Euler's method, while Equation 4.16 uses forward Euler's method for next state value prediction. In Equation 4.16, the terms z(k) and z(k+1) represent current and future state variables, respectively.

$$\frac{dz}{dt} \cong \frac{z(k+1) - z(k)}{T_s} \tag{4.16}$$

The earlier mentioned approximations find application in representing the discrete model of a PMSM during FCSMPC operation. By incorporating Equations 4.13 and 4.14, the discretized model of the SPMSM in the stationary reference frame is presented through Equations 4.17, 4.18, and 4.19. This analytical approach is equally applicable to the IPMSM.

$$\Psi_s(k) = (V_s - R_s i_s)T_s + \Psi_s(k-1) \tag{4.17}$$

$$i_s(k) = \left(\mathrm{V}_s - j\Psi_m \omega_e e^{j\theta_e}\right)\frac{T_s}{L_s} + i_s\left(1 - \frac{T_s R_s}{L_s}\right) \tag{4.18}$$

$$T_m(k)_n = (1.5p) \times imag\left(\overline{\Psi}_s(k)_n \times i_s(k)_n\right) \tag{4.19}$$

4.2.2 Mathematical Model of Two-Level VSI

This part describes the schematic view of a two-level voltage source inverter (VSI) supplied from a DC voltage source (V_{dc}), shown in Figure 4.6. Each leg of the VSI has two power electronic switches. The switches are grouped into top and bottom pairs, denoted as (Sa^+, Sb^+, Sc^+) and (Sa^-, Sb^-, Sc^-). Each leg has pole voltages (V_{ao}, V_{bo}, and V_{co}) that are 120° apart electrically.

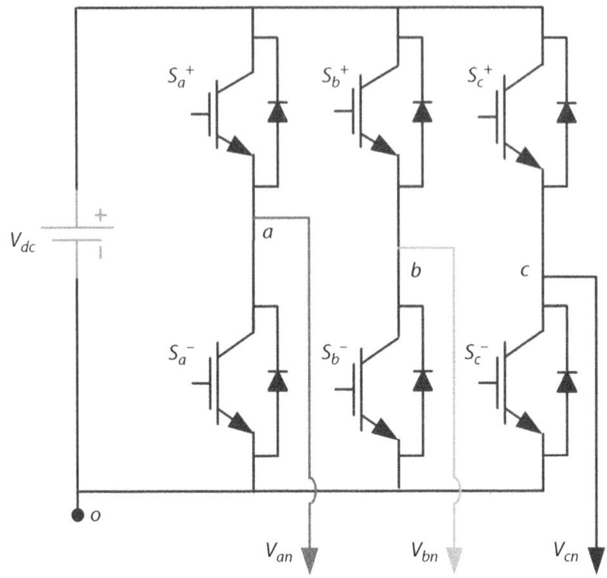

Figure 4.6 Schematic view of two-level VSI.

By turning on the top switch of a leg, a positive pole voltage of $+V_{dc}$ is achieved, and by turning on the bottom switch, a zero-pole voltage is obtained. This results in two possible switching states for each leg, leading to a total of 2^3 (or 2 cubed) switching combinations for the entire two-level VSI operation [27].

The mathematical equation expressing the pole voltages in relation to the switching states is given by Equation 4.20.

$$\begin{pmatrix} V_{ao} \\ V_{bo} \\ V_{co} \end{pmatrix} = \begin{pmatrix} S_a^+ \\ S_b^+ \\ S_c^+ \end{pmatrix} V_{dc} \tag{4.20}$$

In Equation 4.20, the switching states Sa^+, Sb^+, and Sc^+ can take on values of '1' or '0' based on whether they are in a switching turn-on or turn-off condition. The phase voltages are expressed as functions of the pole and common mode voltage (V_z), and these relationships are defined by Equations 4.21 and 4.22.

$$\begin{pmatrix} V_{an} \\ V_{bn} \\ V_{cn} \end{pmatrix} = \begin{pmatrix} V_{ao} \\ V_{bo} \\ V_{co} \end{pmatrix} - \begin{pmatrix} V_z \\ V_z \\ V_z \end{pmatrix} \tag{4.21}$$

$$\begin{pmatrix} V_{an} \\ V_{bn} \\ V_{cn} \end{pmatrix} = \begin{pmatrix} 2/3 & -1/3 & -1/3 \\ -1/3 & 2/3 & -1/3 \\ -1/3 & -1/3 & 2/3 \end{pmatrix} \begin{pmatrix} V_{ao} \\ V_{bo} \\ V_{co} \end{pmatrix} \tag{4.22}$$

In Equation 4.21, the common mode voltage is obtained by summing all the pole voltages in Equation 4.23 in the context of a balanced three-phase load.

$$V_z = \frac{V_{ao} + V_{bo} + V_{co}}{3} \tag{4.23}$$

From the above analysis, mathematical modeling of two-level VSI is drawn for PMSM drive operation.

This section concludes the dynamic mathematical modeling of PMSM in two-phase rotating and stationary reference frames. Along with this, considering SPMSM, the discretized equations of state variables are presented for the sake of FCSMPC analysis. The same methodology can be applicable for IPMSM and also for the representation of discretized machine equations considering Equations 4.5 to 4.8. The regulated supply to PMSM is considered from two-level VSI whose mathematical modeling is also discussed in detail.

4.3 Optimal Control Concepts of Finite Control Set Model Predictive Control (FCSMPC)

Presently, the FCSMPC scheme is emerging in the field of electric motor drives by highlighting its key characteristics such as its intuitive nature, simplicity, and ease of incorporating multiple objectives. The application of FCSMPC, particularly in the PMSM drive, has several benefits to yield performance improvements compared to the conventional DTC approach. The FCSMPC is categorized into two sub-schemes [18]: predictive torque control (PTC) and predictive current control (PCC). This section deliberates the detailed principles of two sub-FCSMPC schemes (PTC and PCC).

4.3.1 PTC Operating Principle

Figure 4.7 illustrates the operational block diagram of the FCSPTC scheme. The torque reference is derived from the PI controller. The various input variables are provided to operate the control algorithm such as motor speed, measured current, torque reference, and flux reference magnitude. Thus, all these inputs are utilized for the control process and finally generate optimal switching pulses for VSI to supply regulated power to the PMSM drive. The FCSPTC algorithm incorporates a mathematical model of the system and this model is represented in a discrete form. The derivative model of state variables such as stator current and flux for PMSM drive is indicated by Equations 4.13 and 4.14, as given in section 4.2. Based on these equations, the mathematical model of the PMSM is discretized using Euler's method and it is denoted by Equations 4.17 and 4.18 in section 4.2. The operation of the FCSPTC algorithm is structured into three key stages as discussed in detail below [20].

4.3.1.1 Measurements and Estimation

The initial stage of FCSPTC operation requires direct measurement of motor speed, DC link voltage of VSI, and phase currents for generating stator flux and current space vector. The measured instantaneous values of motor speed, phase currents, and DC link voltage (V_{dc}) are directly input into the PTC algorithm as these inputs are crucial parameters to make a real-time prediction and control optimization.

4.3.1.2 Stator Current Measurement

The measurements of three-phase stator currents i_a, i_b, and i_c are directly attained from the motor phases *a*, *b*, and *c* through the placement of suitable current sensors. Furthermore, the stator currents undergo Clarke's transformation (αβ-stationary reference frame) to represent the current space vector in the two-coordinate frame as i_s $(= i_\alpha + j\, i_\beta)$.

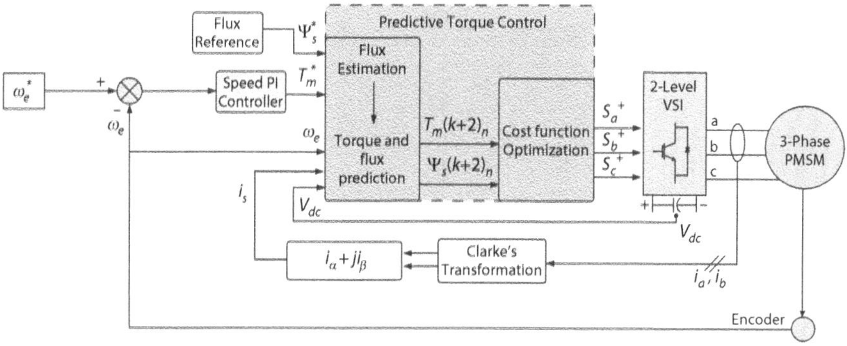

Figure 4.7 FCSPTC for PMSM drive fed by two-level VSI.

4.3.1.3 *Stator Flux (Ψ_s) Estimation*

The stator flux is an important parameter in the analysis and control of electrical machines. It cannot be measured directly from the machine. Thus, Euler's mathematical analysis is applied to estimate the machine stator flux from the stator voltage space vector, as described in Equation 4.9 of section 4.2. Thus, stator flux at the present k^{th} instant is estimated using data from the previous sample instant (k-1) as denoted by Equation 4.24. Here, V_s is the voltage space vector applied in the previous sample state and T_s is the sample time.

$$\Psi_s(k) = (V_s - R_s i_s)T_s + \Psi_s(k-1) \tag{4.24}$$

4.3.1.4 *Predictions*

For the two-level VSI, the generation of net voltage space vector (V_s) is represented by Equation 4.25. This generalized representation allows for recognizing possible voltage space vectors (VVs) based on different switching combinations of a two-level VSI. Table 4.1 provides a comprehensive list of these VVs, and their positions in the α-β voltage space vector plane are illustrated in Figure 4.8. Among the total of eight VVs, six belong to the active VV group, while the remaining two are categorized as null VVs.

$$V_s = \left(S_a^+ + S_b^+ e^{j(2\pi/3)} + S_c^+ e^{j(4\pi/3)}\right)\left(\frac{2}{3}\right)V_{dc} \tag{4.25}$$

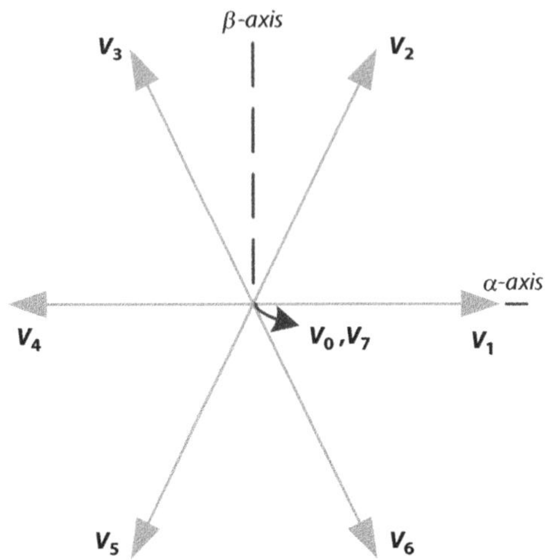

Figure 4.8 Voltage space vectors of two-level VSI operation.

The second stage involves predicting stator flux, current, and torque. These predictions are gained by switching VVs associated with a given inverter. When coming to the case of a two-level VSI, possibly '8' switching vectors are outlined in Table 4.1 in the stationary reference frame. Providing the information regarding stator flux and current at the previous sample instant, the stator flux, current, and torque are estimated in advance, taking into account all eight predictive VVs. The computation

Table 4.1 Possible prediction VVs for two-level VSI operation.

Switching combinations of VSI					
S_a^+	S_b^+	S_c^+	Net VV V_s	$V_{s\alpha}$	$V_{s\beta}$
1	0	0	V_1	$2V_{dc}/3$	0
1	1	0	V_2	$V_{dc}/3$	$V_{dc}/1.732$
0	1	0	V_3	$-V_{dc}/3$	$V_{dc}/1.732$
0	1	1	V_4	$-2V_{dc}/3$	0
0	0	1	V_5	$-V_{dc}/3$	$-V_{dc}/1.732$
1	0	1	V_6	$V_{dc}/3$	$-V_{dc}/1.732$
1	1	1	V_7	0	0
0	0	0	V_0	0	0

process for stator flux, current, and torque predictions are as follows by Equations 4.26 to 4.28.

Stator flux prediction:

$$\Psi_s(k+1)_n = \left((V_s)_n - R_s i_s(k)\right)T_s + \Psi_s(k) \tag{4.26}$$

Stator current prediction:

$$i_s(k+1)_n = \left((V_s)_n - j\Psi_m \omega_e e^{j\theta_e}\right)\frac{T_s}{L_s} + i_s(k)\left(1 - \frac{T_s R_s}{L_s}\right) \tag{4.27}$$

Finally, the motor torque prediction using Equations 4.26 and 4.27 is summarized in Equation 4.28 as follows:

$$T_m(k+1)_n = (1.5p) \times \text{imag}\left(\overline{\Psi}_s(k+1)_n \times i_s(k+1)_n\right) \tag{4.28}$$

In the above-discretized prediction equations, the subscript n indicates the VV number, i.e., zero to seven.

4.3.1.5 *Cost Function Assessment*

The cost function comprises multiple terms representing various control objectives that need to be assessed. In view of all switching states, the predicted torque and flux control values are compared with the reference values of their respective. The optimal switching state is determined by selecting the VV that achieves the closest proximity between the predicted and reference values of the control variable. Upon considering the control objectives (torque and flux), a fundamental expression of the cost function is outlined in Equation 4.29.

$$G_n = \left|T_m^* - T_m(k+1)_n\right| + W\left\|\Psi_s^*\right| - \left|\Psi_s(k+1)_n\right\| \tag{4.29}$$

In this context, W stands for the "flux weighting factor," indicating the relativeness between the torque and stator flux control objectives. The torque reference is shown as T_m^* which is generated by the PI controller, and ψ_s^* is the stator flux reference.

Due to the computation delay problem, there is a negative impact on steady-state performance with one-step prediction horizon-based FCSMPC. The enhancement in steady-state performance is achieved through a long prediction horizon. However, this approach again imposes complexity on the control process. From this discussion, two-step-ahead prediction (k+2) is considered the optimal solution to compensate delay problem (see Equations 4.30 to 4.32)

$$\Psi_s(k+2)_n = \left((V_s)_n - R_s i_s(k)\right)T_s + \Psi_s(k+1) \tag{4.30}$$

$$i_s(k+2)_n = \left((V_s)_n - j\Psi_m \omega_e e^{j\theta_e}\right)\frac{T_s}{L_s} + i_s(k+1)\left(1 - \frac{T_s R_s}{L_s}\right) \tag{4.31}$$

$$T_m(k+2)_n = (1.5p) \times \text{imag}\left(\overline{\Psi}_s(k+2)_n \times i_s(k+2)_n\right) \tag{4.32}$$

With the application of Equation 4.30 to 4.32, the cost function is redeveloped for two-step optimization and is given by Equation 4.33.

$$G_n = \left|T_m^* - T_m(k+2)_n\right| + W\left\|\Psi_s^*\right| - \left|\Psi_s(k+2)_n\right\| \tag{4.33}$$

In Equation 4.33, the flux weight factor is pointed as the fraction of nominal values of motor torque and flux which is signified by Equation 4.34. However, this value of weight factor (W) is advised only at the initial stage, i.e., during the start of operation and it may differ in real-time execution for proper tuning.

$$W = T_m^{nom} \big/ \Psi_s^{nom} \tag{4.34}$$

4.3.2 PCC Operating Principle

As mentioned earlier, the PCC operation also involves three key stages [25]: i. measurement of currents, ii. prediction, and iii. cost function optimization. In the rotating (dq) reference frame, the flux and torque are indirectly regulated through stator currents. Therefore, the stator current phasor is the fundamental control variable. The functional block diagram is illustrated in Figure 4.9. The stator current along with the d-axis is denoted as i_{ds}^* and its value is set to zero since the PMSM has inbuilt rotor flux, and demagnetization of the rotor flux is only created for above-rated speed conditions. Meanwhile, the q-axis of the stator current is denoted as i_{qs}^* which is established by the speed of the PI controller. The development of i_{qs}^* helps to maintain the desired speed. Since the VVs of two-level VSI are derived in stationary (αβ) reference frame, the input reference dq currents to the controller are also converter to the same αβ reference frame. The main purpose of this controller is to reduce the error between reference and actual stator phase currents. In the beginning stage of the operation, the current phasor which represents the magnitude and phase angle at sample (k +1) is projected as given by Equation 4.35. In the second stage, the current prediction at sample (k+2) is established based on eight possible VVs of two-level VSI. The equation governing this two-step-ahead prediction of stator phase current is referred to by Equation 4.36.

$$i_s(k+1) = i_s(k) + \frac{T_s}{L_s}\left(V_S(k) - i_s(k)R_s - j\Psi_m\omega_e e^{j\theta_e}\right) \quad (4.35)$$

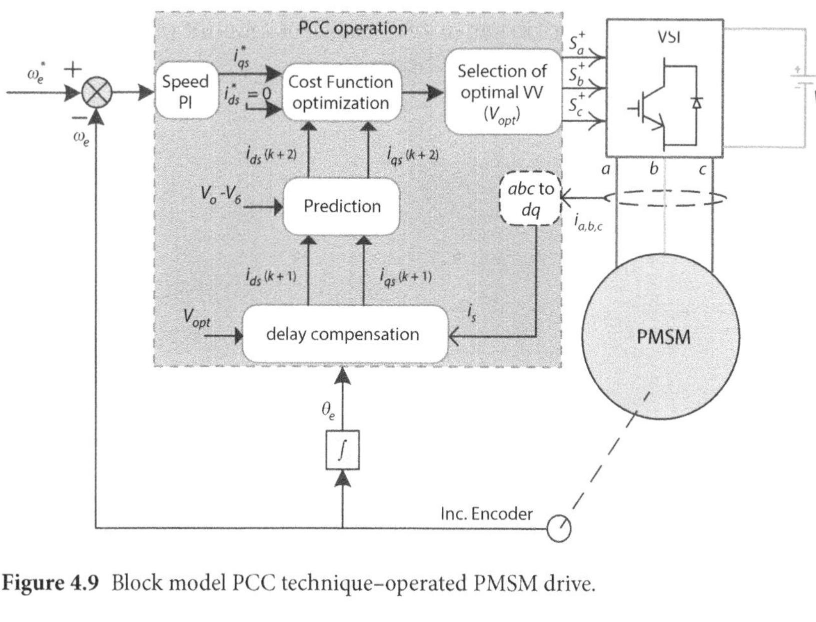

Figure 4.9 Block model PCC technique–operated PMSM drive.

Prediction of control variable:

$$i_s(k+2) = i_s(k+1) + \frac{T_s}{L_s}\left((V_s)_n - i_s(k+1)R_s - j\Psi_m\omega_e e^{j\theta_e}\right) \quad (4.36)$$

With the obtained two-step predicted current space vector Equation 4.36, the cost function can be formulated and is given by Equation 4.37. Thus, the optimal switching state is determined by selecting the VV that achieves the closest proximity between the predicted and reference values of the current.

$$G_n = \left|i_s^* - i_s(k+2)_n\right| \quad (4.37)$$

4.3.3 Simulation Analysis

The simulations have been performed on a PMSM drive using three different control techniques: DTC, PTC, and PCC. The operating condition of the PMSM drive is at a load torque of 12.5 N-m, at the rated speed of 1500 rpm. The responses of the PMSM drive for each control technique are represented in Figure 4.10. The PCC and PTC demonstrate better control response overall compared to the DTC. The fluctuation (standard deviation) in flux and torque is represented in Table 4.2. As a part of this analysis, PTC yields better performance with the least amount of torque ripple as compared with DTC and PCC schemes.

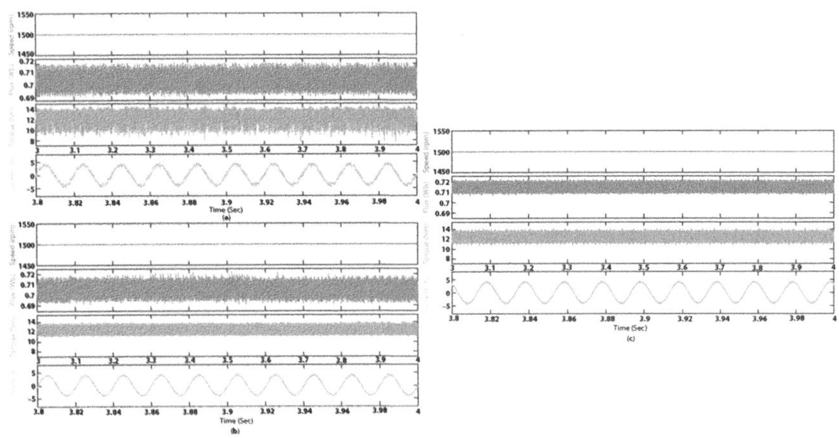

Figure 4.10 Response of PMSM drives during loaded condition at 1500-rpm (a) DTC, (b) PTC, and (c) PCC.

Table 4.2 Control techniques comparison analysis.

	Control techniques		
Parameters	**DTC**	**PTC**	**PCC**
Flux ripple (Wb)	High (0.007)	Low (0.005)	Lower (0.003)
Torque ripple (Nm)	High (1.207)	Lower (0.615)	Low (0.681)
Current THD (%)	High (15.2)	Low (7.4)	Lower (6.7)

In conclusion, the operation of the PMSM drive has been elucidated using FCSMPC techniques, specifically PTC and PCC. Furthermore, MATLAB Simulink investigations have been conducted to assess the performance of these control techniques.

4.4 Constraints in FCSMPC Operation

Currently, there is a trend for MPC application in the power electronics domain since it has been verified both in theory and real time. However, the use of MPC in power electronics and drives has raised some limitations. The in-detail key constraints [18] in MPC operation are provided below.

4.4.1 Weighting Factors Calculation

Research has been conducted up to now investigating various problems in MPC operation, specifically addressing the challenge of selection of weighting factors in the cost function [20–22]. The weighting factor, also called scalar factor selection, is attained with the formal branch and bound algorithm. In situations where multi-objectives are involved in the cost function, the choice of scalar factor becomes more cumbersome and imposes a computational burden. This complexity is more prominent than tuning the PI coefficients. The use of a decision-making strategy is one of the alternatives to solve this problem in MPC operation. So, this way, the selection of multiple objective controls is retained. As a result, it can form objective formulation and optimization.

4.4.2 Steady-State Error Issues

As explained in section 4.3, the MPC algorithm needs to operate in a discrete-time nature at each sampling time. Therefore, the control performance is predicted over a sample, and control input as a VV is applied for each sample instant. The application of single control input in a sample instant results in an increment of steady-state ripples. Therefore, the advancement in the MPC algorithm is mainly focused on maximizing the control inputs during one sampling period in view of system performance [19, 24]. To maximize the control inputs, several research studies focus on space vector modulation techniques and duty ratio control.

4.4.3 Long Prediction Horizon

The FCSMPC functioning is categorized into two types based on the length of the prediction horizon: short and long prediction horizon. With the long prediction horizon, improved performance and efficiency of the MPC scheme are attained. Several research studies are presented to develop a steady-state performance with the application of a long prediction horizon in motor drives. However, in situations where the multilevel inverters operate with MPC and involve long prediction horizons, the computational burden for controller execution develops. Hence, there is a compromise between computational burden and prediction horizon length. The development of new techniques to enhance computational efficiency might require greater attention [24–27]. Also, the enhancement in computation power offered through new digital platforms resolves the development of a long prediction horizon in MPC techniques.

Finally, this section summarizes the present constraints of the FCSMPC scheme for power electronics and drive applications. The key ongoing research studies to advance and mitigate constraints of basic FCSMPC are discussed. Therefore, the entire discussion justifies the future strength of FCSMPC in various applications of power electronics and drives.

4.5 Real-Time Implementation of FCSMPC for PMSM Drive

In this section, a detailed description of the experimental test setup used for the execution of control schemes for the PMSM drive is presented. The block model of the experimental setup photograph for a two-level VSI-fed

PMSM drive is shown in Figure 4.11, in which the hardware requirements for experimentation are presented.

The existing PMSM specifications are listed in Table 4.3. The PMSM shaft is coupled with a DC machine and its ratings are listed in Table 4.3. The DC machine as a generator acts in the form of load for PMSM when its armature winding is connected to a resistive load bank. By varying resistive load tapings, the different load torques on the PMSM shaft can be applied. The PMSM is supplied by VSI. The three-phase VSI specifications with the IGBT module are listed in Table 4.3.

Using the digital signal processing and control engineering (dSPACE-DS1104) platform, all the experimentations are conducted. For this, the Peripheral Component Interconnect (PCI) slot in the host Personal Computer (PC) is interfaced with DS1104. While installing dSPACE software, the existing MATLAB can be associated with it. Every time, when MATLAB starts, the dSPACE software gets configured and displays Real-Time Interface (RTI) DS1104 platform activation in the MATLAB command window. The control schemes for the PMSM drive are implemented in MATLAB/Simulink and built in the dSPACE-1104 model which is interfaced with the host PC.

The dSPACE Input Output (I/O) interface facilitates the required inputs for the execution of control schemes and the control outputs from control schemes to drive PMSM. The essential input variables for the control

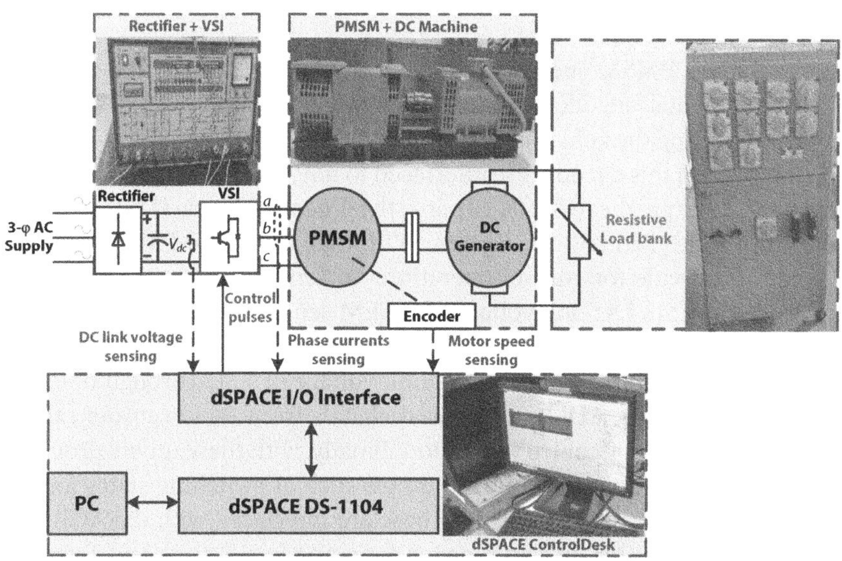

Figure 4.11 Block diagram of PMSM drive experimental test bench setup.

Table 4.3 Experimental setup specifications.

PMSM specifications		**Load (DC machine) specifications**	
Parameter	**Value**	**Parameter**	**Value**
Stator Resistance (R_s)	1.12 Ω	Rated power	3.7 kW
Stator Inductance (L_s)	10.5 mH	Rated speed	1500 RPM
Rated speed (N_{rated})	1500 RPM	Rated torque	23.5 Nm
Rated power	3.7 kW	**VSI specifications**	
Poles (P)	4	Switch model	IGBT- SKM75gb12t4
Rated voltage (Line-Line)	415 V	Voltage rating	1200 V
Rated current (Line)	7.54 A	Current rating	75 A
Inertia (J)	0.031 kg-m^2	DC link capacitor	4700 μF/ 2 × 450 V
Rated flux	0.705 Wb		

operation are PMSM speed, phase currents, and DC link voltage of VSI. The incremental encoder having Pulses per Revolution (PPR)- 1024 is used for sensing the actual speed of PMSM. The output signals (A-, B-, and Z-index) from this encoder are interfaced to dSPACE Incremental Encoder channel 1. From the DS1104 library, the Encoder block (DS1104ENC_POS_C1, ENC_SETUP) is accessed. With this, the actual PMSM speed is gained to provide for control operation. To sense PMSM phase currents (i_a & i_b) and VSI's DC link voltage, the LEM sensors are used. For sensing currents, the sensor model LA 25-NP is used. The voltage sensor of model LV 25-P is used for sensing the DC link voltage of VSI. Through dSPACE RTI blocks such as ADCs, the sensed signals from these sensors can be gained and given for control operation. Finally, with these inputs provided for control operation, the control outputs such as switching states for the VSI-fed PMSM drive are derived. These are interfaced with dSPACE RTI blocks named Master Bit Out pins for accessing in real time and connecting to the hardware setup.

After the successful setting up of the control algorithm and dSPACE interfacing in Simulink, the entire model is built. With this, the code generation and compilation start for the build model. Finally, the file name with the .sdf extension (system description file) is loaded to the DS1104 platform. This completes the building process. The other software linked with the dSPACE experiment is ControlDesk, which facilitates measurement and adjusting variables in real time. For this, the generated variable description file (.sdf) is imported to ControlDesk. Therefore, the entire control operation of VSI-fed PMSM drive can be monitored. Additionally, the reference speed variable can be altered in real time to check the speed dynamics for control operation. This is about hardware information for the provided experimentations.

This section focuses on the FCSPTC a part of MPC and it is applied to control the PMSM drive. The operation and the performance of FCSPTC are also compared with the existing DTC method. It is difficult to achieve an optimal response in the PMSM drive when using a basic PTC due to the issue of weight factors (WF) tuning. The WF plays an important role in the control algorithm, and tuning of it is important for achieving desired performance. Preliminary tests are carried out on the basic PTC to identify the optimal weight value. The weight value is a parameter in the control algorithm and it is determined to be 150 for achieving the desired response of the PMSM drive as presented in the accompanying Figure 4.12.

Figures 4.13 and 4.16 illustrate the comparative analysis of the performance of two control methods, traditional DTC and FCSPTC applied to the PMSM drive. The experimental analysis was conducted across different operating speed ranges and load torque during steady-state conditions. Figure 4.13 represents the motor response under specific operating conditions at a low speed of 20 revolutions per minute (rpm) and a load torque of 12 Newton-meters (N-m). Figure 4.14 shows the response of the

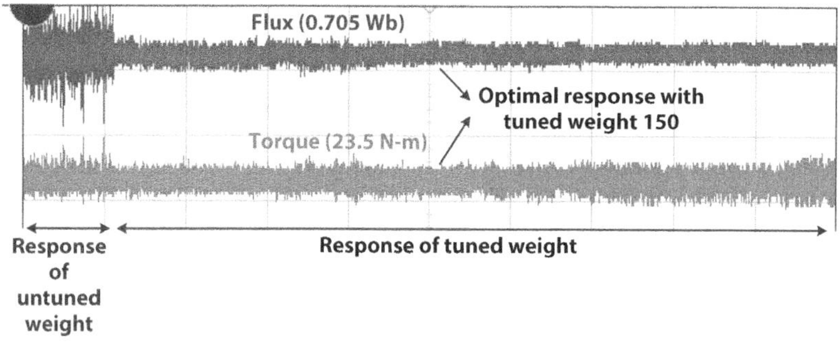

Figure 4.12 Experimental preliminary tests on basic PTC to determine optimal response.

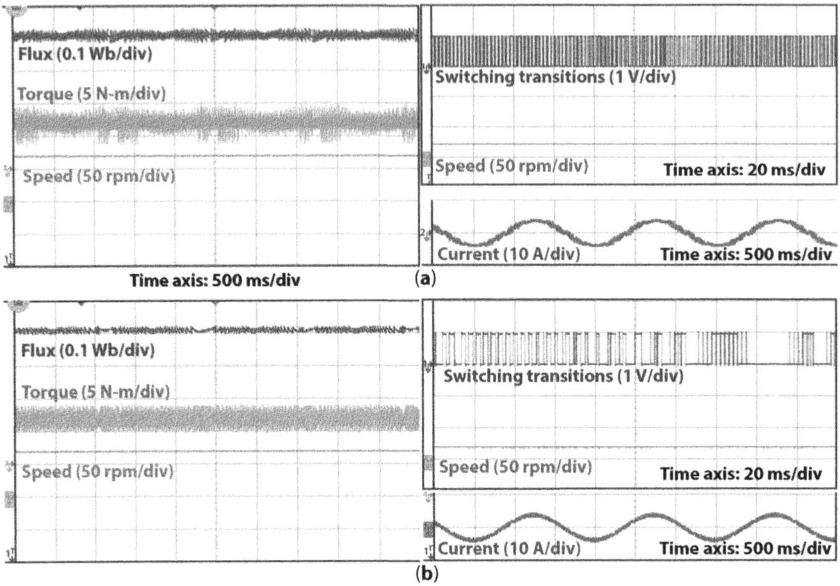

Figure 4.13 PMSM drive response during low-speed of 20 rpm at load torque of 12 N-m. (a) Traditional DTC and (b) FCSPTC.

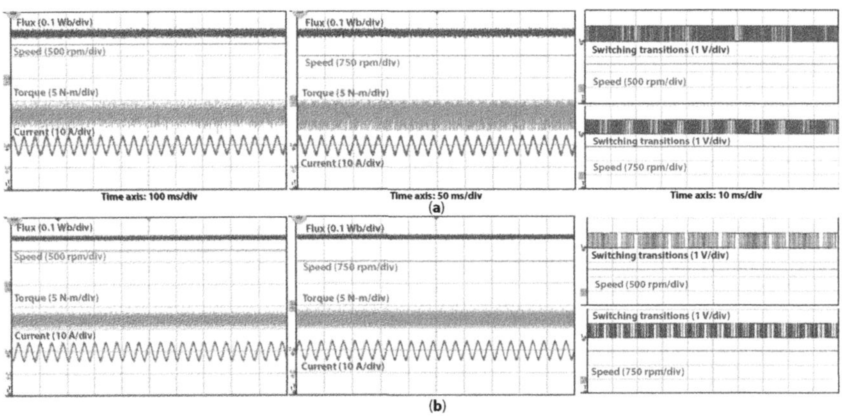

Figure 4.14 PMSM drive response during medium (750 rpm) and high (1500 rpm) speed at load torque of 12 N-m. (a) Traditional DTC and (b) FCSPTC.

PMSM drive under the condition of maintaining a consistent load torque of 12 N-m and speed evaluation from medium to high range (750 rpm and 1500 rpm) with traditional DTC and FCSPTC. For traditional DTC and FCSPTC, to understand the motor performance at its rated operating parameters of 1500-rpm speed and 23.5 N-m rated torque, the response is shown in Figure 4.15 (left part).

Therefore, all steady-state analyses represent that the FCSPTC method enhances the responsiveness of PMSM drives in terms of reducing the ripple of flux and torque, along with improved response in current. In Table 4.4, quantitative response derivation (ripple) for flux and torque, and total harmonic distortion values for current are listed. The response of the traditional DTC method for flux and torque exhibits higher ripple content due to hysteresis controllers and incorrect selection of VV during operation. Therefore, these issues are resolved in FCSPTC. As a result, it asserts the validity and advantage of the FCSPTC over the traditional DTC approach.

The FCSPTC also reduces the switching frequency in a control system. Therefore, the switching transitions of the control system are showcased at different operational speeds of a PMSM drive, including 20 rpm, 750 rpm, and 1500 rpm. Figures 4.13 and 4.14 indicate that FCSPTC exhibits a reduced number of switching state transitions in comparison to traditional DTC. Therefore, FCSPTC achieves a reduction in the switching frequency.

It is crucial to understand the behavior of the dynamic response of the PMSM drive. Therefore, dynamic analysis is subjected to varying operating conditions of speed and load torque. The utilization of the dSPACE control-Desk platform ensures the application of reference speed

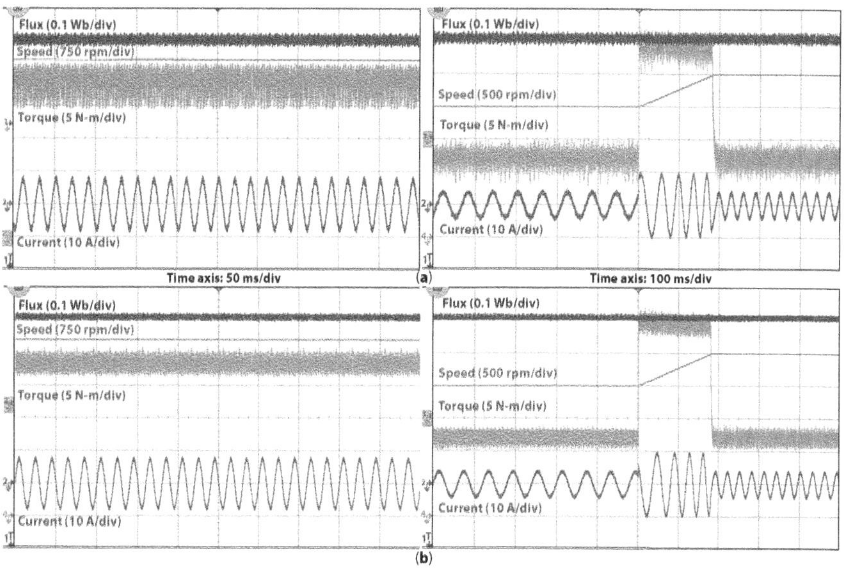

Figure 4.15 PMSM drive response during rated speed and torque condition, speed variations from 500 rpm to 1000 rpm. (a) Traditional DTC and (b) FCSPTC.

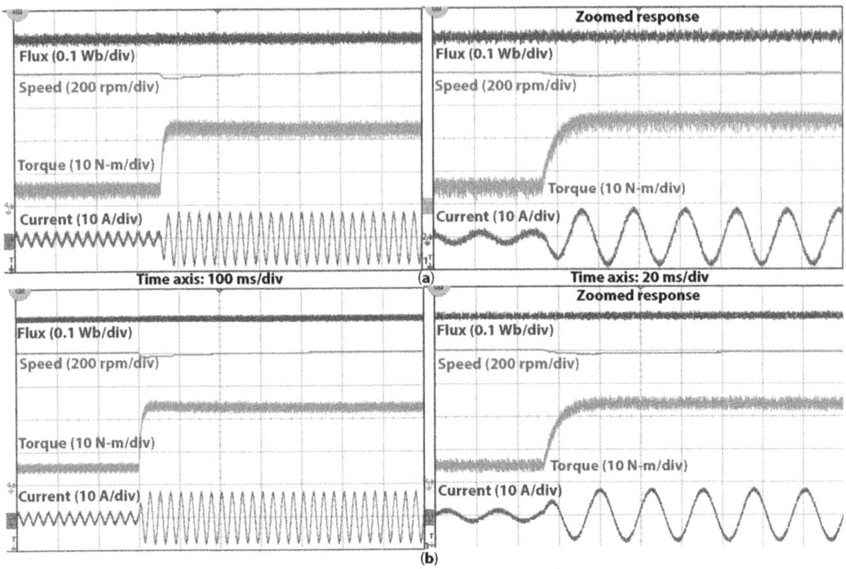

Figure 4.16 PMSM drive response during load torque variations from 5 N-m to 23.5 N-m. (a) Traditional DTC and (b) FCSPTC.

disturbances to the PMSM drive in real time. Figure 4.15 (right part) shows the dynamic response results of the PMSM drive under speed disturbances ranging from 500 to 1000 rpm while maintaining a constant load torque of 12 N-m. The load torque is deployed by incorporating a DC generator coupled to the PMSM drive. Additionally, various electrical loads are switched on the DC generator to introduce dynamic load variation. Changes in electrical load impact the DC generator, leading to equivalent disturbances in load torque on the PMSM drive. Therefore, this setup establish a connection between DC generator's electrical load variation and motor's load torque performance. The load torque disturbances are applied specifically during PMSM drive operation at a constant speed of 1200 rpm and load torque is varied systematically, ranging from 5 N-m to 23.5 N-m. Overall, the experiment allows a detailed analysis of the motor's response to load variations at a specific operating speed. Therefore, the obtained dynamic results are shown in Figure 4.16.

The right part of Figures 4.15 and 4.16 shows the observation of the dynamic response of both traditional DTC and FCSPTC methods. In Figure 4.15, dynamic analysis is focused for a time of 0.18 seconds. During this time, the speed of the PMSM drive was varied. In Figure 4.16, dynamic analysis is observed over a time of 0.2 seconds. During this time, the PMSM drive is undergoing load variations. Therefore, similarity in dynamic responses and improvement in the steady-state response is observed in overall experimentations of FCSPTC in comparison with DTC.

The above results indicate that the FCSPTC performs well in maintaining a constructive response in different steady-state conditions. Also, this control strategy is effective in responding quickly to changes. Besides this, there is an observation in the reduction of switching frequency which leads to low switching loss.

Finally, this section summarizes the complete experimental setup design of the PMSM drive. The real-time implemented PMSM drive operational results under FCSPTC and DTC are exhibited and compared.

Table 4.4 PMSM drive performance.

Operating speeds (rpm):	**20**	**750**	**1500**	**Rated speed (1500 rpm)**
	Half load (12 N-m)			**Rated torque (23.5 N-m)**
Traditional DTC steady-state response				
Torque ripple (N-m)	1.15	1.456	1.667	1.723
Flux ripple (mWb)	9.23	9.451	8.813	8.614
Average switching frequency (kHz)	2.68	5.04	4.33	4.25
Current THD (%)	18.3	20.2	20.3	10.3
FCSPTC steady-state response				
Torque ripple (N-m)	**1.083**	**1.058**	**1.033**	**1.065**
Flux ripple (mWb)	**5.100**	**4.965**	**5.396**	**5.058**
Average switching frequency (kHz)	**0.81**	**2.97**	**3.96**	**3.68**
Current THD (%)	**12.5**	**12.2**	**12.4**	**6.2**

Acknowledgements

This book chapter is supported by the **"Science and Engineering Research Board-Department of Science and Technology, India"** with **SERB File No: SUR/2022/000149** and **TiHAN/IITH/03/2022-23/185. The authors are thankful to these research funding agencies.**

References

1. Madichetty, S., Mishra, S., Basu, M., New trends in electric motors and selection for electric vehicle propulsion systems. *IET Electr. Syst. Transp.*, *11*, 3, 186–199, 2021.
2. De Klerk, M.L. and Saha, A.K., A comprehensive review of advanced traction motor control techniques suitable for electric vehicle applications. *IEEE Access*, *9*, 125080–125108, 2021.
3. Wang, Z., Ching, T.W., Huang, S., Wang, H., Xu, T., Challenges faced by electric vehicle motors and their solutions. *IEEE Access*, *9*, 5228–5249, 2020.
4. Cui, W., Ren, L., Zhou, J., Zhang, Q., A new IPMSM with hybrid rotor structure for electrical vehicle with reduced magnet loss. *IEEE Trans. Magn.*, *58*, 2, 1–6, 2021.
5. Du Plessis, F., Pastellides, S., Wang, R.J., Design and Comparison of Three Surface-Mounted PM Motors for a Light Electric Vehicle, in: *2021 Southern African Universities Power Engineering Conference/Robotics and Mechatronics/Pattern Recognition Association of South Africa (SAUPEC/RobMech/PRASA)*, 2021, January, IEEE, pp. 1–6.
6. Sahu, S., Nayak, B., Dash, R.N., Analysis of Five-Phase Surface PMSM for Application in Electric Vehicles, in: *Sustainable Energy and Technological Advancements: Proceedings of ISSETA 2021*, Springer Singapore, Singapore, pp. 587–595, 2022.
7. Aiso, K. and Akatsu, K., Performance comparison of high-speed motors for electric vehicle. *World Electr. Veh. J.*, *13*, 4, 57, 2022.
8. Chinthala, A. and Vuddanti, S., Performance Analysis of Induction Motor and PMSM for Electrical Vehicle Traction Application, in: *2022 IEEE International Conference on Distributed Computing and Electrical Circuits and Electronics (ICDCECE)*, 2022, April, IEEE, pp. 1–6.
9. Lu, Y., Jiang, Z., Chen, C., Zhuang, Y., Energy efficiency optimization of field-oriented control for PMSM in all electric system. *Sustain. Energy Technol. Assess.*, *48*, 101575, 2021.
10. Furmanik, M., Gorel, L., Konvičný, D., Rafajdus, P., Comparative study and overview of field-oriented control techniques for six-phase PMSMs. *Appl. Sci.*, *11*, 17, 7841, 2021.

11. Koç, M., Unified Field Oriented Controlled Drive System for All Types of PMSMs Considering System Nonlinearities. *IEEE Access, 10*, 56773–56784, 2022.
12. Chen, Z., Shi, T., Lin, Z., Wang, Z., Gu, X., Analysis and control of current harmonic in IPMSM field-oriented control system. *IEEE Trans. Power Electron., 37*, 8, 9571–9585, 2022.
13. Meesala, R.E.K. and Thippiripati, V.K., An improved direct torque control of three-level dual inverter fed open-ended winding induction motor drive based on modified look-up table. *IEEE Trans. Power Electron., 35*, 4, 3906–3917, 2019.
14. Wang, M., Sun, D., Ke, W., Nian, H., A universal lookup table-based direct torque control for OW-PMSM drives. *IEEE Trans. Power Electron., 36*, 6, 6188–6191, 2020.
15. Rehman, A.U., Choi, H.H., Jung, J.W., An optimal direct torque control strategy for surface-mounted permanent magnet synchronous motor drives. *IEEE Trans. Ind. Inf., 17*, 11, 7390–7400, 2021.
16. Nasr, A., Gu, C., Wang, X., Buticchi, G., Bozhko, S., Gerada, C., Torque-performance improvement for direct torque-controlled PMSM drives based on duty-ratio regulation. *IEEE Trans. Power Electron., 37*, 1, 749–760, 2021.
17. Lemma, B.D. and Pradabane, S., Control of PMSM Drive Using Lookup Table Based Compensated Duty Ratio Optimized Direct Torque Control (DTC). *IEEE Access, 11*, 19863–19875, 2023.
18. Li, T., Sun, X., Lei, G., Guo, Y., Yang, Z., Zhu, J., Finite-control-set model predictive control of permanent magnet synchronous motor drive systems—an overview. *IEEE/CAA J. Autom. Sin., 9*, 12, 2087–2105, 2022.
19. Deng, W., Zhang, Q., Xie, W., An Enhanced Virtual Vector-Based Model Predictive Control for PMSM Drives to Reduce Common-Mode Voltage Considering Dead Time Effect. *IEEE J. Emerging Sel. Top. Power Electron.*, 11, 5, 4659–4668, 2023, (Early access, doi: 10.1109/JESTPE.2023.3298666).
20. Kusuma, E., Eswar, K.M.R., Kumar, T.V., An effective predictive torque control scheme for PMSM drive without involvement of weighting factors. *IEEE J. Emerging Sel. Top. Power Electron., 9*, 3, 2685–2697, 2020.
21. Wang, X., Lin, X., Huang, Q., Xie, W., An Improved Parallel Predictive Torque Control for Permanent Magnet Synchronous Motor. *IEEE Access, 11*, 32496–32507, 2023.
22. Bharath Kumar, K. and Praveen Kumar, K.V., Simple predictive torque control of an open-end winding interior permanent magnet synchronous motor drive without weighting factor for electric vehicle applications. *Int. J. Circuit Theory Appl.*, 52, 1, 280–301, 2024 (https://doi.org/10.1002/cta.3744, Early access).
23. Meesala, R.E.K., Udumula, R.R., Nizami, T.K., Hosseinpour, A., Development of enhanced direct torque control for surface-mounted permanent magnet synchronous motor drive operation. *IET Power Electron.*, 16, 11, 1814–1827, 2023.

24. Li, X., Xue, Z., Zhang, L., Hua, W., A low-complexity three-vector-based model predictive torque control for SPMSM. *IEEE Trans. Power Electron.*, *36*, 11, 13002–13012, 2021.
25. Rajanikanth, P., Parvathy, M.L., Thippiripati, V.K., Enhanced Model Predictive Current Control based Five-Phase PMSM Drive. *IEEE J. Emerging Sel. Top. Power Electron.*, 12, 1, 838–848, 2023, (Doi: 10.1109/JESTPE.2023.3342653. Early access).
26. Koç, M. and Özçiflikçi, O.E., Precise torque control for interior mounted permanent magnet synchronous motors with recursive least squares algorithm-based parameter estimations. *Eng. Sci. Technol. Int. J.*, *34*, 101087, 2022.
27. Meesala, R.E.K., Athikkal, S., Aruldavid, R., Improved direct torque controlled PMSM drive for electric vehicles. *J. Inst. Eng. India Ser. B*, *103*, 4, 1177–1188, 2022.

5

Kinematic and Dynamic Modeling of Robots

Suman Lata Tripathi* **and Deepika Ghai**

School of Electronics and Electrical Engineering, Lovely Professional University, Phagwara, Punjab, India

Abstract

This chapter presents a comprehensive investigation into the modeling and simulation of walking robots using MATLAB, with the overarching goal of advancing the understanding and capabilities of dynamic robotic systems. Drawing inspiration from the biomechanics of living organisms, our study focuses on achieving accurate representations of walking robot kinematics and dynamics, leveraging the computational power of MATLAB for simulation and control system design. A successful implementation might result in nimble robots for a variety of applications. The use of MATLAB provides a comprehensive approach that includes mechanical, electrical, and control components. The initiative seeks practical applications in real-world robots as well as scholarly discourse. The convergence of engineering and biology in this work is a step toward more adaptable and flexible robotic systems. Despite obstacles, this chapter also gives predictions about the continued breakthroughs in walking robot development and seeks to encourage creativity in the larger area of robotics. The findings are likely to add to the continuing discussion over the synthesis of biological and artificial locomotion, with implications for the future of robotics.

Keywords: Dynamic modeling, robotic arm, actuator, sensor, MATLAB, controllers

**Corresponding author*: tripathisumanlata78@gmail.com

Arindam Mondal and Souvik Ganguli (eds.) Controller Design for Industrial Applications, (89–104)

5.1 Introduction

The creation and study of walking robots have become critical in the fast-growing area of robotics for a variety of applications ranging from search and rescue missions to industrial automation. Robots' capacity to handle complicated terrains with a stride similar to human or animal movement is critical for their efficacy in real-world circumstances. An unmanned underwater vehicle (UUV)–based robotic model was designed using MATLAB to tackle a tunnel-like environment [1]. In civil construction areas, a cable-driven parallel robot is modeled for three-dimensional (3D) printing using a static kinematic error model to improve motion accuracy [2]. Robotic technology is now popular in the medical sector, specifically in surgical assistance, patient care, and mobile robots [3, 4]. Robots help in healthcare to maintain uninterrupted supply delivery and least disinfections. In a COVID-like pandemic situation, medical robots, robotic assistance, and walking robots can boost patient care in an effective manner. The design of robots depends mainly on their job role, such as controlled movement, interfacing with sensors, and other assistive devices. Also, a foot profile of the robot was modeled to optimize for foot height and minimum foot length using a genetic algorithm in the artificial neural network [5]. So, a detailed model of robots must be designed and simulated before deploying them in the real world.

The goal of this chapter is to model and simulate walking robots using MATLAB, a strong tool for dynamic analysis and control system design. Walking robots, which are frequently influenced by the biomechanics of live beings, provide distinct design and control issues. The use of MATLAB provides a versatile and effective framework for the integration of these systems' mechanical, electrical, and control elements. The chapter deals with the complexities of modeling walking robots taking into account variables such as kinematics, dynamics, and control techniques.

Creating devices with features that are similar to those seen in biologically occurring organisms according to Ramiro and Tomas (2013) [6] takes a lot of money and effort to construct, even for just prototypes, which forces the development of models that enable the desired research to be realized in software, which is MATLAB/Simmechanics [7]. The salient features of biological entities pertinent to research on locomotion ought to be incorporated into these models to aid in the creation of legged robot gaits based on central pattern generators. Thus, it guided the development of the model, which was designed to accommodate various gaits by directly introducing the angular positions of the hip and knee joints.

5.2 Required Block for Robotic Model

The major blocks required for the robotic model are described here:

Actuators: Actuators are any devices that execute any action. Actuators in robotics are motors that cause the manipulator to move. Different types of actuators are utilized for this purpose, with the most prevalent being DC, Stepper, and DC Servo motors. Parameters like torque, speed, and efficiency are used to assess its performance.
Sensors: Sensors are devices that transform electrical signals into physical signals that can be measured. Depending on the situation, several sensors are available for use. Feedback sensors are utilized to determine the present location of the manipulator. Sensors connected to end effectors are utilized to interact with the environment.
Controllers: The fundamental unit of every robot or controlled machine is the controller. PIC microcontroller, Raspberry Pi, Tiva C Launchpad, Arduino, and other controllers can be used for this purpose.
Walking Robots: Humanoid robots, often known as bipedal robots, are devices meant to emulate human-like walking and movement. The intricacy of human mobility, as well as the necessity for stability and balance, makes developing walking robots a difficult undertaking.
Assistive Robotics: Humanoid robots with walking skills can be built to help people with mobility issues by assisting them with activities like carrying goods or moving about in different locations.
Robotic Locomotion: It refers to the transfer or movement of robots from one location to another. It includes a variety of methods and systems that allow robots to explore and traverse their surroundings. The mode of movement chosen depends on the design of the robot, its intended application, and the features of the environment in which it functions.
Software: A manipulator's primary tool is likewise software. A variety of software is available to control the manipulator.

5.3 Robotic Arm Model

The robotic arm model mainly deals with the movement of robotic arm joints and links.

The friction torque (τ_f) and the mass moment of inertia along the motor axis (I) determine the minimum motor torque required, represented as

$$\tau = \tau_f + I\alpha$$

where α is the required rotational acceleration.

The torque from the motor at the joint (on the other side of the transmission) will be a function of the efficiency of the transmission. The actuator torque is expressed by

$$\tau_a = \tau_m \eta_g R_g$$

Here,
τ_m: represents motor torque,
η_g: represents gears efficiency, and
R_g: represents gears ratio.

Degree of Freedom
Figure 5.1 shows the pictorial view of robotic arms. According to Grubler criteria, the number of degrees of freedom can be calculated using the following equation:

$$F = \lambda(n - j - 1) + \sum_{i=1}^{j} f_i$$

where n is the number of links, j is the number of joints, λ is the number of DOF in the space, and fi is the number of DOF permitted by joint Ji.

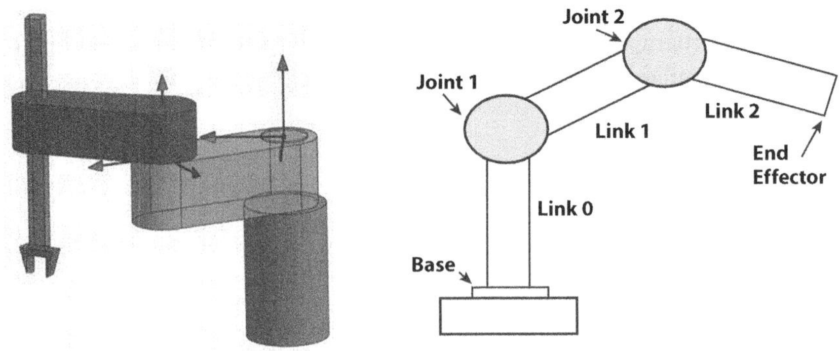

Figure 5.1 Example of the robotic arm.

5.4 Tool Description

In this section, all tools used will be discussed thoroughly including their function. MATLAB is the primary tool in this project, and it is a high-level programming language and interactive environment developed by MathWorks. It is widely used in various fields such as mathematics, engineering, science, and finance for numerical computation, data analysis, visualization, and algorithm development. Here are some key features and uses of MATLAB:

Numeric Computing: MATLAB excels in performing mathematical calculations, matrix manipulations, linear algebra, statistical analysis, and numerical simulations.
Programming Environment: It provides an interactive environment with a command-line interface where users can execute commands, write scripts, and create functions. It also has a graphical user interface (GUI) for easier navigation and visualization.
Data Analysis and Visualization: MATLAB offers powerful tools for importing, cleaning, analyzing, and visualizing data, including plotting 2D and 3D graphs, creating animations, and generating custom plots.
Algorithm Development: It is commonly used for developing and implementing algorithms, simulations, and mathematical models due to its extensive library of functions and toolboxes, covering various areas like signal processing, image processing, control systems, optimization, and more.
Application Development: MATLAB allows users to create standalone applications, interfaces, and graphical components using its App Designer, enabling the development of custom tools and applications.
Integration and Deployment: MATLAB supports integration with other programming languages (C, C++, Python, etc.) and systems, and it provides options for deploying applications to different platforms, including desktop, web, and hardware.

5.5 Methodology/Design Steps

Robot architecture refers to the blueprint that outlines how a robot's hardware and software components are organized and interact. It involves designing the structure of the robot's systems, including its physical parts (like sensors and actuators), software framework (such as control algorithms and decision-making processes), communication interfaces, safety measures, and behavior hierarchies. Essentially, it is the plan that defines how a robot perceives information, processes it, and performs tasks within its environment, tailored to suit its intended function and optimize its performance. Figure 5.2 shows the robotic architecture at the block level. Robotic architectures include controller and robot models as major blocks and submodules as actuators for motion planning and robot mechanics.

The 3D rigid body mechanics of walking robots are built from scratch. The properties are set up correctly, like for example, the dimensions, cross-sections, masses, etc. Figure 5.3 describes a model of a *walking robot in MATLAB Simscape Multibody*.

The next step is adding dynamics to it, which is shown in Figure 5.4. The first one is internal mechanics, and the other one is external mechanics. In internal mechanics, each joint block—whether it is translational or rotational—accounts for initial conditions, mechanical strength, and damping coefficient. The external mechanics are set up through direction and gravity. Similarly, legged robots are modeled contact with respect to the ground.

Actuator dynamics and control in MATLAB refer to the analysis, modeling, and regulation of actuators within systems. It involves creating mathematical models of actuator behavior, designing control systems using MATLAB's tools, simulating actuator responses, and implementing and

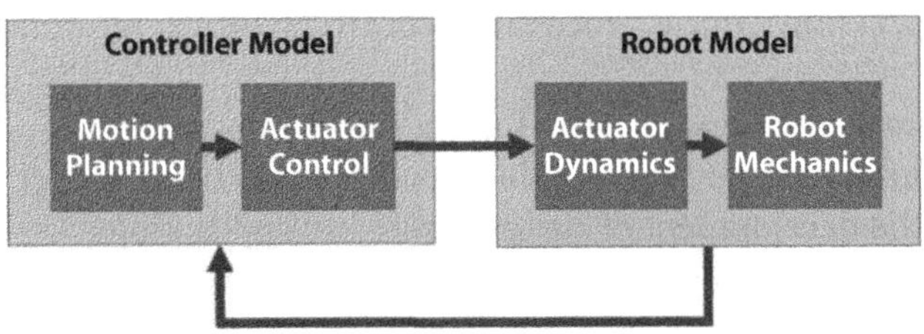

Figure 5.2 Robot architecture.

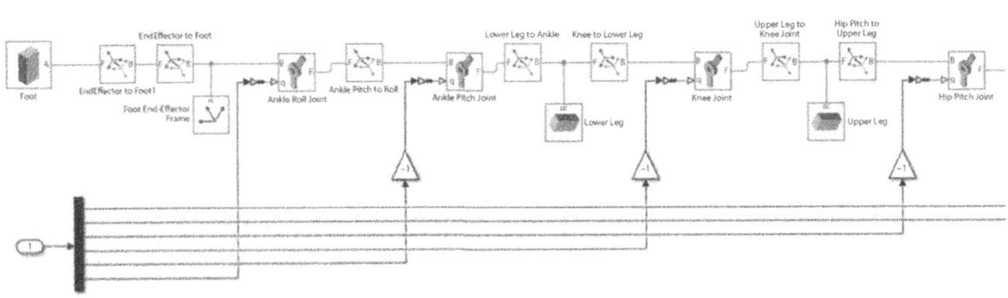

Figure 5.3 Walking robot design in Simscape Multibody.

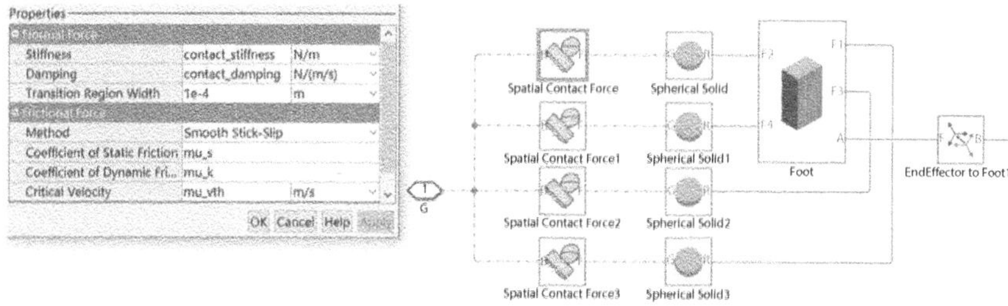

Figure 5.4 Dynamic settings of the walking robot.

testing control strategies. In the architecture diagram earlier, the actuator is the link between the algorithm and the robot, whereas actuator modeling includes the controller and the robot.

Actuator control is responsible for motion to an actuator in Simscape; the first step is to perform actuator sizing as shown in Figure 5.5. This allows you to calculate the power consumed needed for the actuator to perform as desired task. The actuator model designed in Simulink must be tested before deploying it in the target robot.

Actuator dynamics: Simscape can be used to obtain a detailed actuator design that allows the connection of robotic motion in 3D mechanical to other physical domains, like, for example, piston actuators in the fluid domain or motors in the electrical domain. Figure 5.6 shows a MATLAB-based model of actuator dynamics.

The model descriptions are different for different tasks. Based on the available description, the speed range of the simulation will be decided to be faster or slower than in a real-time situation, which is a critical trade-off for the robot design.

Motion planning in MATLAB entails the development and implementation of algorithms that enable efficient pathfinding and movement for robots or systems navigating within physical spaces. This process involves

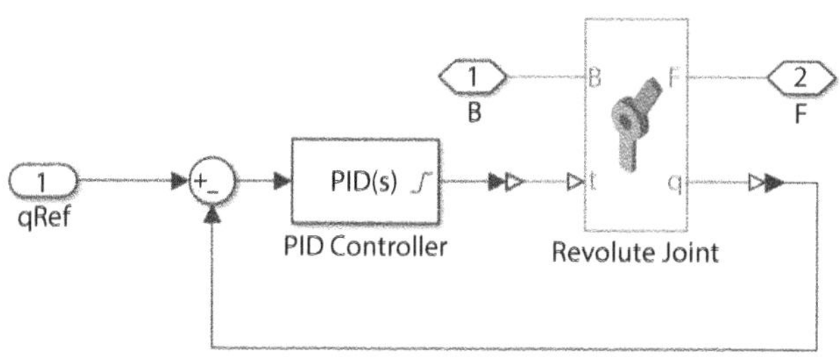

Figure 5.5 Actuator controller of the robot.

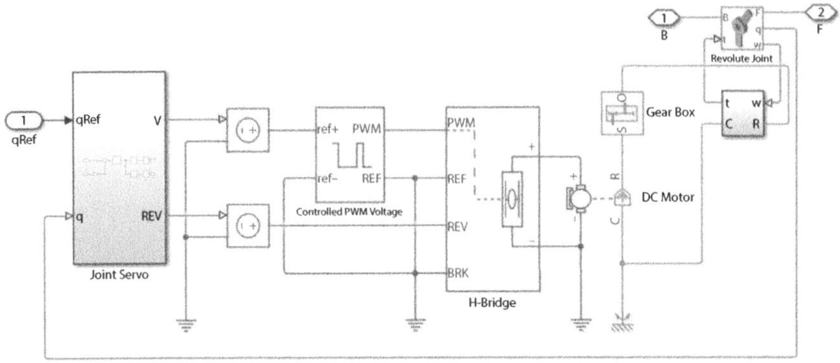

Figure 5.6 Actuator dynamics using Simscape.

creating models of the environment, generating feasible paths considering obstacles and constraints, and designing smooth trajectories for the robot to follow. Engineers leverage MATLAB's diverse toolset to simulate and visualize these planned paths, ensuring their validity and optimizing motion strategies. Finally, these algorithms are integrated into the robot's control system or simulated environments to facilitate autonomous or controlled motion.

The initial open-loop walking pattern is already designed here which makes a smooth robot walk. The walking pattern can be improved with closed-loop components or a reference tracking system. Optimization to refine the walking pattern is a technique that can also be used as listed below.

i) Robot design mainly depends on determining optimal geometry like position, size, cross-sections, or dynamics parameters such as mass/damping/stiffness, or equivalent electrical or fluid actuators.
ii) Robot control also includes rate limits, tuning of gains, and thresholds to meet required performance and safety.
iii) Robot motion planning is to find a motion input sequence that makes satisfactory planning.

5.6 Result and Discussion

Simulation results of the 3D positioning of the walking robot are depicted in Figure 5.7. While the x, y, and z positions of the torso are not identical,

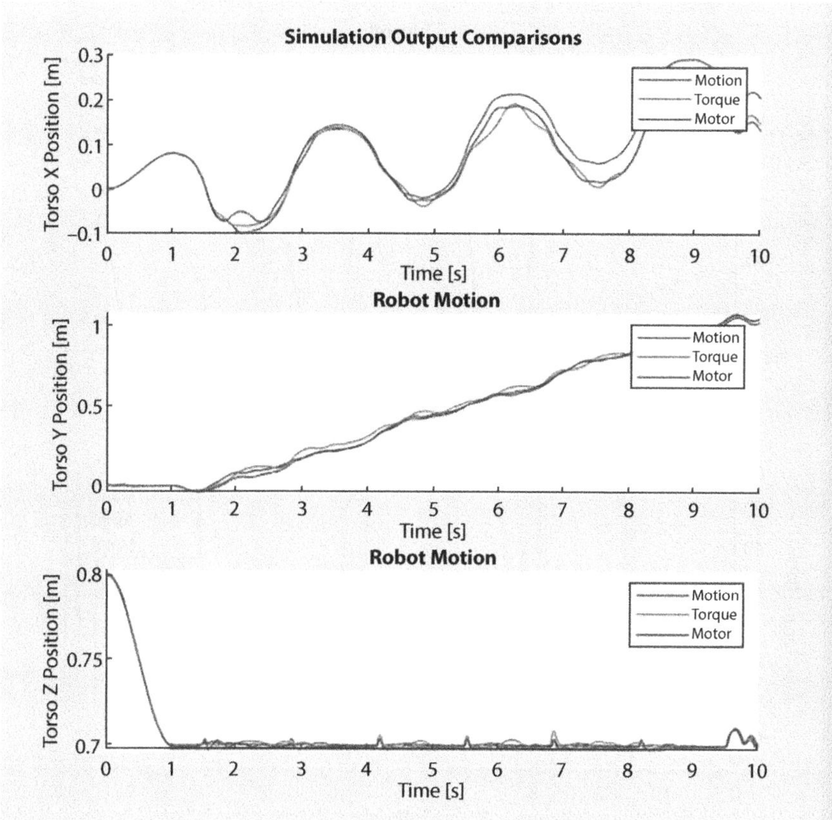

Figure 5.7 3D positioning.

they follow the same overall trajectory shape. The robot responds consistently to joint commands, displaying a similar overall trajectory in different spatial dimensions.

Comparing the hip roll angle the joint angle at the top versus the joint torque at the bottom analysis involves a detailed examination of the hip roll angle's relationship with both joint angle and joint torque. Figure 5.8 shows the characteristics of joint angle and joint torque with respect to time. Zooming in on specific regions allows for a more granular exploration of how these variables interact. Figure 5.9 depicts the zoomed joint angle and joint torque developed in robots.

The comparison between different aspects of a robot's hip roll responses is almost similar, as shown by the blue line (motion actuated), the red line (motor actuated), and the black line (realistic motor model). The

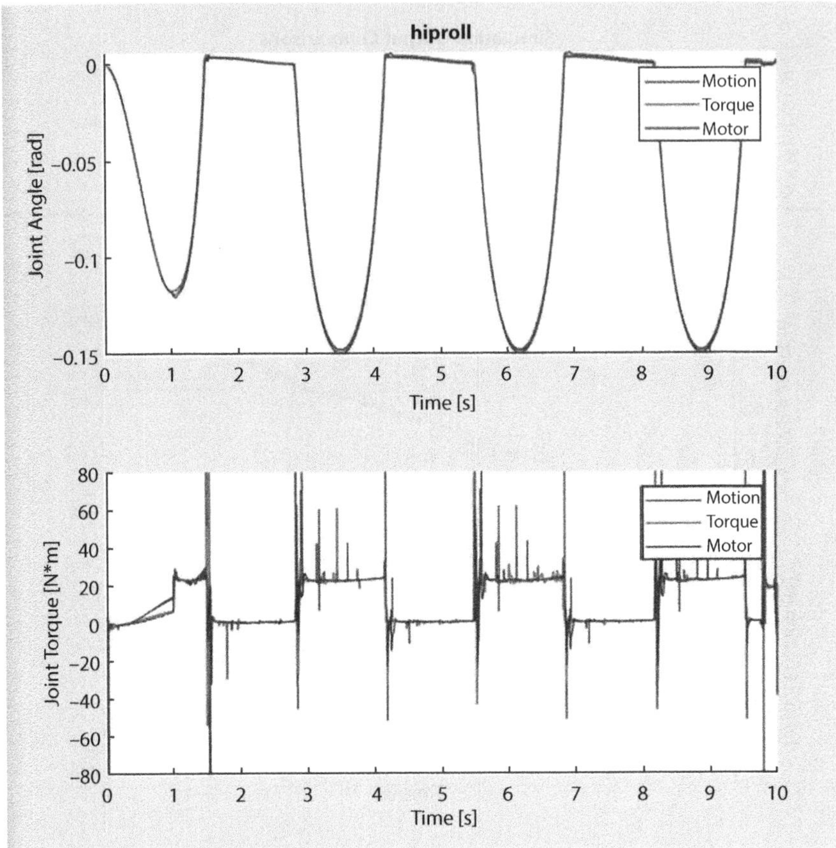

Figure 5.8 Joint angle and joint torque.

motion-actuated line (blue) shows a smooth curve compared to the motor (red) or the torque actuator (black) with more oscillations or blips in the model because of the dynamics. Motion actuation is more idealized. There are these big spikes and torque that are not realizable with a real actuator so you see that as we go from the motion model to the black line to the motor model, the spikes are less exaggerated. The discrepancies observed are due to the differences in the actuation methods, and despite these variations, the robot's walking capability is maintained, showcasing the robustness of the system.

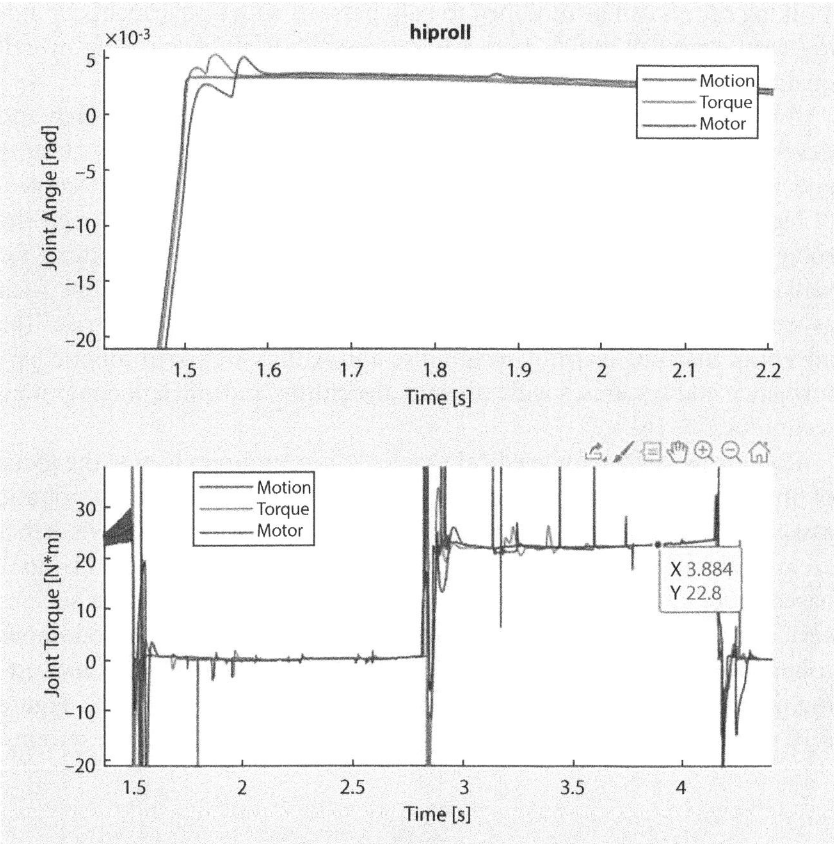

Figure 5.9 Zoomed joint angle and joint torque.

5.7 Applications and Future Scope

Robotic control software, especially for a walking robot, can have a wide range of applications in a variety of sectors and areas. Here are some examples of potential uses. In manufacturing plants, walking robots can be used for assembly line duties such as transporting components from one station to another or constructing complex items. Walking robots might be built to help those with mobility issues by giving companionship and support. They might potentially be used in hospitals to distribute medical supplies.

Walking robots can be modified to help persons with impairments by aiding with everyday duties, increasing accessibility, and improving overall quality of life.

Additionally, it can be used for various purposes such as research and development in robotics, exploring locomotion and gait patterns, testing and improving mechanical designs, studying the dynamics and control of legged systems, simulating and analyzing the interaction between the robot and its environment, and developing and testing algorithms for walking and navigation and potentially even practical applications such as search and rescue, exploration, or assistance in various industries. The advent of machine learning techniques and AI has improved robotic performance and responses with efficient algorithms and efficient computing techniques [8–16].

Robotic assistance for medical practices and surgeries is now the focus of many researchers and medical industries [17–19]. Efficient robotic assistance depends on different sensor modules and integrated devices that are capable of providing accurate and desired performance [20, 21]. IoT-based robotic systems are popular in many real-time applications such as agriculture crop production and precision farming [22–27]. Autonomous robots with human interaction have proven useful in forecasting, harvesting, planting, and packaging different agricultural crops [28–30]. Figure 5.10 presents the major application areas of autonomous robotic systems.

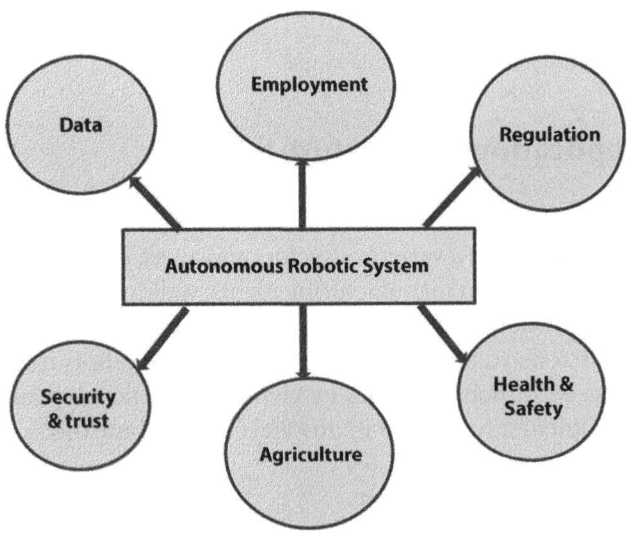

Figure 5.10 Autonomous robotic system.

Robots can improve decision-making, farmer well-being, lifestyle, and safety and perform dangerous tasks.

5.8 Conclusion

In conclusion, the study discusses the process of modeling and controlling a walking robot using robotic control techniques. The input emphasizes the importance of accurately modeling the mechanics of the robot, including rigid bodies, degrees of freedom, and physical contact with the ground. It also highlights the trade-offs between model fidelity and simulation efficiency. The input provides recommendations for creating parameterized models, importing CAD or URDF models, and using Simulink and Simscape Multibody libraries for joint modeling.

The model is then simulated using Simulink, and the Mechanics Explorer provides a visualization of the model's structure. Parameters such as initial conditions, mechanical properties, and actuation methods can be adjusted to affect the simulation. The physical contact between the robot's feet and the ground is implemented using spatial contact force blocks, which model contact between solid geometries. Friction is also considered to enable the robot to move forward during walking.

References

1. Kim, J., Surface Tracking Controls of an Unmanned Underwater Vehicle with Fixed Sonar Ray Measurements in Tunnel-Like Environments. *J. Intell. Rob. Syst.*, 110, 29, 2024, https://doi.org/10.1007/s10846-023-02044-5.
2. Qian, S., Jiang, X., Qian, P., *et al.*, Calibration of Static Errors and Compensation of Dynamic Errors for Cable-driven Parallel 3D Printer. *J. Intell. Rob. Syst.*, 110, 31, 2024, https://doi.org/10.1007/s10846-024-02062-x.
3. Bacik, J., Durovsky, F., Biros, M., Kyslan, K., Perdukova, D., Padmanaban, S., Pathfinder-development of automated guided vehicle for hospital logistics. *IEEE Access*, 5, 26892–26900, 2017, https://doi.org/10.1109/ACCESS.2017.2767899.
4. Bernhard, L., Schwingenschlögl, P., Hofmann, J., *et al.*, Boosting the hospital by integrating mobile robotic assistance systems: a comprehensive classification of the risks to be addressed. *Auton. Rob.*, 48, 1, 2024, https://doi.org/10.1007/s10514-023-10154-0.
5. Venkatagiriyappa, R.B., Annigeri, A.R., Srikantamurthy, J.S., Development of a model for a walking robot made of Desai mechanism using ANN and

regression approach. *J. Eng. Appl. Sci.*, 70, 127, 2023, https://doi.org/10.1186/s44147-023-00296-4.
6. Ramiro, and Tomás, Multilegged walking robot modelling in MATLAB Simmechanics TM and its simulation, 2013.
7. https://www.mathworks.com/videos/model-based-control-of-humanoid-walking-1574399243682.html
8. Goyal, N., Sharma, S., Rana, A.K., Tripathi, S.L. (Eds.), *Internet of Things: Robotic and Drone Technology*, CRC Press, Boca Raton, UK, 2022.
9. Thillaiarasu, N., Tripathi, S.L., Dhinakaran, V. (Eds.), *Artificial Intelligence for Internet of Things: Design Principle, Modernization, and Techniques*, CRC Press, Boca Raton, UK, 2022.
10. Tripathi, S.L., Ganguli, S., Kumar, A., Magradze, T. (Eds.), *Intelligent green technologies for sustainable smart cities*, John Wiley & Sons, USA, 2022, doi: 10.1002/9781119816096.
11. Ghai, D., Tripathi, S.L., Saxena, S., Chanda, M., Alazab, M., Machine Learning Algorithms for Signal and Image Processing, Wiley IEEE Press, USA, 2022, DOI:10.1002/9781119861850.
12. Kaur, J., Ganguli, S., Tripathi, S.L., *The Roles of Artificial Intelligence, Machine Learning, and IoT in Robotic Applications*, p. 14, CRC Press, Taylor & Francis, Boca Raton, UK, 2022.
13. Tripathi, S.L. and Mahmud, M., *Explainable Machine Learning Models and Architecture*, Wiley & Scrivener Publishing LLC, USA, 2023, https://doi.org/10.1002/9781394186570.
14. Tripathi, S.L., Kumar, A., Mahmud, M., FPGA for secured hardware & IP ownership. *2022 IEEE International Conference of Electron Devices Society Kolkata Chapter (EDKCON)*, Kolkata, India, pp. 184–189, 2022, doi: 10.1109/EDKCON56221.2022.10032935.
15. Arora, K., Kumar, V., Prashar, D., Tripathi, S.L. (Eds.), Modern Computational Techniques for Engineering Applications, 1st ed., CRC Press, Boca Raton, UK, 2023, https://doi.org/10.1201/9781003407409.
16. Kumari, C.L., Kamboj, V.K., Bath, S.K., *et al.*, A boosted chimp optimizer for numerical and engineering design optimization challenges. *Eng. Comput.*, 39, 2463–2514, 2023, https://doi.org/10.1007/s00366-021-01591-5.
17. Tripathi, S.L., Prakash, K.B., Balas, V., Mohapatra, S.K., Nayak, J., Electronic Device and Circuit Design Challenges for Biomedical Applications, Amsterdam, Netherlands, Elsevier, 2021, https://doi.org/10.1016/C2020-0-02289-9.
18. Kumar, K., Bhaumik, S., Tripathi, S.L., Health Monitoring System, in: Electronic Device and Circuits Design Challenges to Implement Biomedical Applications, Elsevier, 2021, Doi: https://doi.org/10.1016/B978-0-323-85172-5.00021-6.
19. Lata Tripathi, S., Dhir, K., Ghai, D., Patil, S. (Eds.), Health Informatics and Technological Solutions for Coronavirus (COVID-19), 1st ed., CRC Press, Boca Raton, UK, 2021, https://doi.org/10.1201/9781003161066.

20. Kumar, K., Sharma, A., Tripathi, S.L., Sensors and their Application" Electronic Device and Circuits Design Challenges to Implement Biomedical Applications, pp. 177–195, Amsterdam, Netherlands, Elsevier, 2021, https://doi.org/10.1016/B978-0-323-85172-5.00021-6.
21. Bajpai, S., Sharma, D., Alam, M., Chandel, V.S., Pandey, A.K., Tripathi, S.L., Curvelet Transform Based Compression Algorithm for Low Resource Hyperspectral Image Sensors. *J. Electr. Comput. Eng.*, 2023, 8961271, 1–18, 2023, https://doi.org/10.1155/2023/89612.
22. Srivastava, Y., Virk, S., Ganguli, S., Tripathi, S.L., *Internet of Robotic Things (IoRT) and Its Manifold Applications*, p. 16, CRC Press, Taylor & Francis, Boca Raton, UK, 2022.
23. Dhonde, M., Sahu, K., Murty, V.V.S., The application of solar-driven technologies for the sustainable development of agriculture farming: a comprehensive review. *Rev. Environ. Sci. Bio/Technol.*, 21, 139–167, 2022, https://doi.org/10.1007/s11157-022-09611-6.
24. Yano, A. and Cossu, M., Energy sustainable greenhouse crop cultivation using photovoltaic technologies. *Renew. Sustain. Energy Rev.*, 109, 116–137, 2019, https://doi.org/10.1016/j.rser.2019.04.026.
25. Umapathy, P., Hemanth, J., Khera, S., Inbamani, A., Tripathi, S.L. (Eds.), *Machine Learning and the Internet of Things in Solar Power Generation*, 1st ed., CRC Press, Boca Raton, UK, 2023, https://doi.org/10.1201/9781003302964.
26. Tripathi, S.L., Dubey, M.K., Rishiwal, V., Padmanabhan, S., *Introduction to AI techniques for Renewable Energy System*, Taylor & Francis, Boca Raton, UK, 2021, https://doi.org/10.1201/9781003104445.
27. Tripathi, S.L. and Padmanabhan, S., Green Energy: Fundamentals, Concepts, and Applications, Scrivener Publishing, Wiley, USA, 2020, DOI:10.1002/9781119760801.
28. Rose, D.C., Lyon, J., de Boon, A., *et al.*, Responsible development of autonomous robotics in agriculture. *Nat. Food*, 2, 306–309, 2021, https://doi.org/10.1038/s43016-021-00287-9.
29. Pal, V.C., Tripathi, S.L., Ganguli, S., *Industrial Control Systems*, Wiley, Scrivener Publishing LLC, USA, 2024, DOI:10.1002/9781119829430.
30. Singh, T., Tripathi, S.L., Mahmud, M., Design of a 16-bit 500 MS/s SAR-ADC at 45 nm for low power and high frequency applications. *Eng. Res. Express*, 6, 015306, 2024. DOI: 10.1088/2631-8695/ad2730.

6

Design of FUZZY-(1+PD)-FOPID Controller for Hybrid Two-Area Power System

Susmit Chakraborty[1]* and Arindam Mondal[2]

[1]*Department of Computer Science and Engineering-Cyber Security and Data Science, Brainware University, Barasat, West Bengal, India*
[2]*Department of Electrical Engineering, Dr. B C Roy Engineering College, Durgapur, West Bengal, India*

Abstract

Approach of the fuzzy logic control (FLC) has been one of the most vibrant and productive fields for study in the use of fuzzy set theories during the past couple of decades. Zadeh's influential publications on the linguistic methodology and analysis of the system based on the idea of fuzzy sets served as the inspiration for Mamdani and his coworkers' revolutionary study on FLC. The fundamental tenet of this strategy was to develop the controller while taking into account the "learning" of a human process of operations. A control algorithm is built from a collection of linguistic rules that characterize the operator's controlling approach, with the words described as fuzzy sets. To triumph over the LFC in all two-area power systems, a hybrid FUZZY-(1+PD)-FOPID controller is introduced in this study as a novel expert control approach. The proposed controller settings such as K_{P1}, K_{P2}, K_{D1}, K_{D2}, and K_{I2} as well as additional parameters, such as the integrator-exponential (λ) and differentiator-exponential (μ), are optimized using a recently published population-based, nature-inspired algorithm named tree-seed algorithm (TSA) while using the integral of squared error (ISE) criteria. The recommended FUZZY-(1+PD)-FOPID controller has been suggested to be utilized for controlling a two-area power system that includes thermal and nuclear units along with non-conventional energy sources (NCESs) like solar and ocean thermal. The whole system with the proposed controller is sketched and tested in the MATLAB simulation platform. The simulation results are analyzed using the

**Corresponding author*: susmit.eee@gmail.com

Arindam Mondal and Souvik Ganguli (eds.) Controller Design for Industrial Applications, (105–124)

acquired time-domain parameters such as settling time (ST), overshoot (OS), and undershoot (US). The suggested controller outperforms competing controllers found in the literature, as evidenced by the comparison of the major performance indices and the robustness of the devised controller to load change.

Keywords: Tree-seed algorithm (TSA), load frequency control (LFC), fractional-order PID controller (FOPID), fuzzy logic controller (FLC), power system (PS)

6.1 Introduction

Load frequency control (LFC), which provides high-quality electric power, is an essential component in raising a nation's standard of life in the world of today, which is changing tremendously fast [1]. The mismatch in tie-bar power and frequency between different areas in the PS is the most obvious disadvantage [2] of this interconnected power system (PS). With the help of LFC, this issue can be mitigated or even cured [3]. To manage load frequency, there are already several controllers [4–6] in use; however, they have shown to be ineffective. The typical PI controller [7] is easy to implement, but the frequency variation is too much to accept. To give the best performance throughout a broad range of operations, a self-adjusting gain controller [8] is necessary. The available self-tuning sophisticated control schemes incorporating sliding modes [9, 10], H-infinity [11], fractional calculus method [12–19], artificial neural network approach with FLC [20], adaptive neuro-fuzzy inference system (ANFIS) [21, 22], an observer of nonlinear disturbances [23], disturbance observer optimized fractional-order three-degrees-of-freedom tilt-integral-derivative (FO-3DOF-TID) [24], and fuzzy logic controllers (FLCs) [7, 25–32] are documented in the appropriate field to improve control performance. Fuzzy provides the ability to select various control gain settings based on the frequency variance between interconnected PSs [33]. The logic that underpins fuzzy control, known as fuzzy logic control (FLC), is fundamentally different from conventional logical systems in that it is more evocative of how people think and communicate naturally. In essence, it gives us a useful way to capture the approximate, imperfect aspect of the real world. It has been demonstrated that the cascaded control technique has the ability to outperform the traditional feedback controller, as highlighted by several studies [34]. The cascade control concept is well-known for quickly rejecting noise before propagation to different areas of the PS [34]. By adding a novel cascade controller called FUZZY-(1+PD)-PID, whose efficiency evaluation has not yet been assessed, this chapter revisits the LFC problem.

It is very much interesting to observe the overall result after processing the controlling signal above three units such as FLC, (1+PD), and, finally, PID. The FLC output is delivered into the input terminal of the (1+PD) block, which generates another output and to use as the input of the FOPID block. This Three-layer control strategy is considered here to mitigate the LFC problem.

Following a review of the literature, several nature-inspired optimization strategies like PSO [35], CSA [36], and BFA [37] have been used in LFC research to properly adjust gain and maximize the benefits of the controllers that have been suggested. In this chapter, the tree-seed algorithm (TSA) is preferred for selecting the appropriate values of controller parameters. TSA is a population-oriented, potent, probability-derived, metaheuristic optimization method that originated by nature. To handle continuous optimization difficulties, Kiran [38] invented TSA, which is in accordance with the relationship of trees and seeds. TSA operates on the three-layer controller to tune the parameters such as K_{P1}, K_{P2}, K_{D1}, K_{D2}, and K_{I2} as well as additional settings, such as integrator-exponential (λ) and differentiator-exponential (μ) for a two-area interconnected hybrid PS network comprising conventional (nuclear and thermal) as well as renewable sources (solar and ocean thermal).

6.2 Plant Model

The plant that has to be controlled in this chapter is a two-area linked hybrid PS with two independent units in each. Each area has a mix of renewable (solar or ocean thermal power plants) and traditional energy sources (thermal or nuclear power plants). The first area includes a thermal power plant (TPP) and a solar photovoltaic unit [37, 39], while the second area includes an ocean thermal power unit [40] and a nuclear power unit [41]. The complete PS block diagram is depicted in Figure 6.1.

6.3 Controller Design

In this manuscript, a three-layer cascade controller named FUZZY-(1+PD)-FOPID is utilized to get rid of the tie-bar power and frequency error between different areas of the PS. Three different layers are discussed in the following three subsections and the complete pictorial representation of the hybrid controller is illustrated in Figure 6.2.

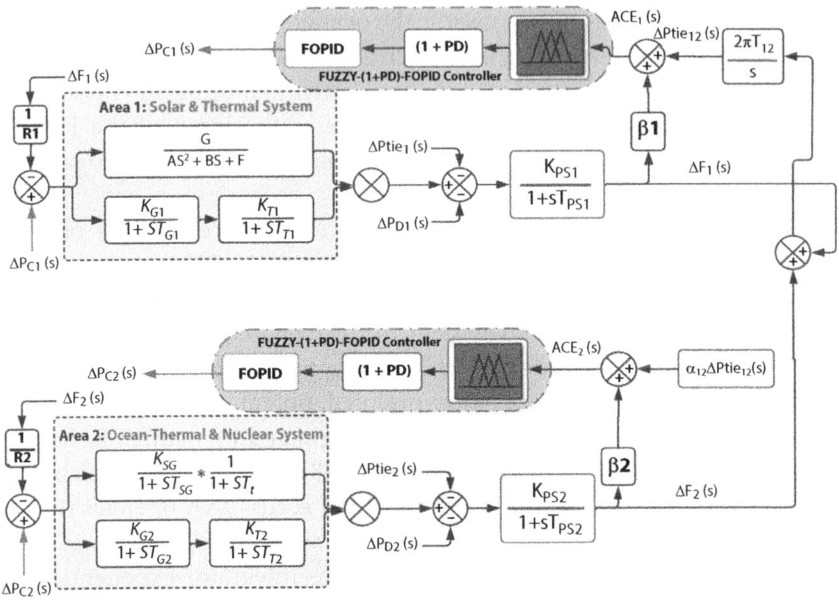

Figure 6.1 Simulation schematic of the hybrid two-area PS network.

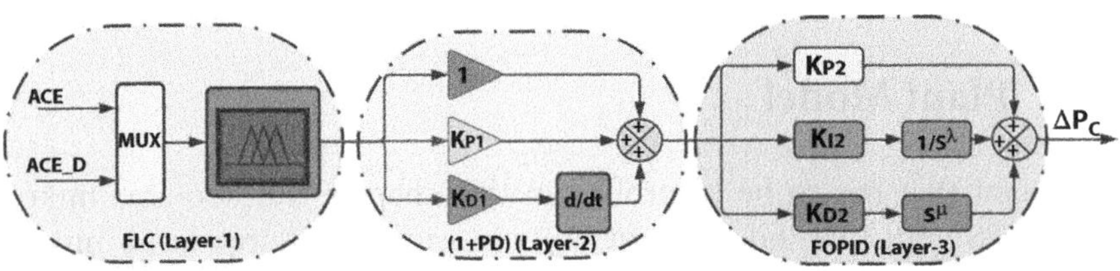

Figure 6.2 Structure of the proposed three-layer controller (FUZZY-(1+PD)-FOPID).

6.3.1 Fuzzy Logic Controller (FLC)

Fuzzy logic is an effective intelligent multivalued logical procedure that handles inaccurate granular data from many data sources. The linguistic information used in this work comes with the help of the formerly developed fuzzy control by Gupta *et al.* [42]. In an FLC, there are four parts: a fuzzifier, a knowledge base, an interference system, and a defuzzifier, as represented in Figure 6.3. Fuzzifier operation, which converts input signals into fuzzy numbers, initiates fuzzy operation. These ambiguous numbers are inputs to the interference system, which acts as a cognitive and value-setting mechanism for the controller's output. It uses a rule basis to make judgments that result in suitable output. The knowledge base, also

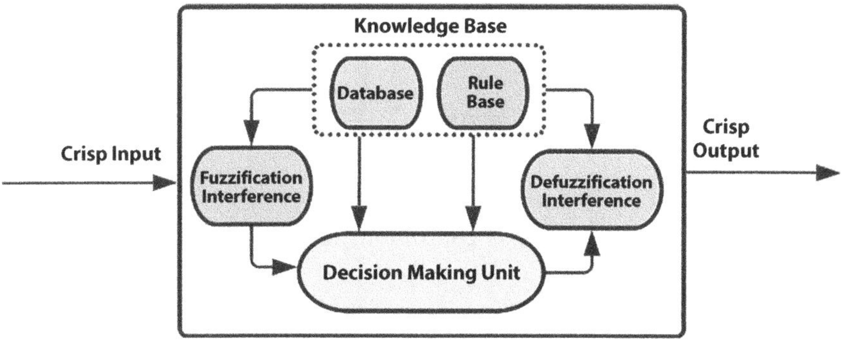

Figure 6.3 Fuzzy inference system (FIS).

known as the rule base, is essentially a collection of data contained within the fuzzy membership function. Following decision-making, the output in the form of fuzzy values is transformed using the defuzzifier procedure into real-world or crisp values. The abrupt diminutive load disturbances that continually interfere with a PS's regular operation are the LFC problem that is being discussed here.

6.3.1.1 *Fuzzy Sets and Fuzzy Logic*

An extension of crisp sets is termed the fuzzy set. Fuzzy sets enable partial membership, whereas crisp sets only allow complete membership or none at all. So, an element could only partially be a part of a set. In a crisp set, a characteristic function $\mu_P(a)$ describes whether an element is a part of the set P or not, where $\mu_P(a)$ can be written as (6.1).

$$\mu_P(a) = \begin{cases} 1; \, if \; a \in P \\ 0 \; if \; a \notin P \end{cases} \tag{6.1}$$

This idea is expanded upon by fuzzy set theory, which introduces partial memberships with values ranging from 0 to 1.

$$\mu_P(a) : A \rightarrow [0,1] \tag{6.2}$$

where A is the universal set stated in the given issue. If this universal set is numerous and limited, then a fuzzy set P in this universe may be derived by listing each member and their degree of membership in the set P as mentioned in Equation 6.3.

$$P = \sum_{i=1}^{n} \{\mu_P(a_i) / a_i\} \tag{6.3}$$

In a similar way when A is continuous, Equation 6.4 may describe a fuzzy set P.

$$P = \int_A \{\mu_P(\mathrm{a}) / a\} \tag{6.4}$$

The membership function is denoted by μ_p(a) as demonstrated in Figure 6.4 to study the LFC problem.

In LFC, there are two inputs in the FLC unit, called area-control error (ACE) along its derivatives (ACE-D) and FLC is the controlled output after defuzzification from the fuzzy unit. Linguistic variables are used to represent fuzzy sets using their corresponding letter representations. The letters Z.R, N.L, N.M, N.S, P.S, P.M, and P.L denote zero, negative-large, medium, small, and positive-small, medium, and large, respectively, and they indicate the linguistic parameters in this work, as listed in Table 6.1. The Mamdani FIS and the center-of-gravity (COG) technique of defuzzification are utilized to generate the FLC's true control output.

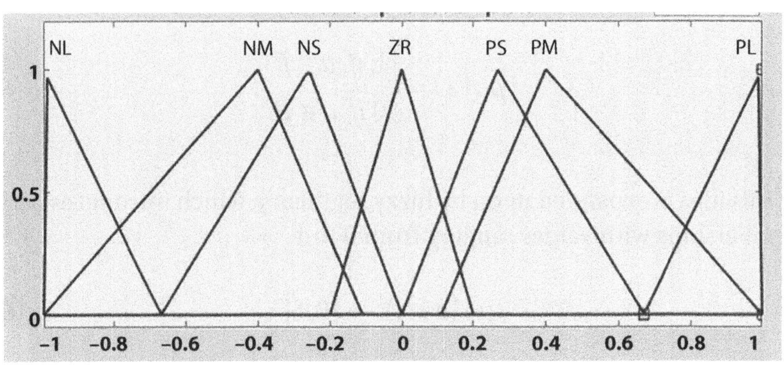

Figure 6.4 Membership distribution for the inputs (ACE and ACE-D) and the output (FLC) signals.

Table 6.1 Rule base for ACE and ACE-D and the output of FLC.

	ACE derivatives						
ACE	N.L	N.M	N.S	Z.R	P.S	P.M	P.L
N.L	N.L	N.L	N.L	N.L	N.M	N.S	Z.R
N.M	N.L	N.L	N.L	N.M	N.S	Z.R	P.S
N.S	N.L	N.L	N.M	N.S	Z.R	P.S	P.M
Z.R	N.L	N.M	N.S	Z.R	P.S	P.M	P.L
P.S	N.M	N.S	Z.R	P.S	P.M	P.L	P.L
P.M	N.S	Z.R	P.S	P.M	P.L	P.L	P.L
P.L	Z.R	P.S	P.M	P.L	P.L	P.L	P.L

From this FLC unit, a conditional time-dependent scaling factor ($\tilde{\bigcap}$) is basically obtained which is further injected as the input to the next layer of the controller.

6.3.2 (1+PD) Controller Unit

Figure 6.5 gives the pictorial demonstration of the (1+PD) unit which is the second layer of this controller.

The FLC output is fed as the input to this layer. The Laplace representation of (1+PD) can be written as Equation 6.5.

$$G_{(1+PD)s} = 1 + K_{P1} + sK_{D1} \tag{6.5}$$

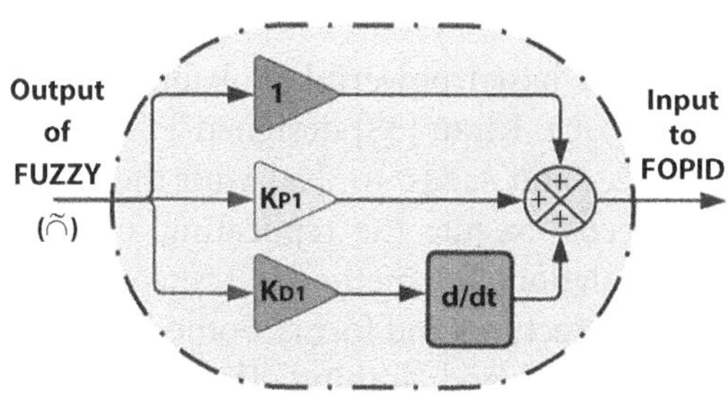

Figure 6.5 (1+PD) controller unit structure.

6.3.3 Fractional-Order PID (FOPID) Controller

The abbreviation FOPID is an acronym that gives fractional-order proportional, integral, derivative controlling action. It was initially put into practice by Kapoor [43]. Due to its greater number dimensionality and superior durable design, the FOPID outperforms the PID controller [44]. The controller settings, however, are more difficult to modify than those of a traditional PID controller. Various design strategies with FOPID control have been documented in the literature [45–48]. Aside from the three parameters K_{P2}, K_{I2}, and K_{D2}, two extra parameters λ and μ must be addressed when representing a FOPID controller. To improve the controller›s dynamic performance, these extra parameters called fractional order (FO) derivative and integrator are used. The Laplace representation of a FOPID controller is noted as Equation 6.6.

$$G_{FOPID} = K_{P2} + S^{\mu}K_{D2} + \frac{K_{I2}}{S^{\lambda}} \tag{6.6}$$

The suggested controller's structure is depicted in Figure 6.5, where ACE and ACE-D are the inputs of the first layer of control (FLC). $\tilde{\cap}$ denotes FLC unit output and also acts as the input to the next layer of the controller ((1+PD)-FOPID), and ΔP_C is the final output of the recommended hybrid controller. The overall control signal of the proposed three-layer hybrid controller is formulated in Equation 6.7.

$$\Delta P_C = \tilde{\cap} x G_{(1+PD)s} x G_{FOPID} = \tilde{\cap} x \left(1 + K_{P1} + sK_{D1}\right) x \left(K_{P2} + S^{\mu}K_{D2} + \frac{K_{I2}}{S^{\lambda}}\right) \tag{6.7}$$

6.4 Tree-Seed Algorithm

It is a natural population-based, powerful, probability-derived, metaheuristic optimization approach. Kiran [38] designed TSA, which is oriented on the interaction of trees and seeds, to deal with the issues of progressing optimization. The seeds execute the replicating and producing mechanisms of the tree population. The method of transportation of seeds varies according to their architectures and forms. Some seeds remain attached to birds' feathers, animal furs, and clothing. They are transported from different trees to various locations. By this process, the seeds germinate in the correct atmosphere and develop into trees. The ground is considered as the search space for optimization issues in TSA. The answers to optimization issues are represented by both seeds and trees. The techniques for exploiting and exploring are critical for the outstanding efficiency of

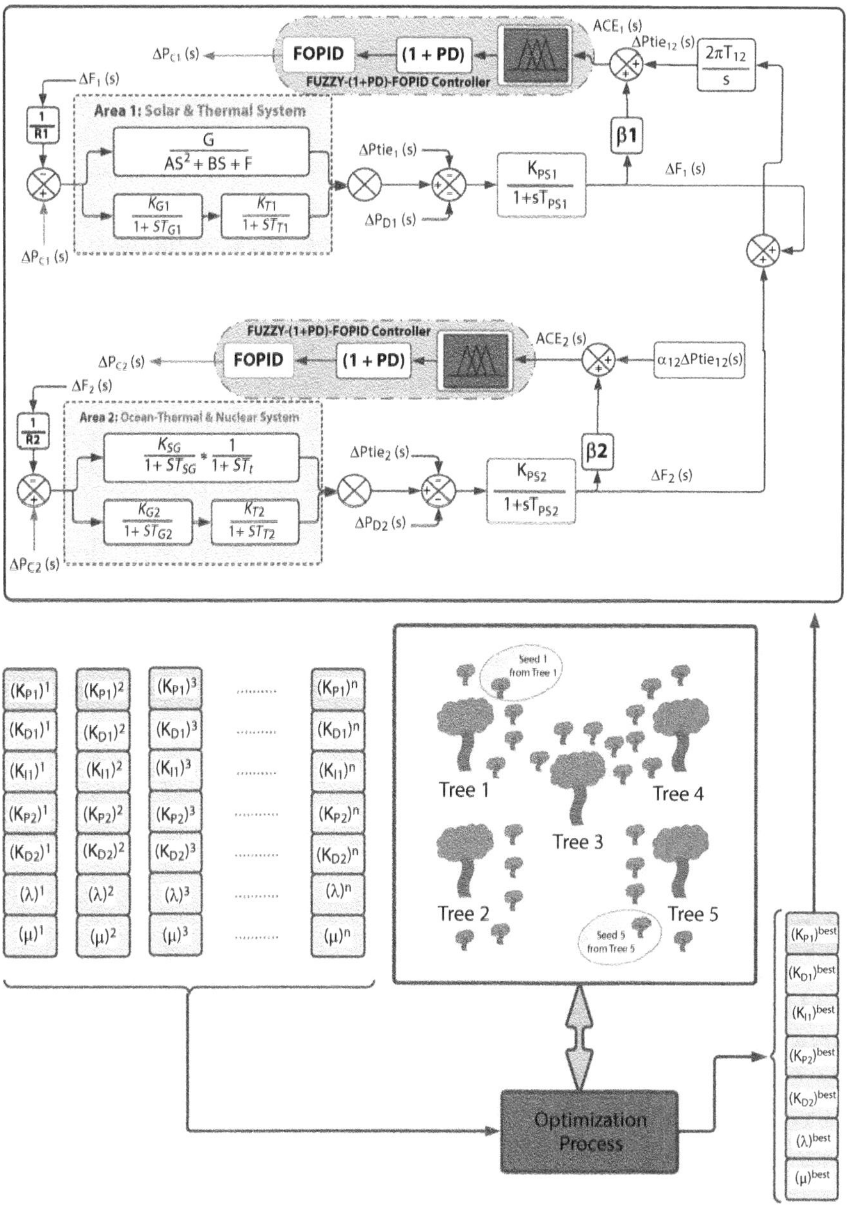

Figure 6.6 Flowchart of TSA optimization process for LFC problem.

the metaheuristic processes and for creating a better-resulting solution. During the exploration process, the trees are randomly placed around the search space, and the seeds with features similar to the trees are employed in the exploitation phase. The seeds were generated using Equations 6.8 and 6.9 as follows:

$$M_{i,j} = R_{i,j} + a_{i,j}x\left(D_j - R_{r,j}\right) \tag{6.8}$$

$$M_{i,j} = R_{i,j} + a_{i,j}x\left(R_{i,j} - R_{r,j}\right) \tag{6.9}$$

where $M_{i,j}$ is the jth dimensional parameters of the seed number 'k' produced from the ith tree, $R_{i,j}$ is the jth dimensional settings of the tree number 'I', D_j is the jth dimensional settings for the best solution of the tree location obtained so far, and $R_{r,j}$ is the jth dimensional settings for the tree number 'r', which is considered randomly and produced in the boundary [-1, 1]. The best-fitted seed is found throughout the search process by making spontaneous choices among the seeds of each tree. The TSA has progressed until the termination criterion is achieved. Figure 6.6 depicts the TSA flowchart in this case.

To properly dampen the frequency and tie-bar power oscillations, an adequate cost function (Q) is required to determine the best control settings. Furthermore, the system performance depends on the choice of a suitable Q. In this case, Q is considered as ISE, which is critical for the system's optimal performance and may be represented as Equation 6.10.

$$\text{Integral of squared error }(\text{ISE}) = \int_0^{\infty}\left\{\Delta F_1^2 + \Delta F_2^2 + \Delta P_{\text{tie12}}^2\right\}dt \tag{6.10}$$

6.5 Result and Analysis

MATLAB is used to simulate a two-area interconnected hybrid PS (picturized in Figure 6.1) having one NCES and one non-renewable type unit in each area. A little change in demand in the first and second areas requires the use of ΔP_{D1} and ΔP_{D2}, respectively. In the relevant areas, frequency errors ΔF_1 and ΔF_2 are observable, as well as a tie-bar power deviation of ΔP_{tie} between the areas. Thermal and solar power plants are considered in

Area-1, whereas nuclear power plants and ocean thermal power plants are located in Area-2.

In this study, the same system is used with two different strategies of control, such as TSA-optimized PID and TSA-tuned FUZZY-(1+PD)-FOPID controllers. Comparisons of the various system results are visualized in Figures 6.7 and 6.8, and the outcomes, such as overshoot (OS), undershoot (US), and settling time (ST), are tabulated in Table 6.2. Step changes are made to both areas, and it is shown that the three observation parameters ΔF_1, ΔF_2, and ΔP_{tie} quickly return to zero. The TSA-optimized

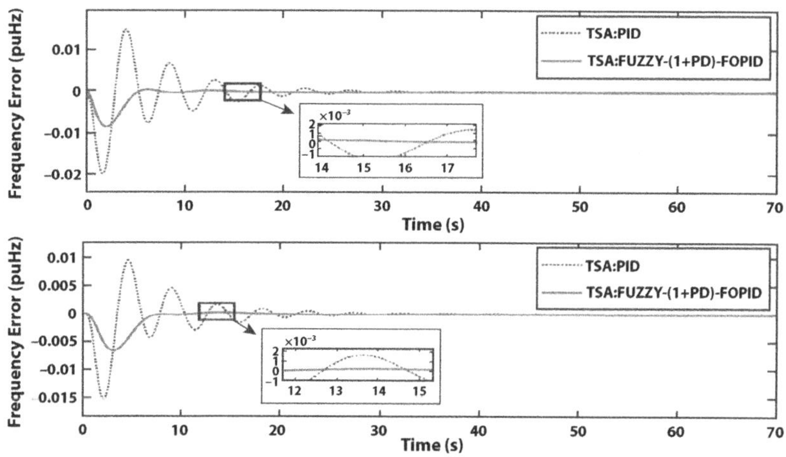

Figure 6.7 Frequency deviation in different areas of the hybrid power system controlled by TSA:PID, TSA:FUZZY-(1+PD)-FOPID controllers.

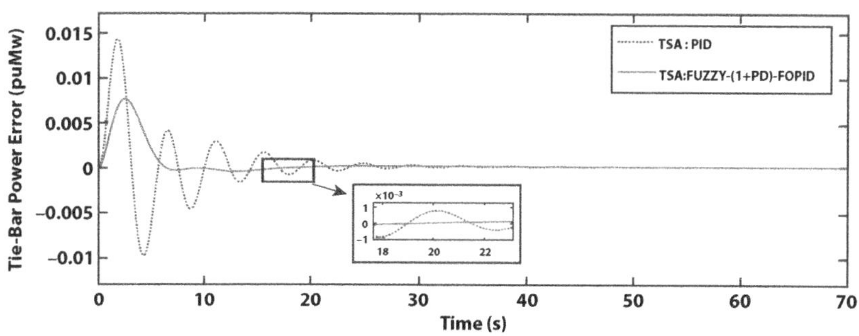

Figure 6.8 Tie-bar power deviation in different areas of the hybrid power system controlled by TSA:PID, TSA:FUZZY-(1+PD)-FOPID controllers.

Table 6.2 Two areas of hybrid power system performance analysis using TSA:FUZZY-(1+PD)-FOPID and TSA:PID controllers.

Functions	Parameters	TSA:PID	TSA:FUZZY-(1+PD)-FOPID
$\Delta F_1(Hz)$	OS	0.0148	0.0007762
	US	-0.0195	-0.008731
	ST	32.17	15.64
$\Delta F_2(Hz)$	OS	0.00912	0.0006628
	US	-0.0151	-0.006423
	ST	31.25	13.57
$\Delta P_{tie}(puMw)$	OS	0.00148	0.00076
	US	-0.009141	-0.0001953
	ST	30.47	18.78

parameters of the proposed three-layer hybrid controller along with the traditional controller such as TSA:PID are listed below in Table 6.3.

The magnitudes of the OS (0.0007762 Hz for ΔF_1, 0.0006628 Hz for ΔF_2 and 0.00076 puMw for ΔP_{tie}), US (-0.008731 Hz for ΔF_1, -0.006423 Hz for ΔF_2-0.0001953 puMw for ΔP_{tie}), and ST (15.64 s for ΔF_1, 13.57 s for ΔF_2 18.78 s for ΔP_{tie}) of the entire system outcomes are obtained from Figures 6.7 and 6.8 and provided in Table 6.2 as time-domain performance indicators. A critical remark in Table 6.2 reveals that our solution achieves the smallest value of different parameters with 92% to 95% lower OS, 55% to 57% lower US, and 51% to 57% lower ST for both the frequency error and 49%, 98%, and 38% lower OS, US, and ST, respectively, for tie-bar power error between the areas, as well as a 67% lower Q in comparison to the conventional approach. Overall, the proposed TSA-tuned three-layer hybrid cascade controller can give the fastest (lowest ST), most stable (lowest US and OS), and most accurate (oscillation-free) step response for the PS under investigation.

In this study, authors have also compared the efficacy of the TSA method with one traditional algorithm (PSO) by considering the same pair of PS and controller. Figures 6.9 and 6.10 illustrate the time-domain results of the same PS using PSO:FUZZY-(1+PD)-FOPID and TSA:FUZZY-(1+PD)-FOPID algorithms. The time-domain parameters are depicted in

Table 6.3 TSA:PID and TSA:FUZZY-(1+PD)-FOPID controller parameters.

Area	Parameters	TSA:PID	TSA:FUZZY-(1+PD)-FOPID
1st	K_{P1}	0.732	7.852
	K_{P2}	1.448	0.710
	K_{D1}	0.677	3.564
	K_{D2}	0.238	0.155
	K_{I2}	0.035	0.013
	μ	--	0.731
	λ	--	0.340
2nd	K_{P1}	0.997	8.125
	K_{P2}	3.143	0.971
	K_{D1}	0.978	3.642
	K_{D2}	1.382	0.415
	K_{I2}	0.010	0.010
	μ	--	0.618
	λ	--	0.542

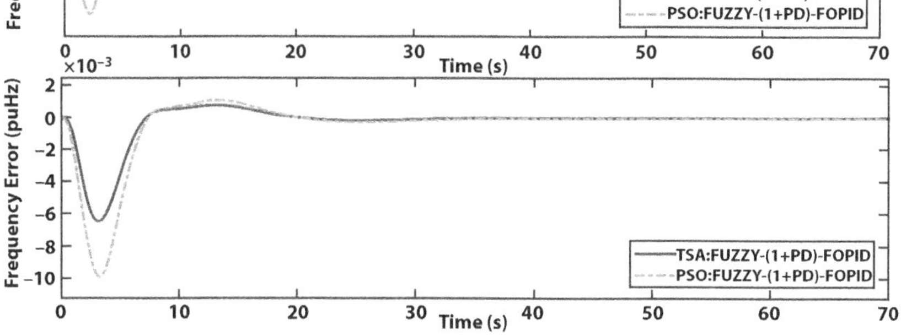

Figure 6.9 Frequency deviation in different areas of the hybrid power system controlled by PSO:FUZZY-(1+PD)-FOPID controllers, TSA:FUZZY-(1+PD)-FOPID controllers.

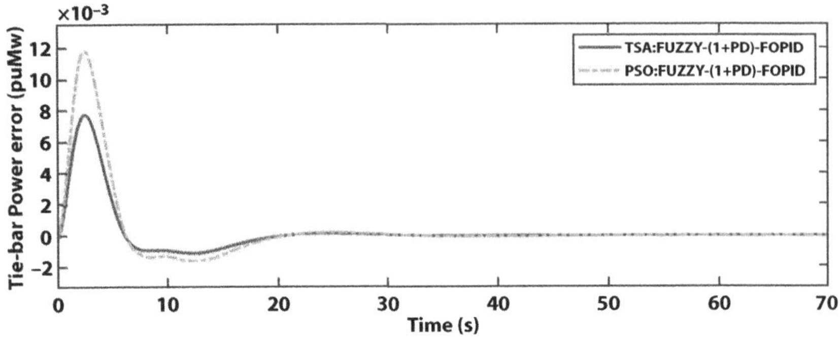

Figure 6.10 Tie-bar power deviation between different areas of the hybrid power system controlled by PSO:FUZZY-(1+PD)-FOPID controllers, TSA:FUZZY-(1+PD)-FOPID controllers.

Table 6.4. The proposed controller (FUZZY-(1+PD)-FOPID) parameters are presented in Table 6.5. Figure 6.9 visualizes the lucid difference between the performance of TSA and PSO over the proposed controller, where TSA:FUZZY-(1+PD)-FOPID outperforms the PSO:FUZZY-(1+PD)-FOPID with 37%, 18%, and 36% lower OS, 66%, 35%, and 9% lower US for $\Delta F_1, \Delta F_2$ & ΔP_{tie}, respectively, and with approximately 1% lower ST.

Table 6.4 Performance indicators of the PS using TSA:PID and TSA:FUZZY-(1+PD)-FOPID controllers.

Functions	Parameters	PSO:FUZZY-(1+PD)-FOPID	TSA:FUZZY-(1+PD)-FOPID
$\Delta F_1(Hz)$	OS	0.00124	0.0007762
	US	-0.0256	-0.008731
	ST	15.87	15.64
$\Delta F_2(Hz)$	OS	0.000812	0.0006628
	US	-0.00991	-0.006423
	ST	13.65	13.57
$\Delta P_{tie}(puMw)$	OS	0.00118	0.00076
	US	-0.0002135	-0.0001953
	ST	18.81	18.78

Table 6.5 PSO:FUZZY-(1+PD)-FOPID and TSA:FUZZY-(1+PD)-FOPID controller parameters.

Area	Parameters	TSA:FUZZY-(1+PD)-FOPID	TSA:FUZZY-(1+PD)-FOPID
1st	K_{P1}	11.724	7.852
	K_{P2}	4.648	0.710
	K_{D1}	1.787	3.564
	K_{D2}	0.985	0.155
	K_{I2}	0.351	0.013
	μ	0.842	0.731
	λ	0.553	0.340
2nd	K_{P1}	12.284	8.125
	K_{P2}	4.998	0.971
	K_{D1}	2.170	3.642
	K_{D2}	1.588	0.415
	K_{I2}	0.031	0.010
	μ	0.627	0.618
	λ	0.543	0.542

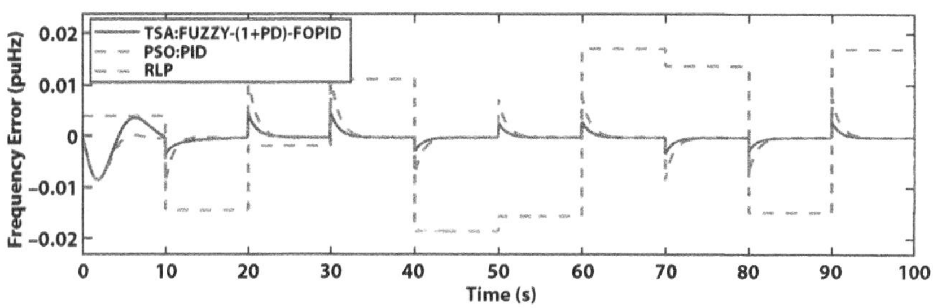

Figure 6.11 Under ±0.02-puMw RLP scenario load frequency fluctuation in the ocean thermal unit.

To test the resilience of the suggested FUZZY-(1+PD)-FOPID controller, a random load type (RLP) noise is offered to Area-2 of the hybrid PS (Figure 6.1), which includes the ocean thermal PS as a power plant. The disturbance's magnitude is calculated to be ±0.02 puMw. A comparison between resilient behavior and a conventional controller, such as a PSO-adjusted PID, is shown in Figure 6.11. The red line in Figure 6.11 denotes the disturbance.

6.6 Conclusion

In this chapter, a three-layer cascade hybrid FUZZY-(1+PD)-FOPID controller application is analyzed in the context of an interconnected hybrid PS network. Initially, only PID controllers were used to regulate this sort of PS network. From the findings, as shown in Figures 6.7 and 6.8, it can be seen that the response graph exhibits strong oscillations. The oscillations are significantly minimized and the system becomes more stable with the skillful incorporation of fuzzy logic and (1+PD) unit with an FOPID controller. As shown in Table 6.2, the suggested controller operates effectively under a variety of settings and, in certain circumstances, outperforms the performance of the traditional controlling strategy (PID). The robustness of the developed controller is examined, and as can be shown in Figure 6.11, this analysis has shown its efficacy. As a result, the control strategy developed utilizing the TSA optimization approach and controller structure proposed in this study may provide an effective substitute for the current method in LFC investigations of interconnected PS networks containing renewable energy sources.

Appendix

- For Thermal Power unit:

$$T_{G1} = 0.08s,\ T_{T1} = 0.3s,\ K_{G1} = 1,\ K_{T1} = 1$$

- For Solar Unit:

$$A = 0.08,\ B = 0.8,\ G = 1,\ F = 0.1$$

- For Ocean Thermal Power Plant:

$$K_{SG} = 1.05,\ T_{SG} = 0.06s,\ T_t = 1.2s$$

- For Nuclear Power Plant:

$$T_{G2} = 0.08s,\ T_{T2} = 0.3s,\ K_{G2} = 1,\ K_{T2} = 1$$

- For Other Parameters:

$$K_{PSi} = 120\frac{Hz}{puMW},\ T_{PSi} = 20s,\ R_1 = R_2 = \frac{2.4\ Hz}{pu},\ \beta_1 = \beta_2 = 0.425\frac{puMW}{Hz}$$

References

1. Kristov, L., The Bottom-Up (R)Evolution of the Electric Power System: The Pathway to the Integrated-Decentralized System. *IEEE Power Energy Mag.*, 17, 2, 42–49, March-April 2019, doi: 10.1109/MPE.2018.2885204.
2. Xu, S., Xue, Y., Chang, L., Review of power system support functions for inverter-based distributed energy resources-standards, control algorithms, and trends. *IEEE Open J. Power Electron.*, 2, 88–105, 2021.
3. Shakibjoo, A.D., Moradzadeh, M., Din, S.U., Mohammadzadeh, A., Mosavi, A.H., Vandevelde, L., Optimized Type-2 Fuzzy Frequency Control for Multi-Area Power Systems. *IEEE Access*, 10, 6989–7002, 2022, doi: 10.1109/ACCESS.2021.3139259.
4. Hidalgo, D., Castillo, O., Melin, P., Type-1 and type-2 fuzzy inference systems as integration methods in modular neural networks for multimodal biometry and its optimization with genetic algorithms. *Inf. Sci.*, 179, 13, 2123–2145, 2009.
5. Zribi, M., Al-Rashed, M., Alrifai, M., Adaptive decentralized load frequency control of multi-area power systems. *Int. J. Electr. Power Energy Syst.*, 27, 8, 575–583, 2005.
6. Panda, G., Panda, S., Ardil, C., Hybrid neuro fuzzy approach for automatic generation control of two–area interconnected power system. *Int. J. Comput. Intell.*, 5, 1, 80–84, 2009.
7. Sahu, R.K., Panda, S., Chandra Sekhar, G.T., A novel hybrid PSO-PS optimized fuzzy PI controller for AGC in multi area interconnected power systems. *Int. J. Electr. Power Energy Syst.*, 64, 880–893, 2015.
8. Rodríguez-Abreo, O., *et al.*, Self-tuning neural network PID with dynamic response control. *IEEE Access*, 9, 65206–65215, 2021.

9. Vrdoljak, K., Perić, N., Petrović, I., Sliding mode based load-frequency control in power systems. *Electr. Power Syst. Res.*, 80, 5, 514–527, 2010.
10. Dahiya, P., Sharma, V., Naresh, R., Optimal sliding mode control for frequency regulation in deregulated power systems with DFIG-based wind turbine and TCSC–SMES. *Neural Comput. Appl.*, 31, 7, 3039–3056, 2019.
11. Rosaline, A.D. and Somarajan, U., Structured H-Infinity controller for an uncertain deregulated power system. *IEEE Trans. Ind. Appl.*, 55, 1, 892–906, 2018.
12. Sondhi, S. and Hote, Y.V., Fractional order PID controller for load frequency control. *Energy Convers. Manage.*, 85, 343–353, 2014.
13. Arya, Y. and Kumar, N., BFOA-scaled fractional order fuzzy PID controller applied to AGC of multi-area multi-source electric power generating systems. *Swarm Evol. Comput.*, 32, 202–218, 2017.
14. Zamani, A., Masoud Barakati, S., Yousofi-Darmian, S., Design of a fractional order PID controller using GBMO algorithm for load–frequency control with governor saturation consideration. *ISA Trans.*, 64, 56–66, 2016.
16. Kumar, N., Kumar, V., Tyagi, B., Multi area AGC scheme using imperialist competition algorithm in restructured power system. *Appl. Soft Comput.*, 48, 160–168, 2016.
17. Ebrahim, M.A., Becherif, M., Abdelaziz, A.Y., PID-/FOPID-based frequency control of zero-carbon multisources-based interconnected power systems under deregulated scenarios. *Int. Trans. Electr. Energy Syst.*, 31, 2, e12712, 2021.
18. Arya, Y., *et al.*, AGC performance amelioration in multi-area interconnected thermal and thermal-hydro-gas power systems using a novel controller. *Eng. Sci. Technol. Int. J.*, 24, 2, 384–396, 2021.
19. Arya, Y., *et al.*, Cascade-IλDμN controller design for AGC of thermal and hydro-thermal power systems integrated with renewable energy sources. *IET Renewable Power Gener.*, 15, 3, 504–520, 2021.
20. Saikia, L.C., *et al.*, Automatic generation control of a multi area hydrothermal system using reinforced learning neural network controller. *Int. J. Electr. Power Energy Syst.*, 33, 4, 1101–1108, 2011.
21. Khuntia, S.R. and Panda, S., Simulation study for automatic generation control of a multi-area power system by ANFIS approach. *Appl. Soft Comput.*, 12, 1, 333–341, 2012.
22. Fathy, A. and Kassem, A.M., Antlion optimizer-ANFIS load frequency control for multi-interconnected plants comprising photovoltaic and wind turbine. *ISA Trans.*, 87, 282–296, 2019.
23. Sharma, D. and Mishra, S., Non-linear disturbance observer-based improved frequency and tie-line power control of modern interconnected power systems. *IET Gener. Transm. Distrib.*, 13, 16, 3564–3573, 2019.
24. Guha, D., Roy, P.K., Banerjee, S., Disturbance observer aided optimised fractional-order three-degree-of-freedom tilt-integral-derivative controller for

load frequency control of power systems. *IET Gener. Transm. Distrib.*, 15, 4, 716–736, 2021.
25. Sahu, B.K., *et al.*, A novel hybrid LUS–TLBO optimized fuzzy-PID controller for load frequency control of multi-source power system. *Int. J. Electr. Power Energy Syst.*, 74, 58–69, 2016.
26. Nayak, J.R., Shaw, B., Sahu, B.K., Application of adaptive-SOS (ASOS) algorithm based interval type-2 fuzzy-PID controller with derivative filter for automatic generation control of an interconnected power system. *Eng. Sci. Technol. Int. J.*, 21, 3, 465–485, 2018.
27. Haroun, Gomaa, A.H., Li, Y.-Y., A novel optimized hybrid fuzzy logic intelligent PID controller for an interconnected multi-area power system with physical constraints and boiler dynamics. *ISA Trans.*, 71, 364–379, 2017.
28. Fathy, A., Kassem, A.M., Abdelaziz, A.Y., Optimal design of fuzzy PID controller for deregulated LFC of multi-area power system via mine blast algorithm. *Neural Comput. Appl.*, 32, 4531–4551, 2020.
29. Nayak, N., *et al.*, Application of modified sine cosine algorithm to optimally design PID/fuzzy-PID controllers to deal with AGC issues in deregulated power system. *IET Gener. Transm. Distrib.*, 13, 12, 2474–2487, 2019.
30. Nayak, P.C., *et al.*, Performance analysis of hydrogen aqua equaliser fuel-cell on AGC of wind-hydro-thermal power systems with sunflower algorithm optimised fuzzy-PDFPI controller. *Int. J. Ambient Energy*, 43, 1, 3454–3467, 2022.
31. Arya, Y., A new optimized fuzzy FOPI-FOPD controller for automatic generation control of electric power systems. *J. Franklin Inst.*, 356, 11, 5611–5629, 2019.
32. Sudha, K.R. and Vijaya Santhi, R., Robust decentralized load frequency control of interconnected power system with generation rate constraint using type-2 fuzzy approach. *Int. J. Electr. Power Energy Syst.*, 33, 3, 699–707, 2011.
33. Kocaarslan, I., Akalın, G., Erfidan, T.A.R.I.K., Application of fuzzy reasoning to load-frequency control in two area power system. *1999 European Control Conference (ECC)*, IEEE, 1999.
34. Chandran, K., *et al.*, Modified cascade controller design for unstable processes with large dead time. *IEEE Access*, 8, 157022–157036, 2020.
35. Sanki, P. and Basu, M., New approach in two-area interconnected AGC including various renewable energy sources using PSO. *Turk. J. Electr. Eng. Comput. Sci.*, 26, 3, 1491–1504, 2018.
36. Chaine, S. and Tripathy, M., Performance of CSA optimized controllers of DFIGs and AGC to improve frequency regulation of a wind integrated hydrothermal power system. *Alexandria Eng. J.*, 58, 2, 579–590, 2019.
37. Hakimuddin, N., Nasiruddin, I., Bhatti, T.S., Generation-based automatic generation control with multisources power system using bacterial foraging algorithm. *Eng. Rep.*, 2, 8, e12191, 2020.
38. Kiran, M.S., TSA: Tree-seed algorithm for continuous optimization. *Expert Syst. Appl.*, 42, 19, 6686–6698, 2015.

39. Biswas, S., Roy, P.K., Chatterjee, K., Renewable energy-based multi-source system under deregulated environment using cokha algorithm. *IETE J. Res.*, 69, 3, 1–19, 2021.
40. Hernández-Romero, I.M., *et al.*, Optimal design of the ocean thermal energy conversion systems involving weather and energy demand variations. *Chem. Eng. Process.-Process Intensif.*, 157, 108114, 2020.
41. Hasan, N. and Farooq, S., Real time simulation of automatic generation control for interconnected power system. *Int. J. Electr. Eng. Inform.*, 4, 1, 40, 2012.
42. Gupta, N. and Garg, R., Tuning of asymmetrical fuzzy logic control algorithm for SPV system connected to grid. *Int. J. Hydrogen Energy*, 42, 26, 16375–16385, 2017.
43. Kapoor, S., Chaturvedi, M., Juneja, P.K., Design of fractional order PID controller for a SOPDTprocess model. *2017 Fourth International Conference on Image Information Processing (ICIIP)*, IEEE, 2017.
44. Nayak, J.R., Shaw, B., Sahu, B.K., Implementation of hybrid SSA–SA based three-degree-of-freedom fractional-order PID controller for AGC of a two-area power system integrated with small hydro plants. *IET Gener. Transm. Distrib.*, 14, 13, 2430–2440, 2020.
45. Abraham, A., *et al.*, Design of fractional order PIλDμ controllers with an improved differential evolution. *Proceedings of the 10th annual conference on Genetic and evolutionary computation*, 2008.
46. Zamani, M., *et al.*, Design of a fractional order PID controller for an AVR using particle swarm optimization. *Control Eng. Pract.*, 17, 12, 1380–1387, 2009.
47. Monje, C.A., *et al.*, On fractional PI λ controllers: some tuning rules for robustness to plant uncertainties. *Nonlinear Dyn.*, 38, 369–381, 2004.
48. Barbosa, R.S., Tenreiro Machado, J.A., Ferreira, I.M., Tuning of PID controllers based on Bode's ideal transfer function. *Nonlinear Dyn.*, 38, 305–321, 2004.

7

Design of MPC-TSA Controller for Hybrid Two-Area Power System

Susmit Chakraborty[1]* and Arindam Mondal[2]

[1]*Department of Computer Science and Engineering-Cyber Security and Data Science, Brainware University, Barasat, West Bengal, India*
[2]*Department of Electrical Engineering, Dr. B C Roy Engineering College, Durgapur, West Bengal, India*

Abstract

In the past several decades, the discipline of control has seen significant advancements in model predictive control (MPC). It offers several benefits, including quick reaction times, steadiness against nonlinear behaviors, and parametric uncertainties. The control input for MPC, which is also termed as control of the receding horizon, is derived by evaluating a discrete-time optimum control problem over a predetermined horizon. This chapter proposes the tree-seed algorithm (TSA) for most ideal MPC controller settings for a two-area interconnected hybrid power system (2-AIHPS) that includes thermal and nuclear units along with non-conventional energy sources (NCES) like ocean thermal power plant (OTPP) and solar plant to reduce power system pulsations. TSA is used to find the best controller settings by minimizing a time-domain–based fitness function such as the integral-time multiplied squared error (ITSE). The performance of the suggested controller may be assessed by contrasting it with TSA-based PID and conventional control approaches like PSO-based PID. In the MATLAB simulation platform, the entire system (2-AIHPS), including the controllers, is sketched and modeled. Simulation outcomes such as settling time (ST), overshoot (OS), and undershoot (US) on a 2-AIHPS are analyzed for all controlling methods discussed above. The suggested controller outperforms other controllers found in the literature on account of its robustness to load variations.

**Corresponding author*: susmit.eee@gmail.com

Arindam Mondal and Souvik Ganguli (eds.) Controller Design for Industrial Applications, (125–140)

***Keywords*:** Load frequency control (LFC), tree-seed algorithm (TSA), model predictive control (MPC), renewable energy sources, ocean thermal power plant (OTPP), power system (PS)

7.1 Introduction

Controlling of load frequency (LFC) is essential for the reliable functioning of the interconnected hybrid PS (IHPS) as additional distributed renewable generation (DRG) sources are included in the networks [1, 2]. The LFC has been widely used in traditional electric power systems and is capable of maintaining frequency stability using a number of approaches including optimum control and proportional-integral (PI) controlling methods [3–5]. To optimize the dual-mode PI controller, the bat-inspired technique was adopted [6]. Constrained population external optimization (CPEO) has been used to present and optimize a traditional PI nonlinear LFC system [7, 8]. A PID, or proportional-integral-derivative, controller has been built using an artificial bee colony (ABC) [9].

However, PI or PID are unable to provide good controlling performances under intricate systems due to difficulties in adjusting the ideal parameters and in the management of frequency control over an IHPS containing multi-area. To give the best performance over a wide variety of operations, a self-tuning gain controller [10] is necessary. The available self-tuning sophisticated control schemes incorporating sliding modes [11, 12], H-infinity [13], fractional calculus method [14, 15], artificial neural network approach with FLC [16], adaptive neuro-fuzzy inference system (ANFIS) [17, 18], an observer of nonlinear disturbances [19], disturbance observer optimized fractional-order three-degrees-of-freedom tilt-integral-derivative (FO-3DOF-TID) [20], and fuzzy logic controllers (FLCs) [21–23] are documented in the appropriate field to improve control performance.

Model predictive control (MPC) has seen a substantial increase in interest recently. The rapid responsiveness and stability of MPC against nonlinearities, restrictions, and parameter uncertainties are what have sparked its growing popularity [24–26]. The performance of the suggested controllers will be improved by MPC's potent features. Some top findings on the control of MPC technique in LFC were given in [27–32]. This chapter suggests the use of TSA to tune MPC-based LFCs with renewable energy power plant units in a 2-AIHPS to improve the power quality. TSA is a population-based, potent stochastic metaheuristic optimization method that originated by nature. To handle continuous optimization difficulties, Kiran [33] invented TSA, which is in accordance with the relationship of trees and seeds. This combination of controlling methods comprising TSA optimization, the MPC unit, and a system like 2-AIHPS containing ocean thermal as a unit is very unique in the LFC research domain until now.

7.2 Plant Model

In this chapter, the plant that has to be controlled is 2-AIHPS with two separate units in each area. Each area has a combination of RES (solar or ocean thermal power plant) and traditional energy sources (thermal or nuclear power plant). The first area includes a thermal power plant (TPP) and a solar photovoltaic unit [34, 35], while another area includes an ocean thermal power unit [36] and a nuclear power unit [37]. The simulation schematic of the model is picturized in Figure 7.1.

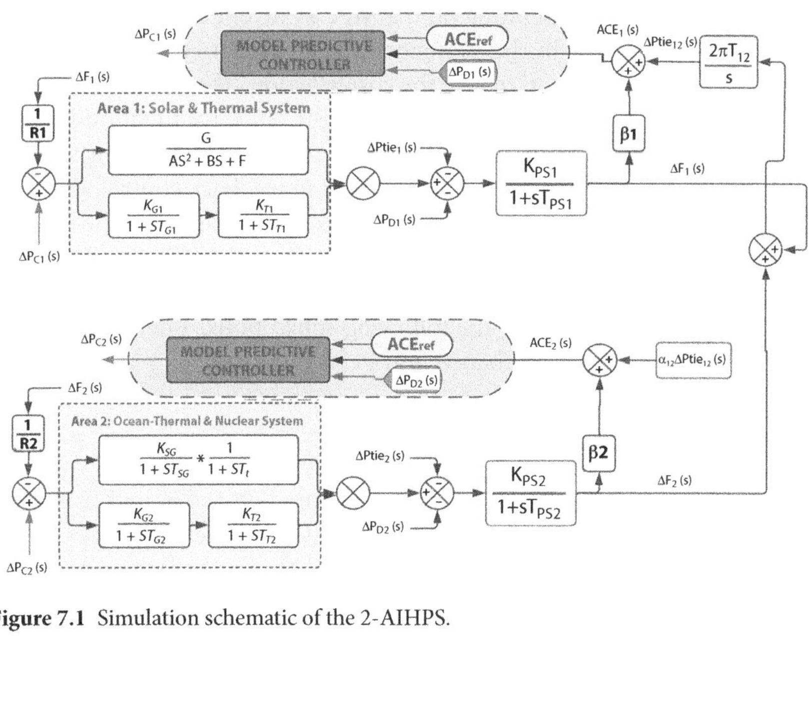

Figure 7.1 Simulation schematic of the 2-AIHPS.

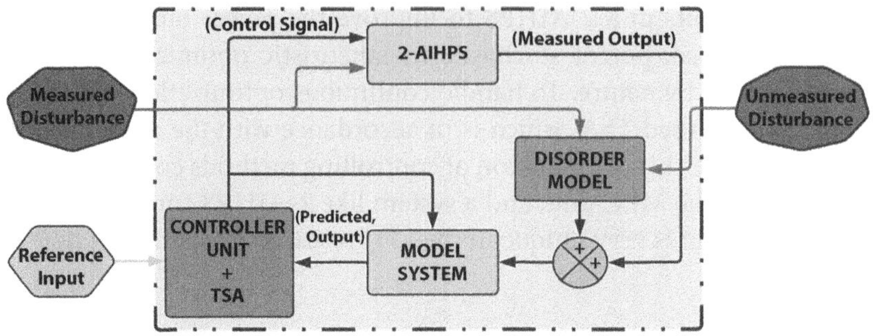

Figure 7.2 General MPC design.

7.3 Model Predictive Controller

MPC is gaining traction as a viable control strategy for stabilizing nonlinear dynamical systems mostly in process control with uncertainty and temporal delays [25, 26, 38]. As illustrated in Figure 7.2, an MPC general unit consists of two fundamental units: the predicting unit (PU) and the controlling unit (CU) [38]. On a bounded predicting horizon, the PU forecasts the system's future behavior depending on the MPC's current output, disturbance, and controlling signal. In the presence of system restrictions, the CU utilizes the expected output to reduce the cost function. There are several MPC approaches that differ in their definitions of the objective function [25, 26]. The measured disturbance (MD) in the MPC can be corrected using the FFC, or feed-forward control approach. In contrast to feedback control, FFC suppresses the observed disturbance before it has any influence on the PS. FFC is used in conjunction with feedback control where the FFC invalidates the majority of the MD, while the feedback control rejects the remainder while also dealing with un-MD. In the following subsections, a detailed study of the models for system, disturbance, and PU is discussed.

7.3.1 Model System

A model system is nothing but a system that predicts the output using past trends and the response generated by a disordered system. The model system is an LTI (linear-time-invariant) system. The estimating and optimization procedures in MPC are preceded by four essential phases, which are as follows: state-space conversion, quantization, delay reduction, and non-dimensional input and output conversion which are represented by Equations 7.1 and 7.2, respectively.

$$x_p(m+1) = D_p\ x_p(m) + F\ J_i\ \mu_p(m) \tag{7.1}$$

$$y_p(m) = J_0^{-1}L\ x_p(m) + J_0^{-1}M\ J_i\ \mu_p(m) \tag{7.2}$$

The diagonal matrices J_i and J_0 are used to scale the input and output of the MPC, respectively. The system mode on a whole is comparable to Equations 7.3 and 7.4, respectively.

$$x_p(m+1) = D_p\ x_p(m) + F_{p\mu}\ \mu_p(m) + F_{p\vartheta}\ \vartheta(m) + F_{p\omega}\ \omega(m) \tag{7.3}$$

$$y_p(m) = L_p\ x_p(m) + M_{p\mu}\ \mu_p(m) + M_{p\vartheta}\ \vartheta(m) + M_{p\omega}\ \omega(m) \tag{7.4}$$

where $L_p = J_0^{-1}L$ and $F_{p\mu}, F_{p\vartheta}$ & $F_{p\omega}$ correspond to the $F\,J_i$ columns. Similarly, comparable columns of $J_0^{-1}M\ J_i$ are $M_{p\mu}, M_{p\vartheta}, M_{p\omega}$. $\vartheta(m)$ and $\omega(m)$ are the measured and unmeasured disturbances, respectively.

7.3.2 Disorder Model

The disorder model performs the scaling of the un-MD using predetermined MD. The disturbance's discrete-time state-space structures are formulated as Equations 7.5 and 7.6.

$$x_\omega(m+1) = D_\omega x_\omega(m) + F_\omega\ \varphi_\omega(m) \tag{7.5}$$

$$\omega(m) = L_\omega x_\omega(m) + M_\omega \varphi_\omega(m) \tag{7.6}$$

where $\omega(m)$ and $\varphi_\omega(m)$ are the dimensionless un-MD and white noise input, respectively.

7.3.3 Predicting Model

This is the main engine of the MPC, where the future value of the controlling signal is predicted using the model of the time-discretized state-space of the measured noise which is structured in Equations 7.7 and 7.8 as follows:

$$x_n(m+1) = D_n x_n(m) + F_n\, \varphi_n(m) \tag{7.7}$$

$$y_n(m) = L_n x_n(m) + M_n \varphi_n(m) \tag{7.8}$$

In this case, $\varphi_n(m)$ is a non-dimensional white noise input. Surprisingly, the MPC operates on intervals that are discrete with unvarying time samples (Ts). If the MPC begins at zero, these intervals are integer multiples of the sampling period, i.e., from 0 to mTs, where m represents the current sampling moment. Figures 7.3 and 7.4 depict the SISO, or serial-input-serial-output–based MPC system that remained operational for a particular duration. The current predicting action μ(m) is computed by resolving the optimization problem for the mth instant of sampling, as listed in Equation 7.9.

$$\min_{\mu_m,\ldots\ldots\ldots,\mu_{m+P-1}} \sum_{i=1}^{P} \left\{ Q\left(r_{m+i} - \bar{y}_{m+i}\right)^2 + R\left(\Delta\mu_{m+i-1}\right)^2 \right\} \tag{7.9}$$

Under consideration, $\mu_{min} \le \mu_{m+i} \le \mu_{max}; y_{min} \le \bar{y}_{m+i} \le y_{max};$

$$\left|\Delta\mu_{m+i}\right| \le \Delta\mu_{max} \tag{7.10}$$

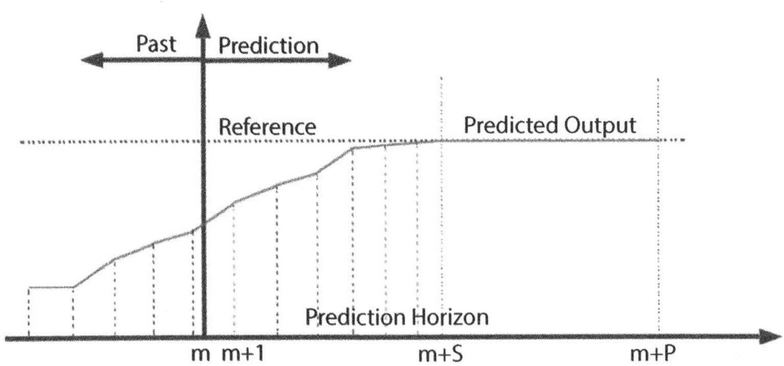

Figure 7.3 Movement of MP-LFC output.

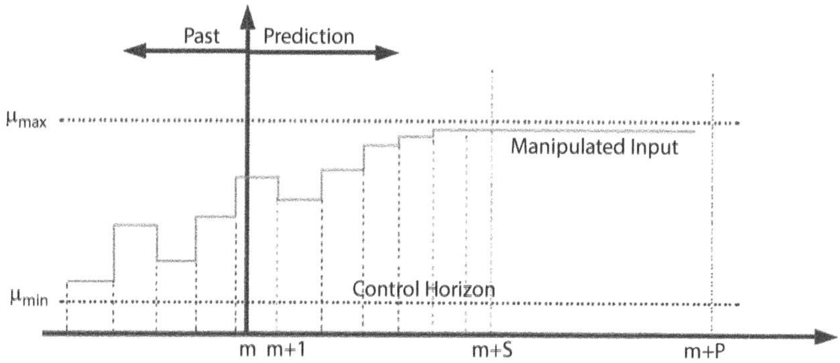

Figure 7.4 Movement of MP-LFC control signal.

where $\Delta\mu_j = \mu_j - \mu_{j-1}$ represents the move alteration at the jth sampling moment, Q & R are non-zero finite weights, and r_{m+i} represents the desired value for reference. Lastly, TSA is used to obtain the best-fitted MPC parameters such as Ts, S, P, Q, and R.

7.3.4 MPC on Load-Frequency-Control (MP-LFC)

The MP controller was utilized to create the controlling signal using area control error signal (ACE_i), unmeasured disturbance (ΔP_{Di}), and the reference value for area control error (ACE_r) as inputs, where ACEr is equal to zero. Figure 7.5 depicts an MP-LFC management technique. In this study, the MP control toolbox from MATLAB Simulink v.2020A was utilized to build the MPC controller.

MPC runs at a rate of $1/(N * T_s)$, where T_s is the period of sampling and N represents how many controlling signals are considered to the PS. In nearly every instance, N is fixed at one. The time span of every single prediction step is determined by T_s. The tracking performance approach is used to pick T_s for this problem. The controller also had an impact on the horizon of prediction P as well as control horizon S. Q and R are the two

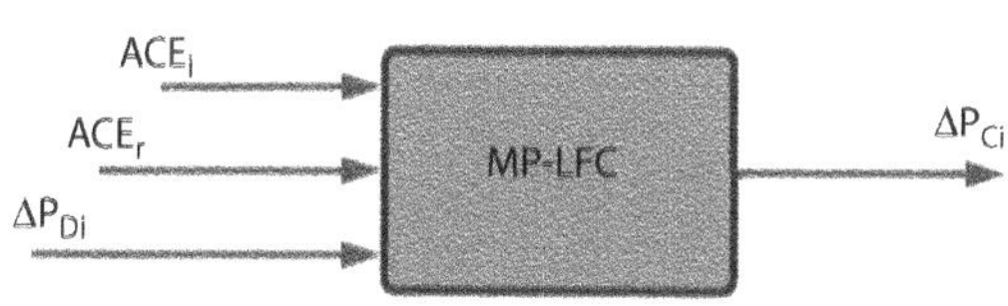

Figure 7.5 A MP-LFC scheme.

weights on the input and output signals of the system that are set to their optimal values. In this study, the optimization method termed TSA or tree-seed algorithm is presented to find the optimum combination of T_s, P, S, Q, and R. In order to properly dampen the frequency oscillations and tie-bar power error in the 2-AIHPS, an adequate fitness function or cost function (J) is required to determine the best control settings. Furthermore, the ability to perform the system is determined by the selection of an appropriate J. In this case, J is considered as ITSE, which is critical for the system's optimal performance and may be represented as (11).

$$\text{Fitness Function }(J)=\int_0^\infty t\left\{\Delta F_1^2+\Delta F_2^2+\Delta P_{tie12}^2\right\}dt \tag{7.11}$$

7.4 Tree-Seed Algorithm

It is a natural population-based, powerful, probability-derived, metaheuristic optimization approach. Kiran [33] designed TSA, which is oriented on the interaction of trees and seeds, to deal with the issues of progressing optimization. The seeds execute the replicating and producing mechanisms of the tree population. The method of transportation of seeds varies according to their architectures and forms. Some seeds remain attached to birds' feathers, animal furs, and clothing. They are transported from different trees to various locations. By this process, the seeds germinate in the correct atmosphere and develop into trees. The ground is considered as the search space for optimization issues in TSA. The answers to optimization issues are represented by both seeds and trees. The techniques for exploiting and exploring are critical for the outstanding efficiency of the metaheuristic processes and for creating a better-resulting solution. During the exploration process, the trees are randomly placed around the search space, and the seeds with features similar to the trees are employed in the exploitation phase. The seeds were generated using Equations 7.12 and 7.13 as follows:

$$M_{i,j}=R_{i,j}+a_{i,j}\ x\left(D_j-R_{r,j}\right) \tag{7.12}$$

$$M_{i,j}=R_{i,j}+a_{i,j}\ x\left(R_{i,j}-R_{r,j}\right) \tag{7.13}$$

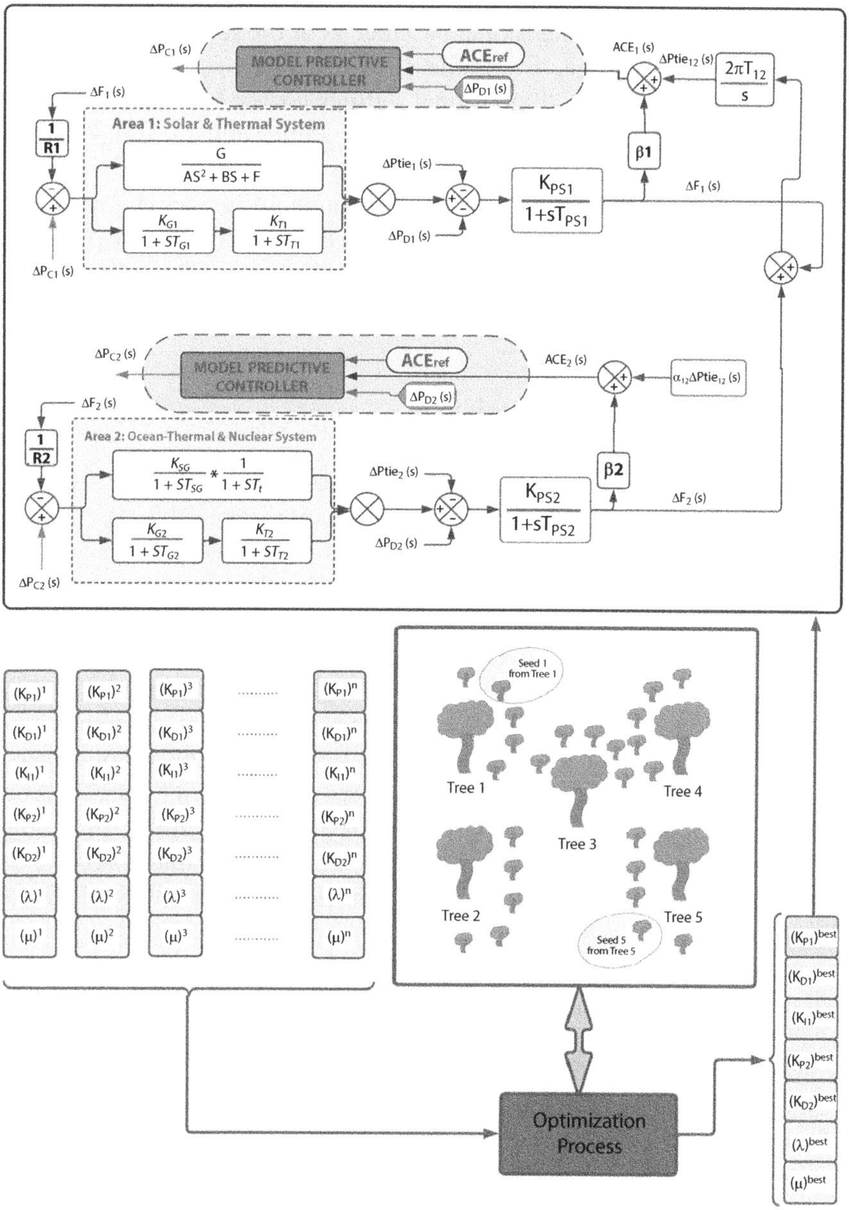

Figure 7.6 Flowchart of TSA optimization process for LFC problem.

where $M_{i,j}$ is the jth dimensional parameters of the seed number 'k' produced from the ith tree, $R_{i,j}$ is the jth dimensional settings of the tree number 'I', D_j is the jth dimensional settings for the best solution of the tree location obtained so far, $R_{r,j}$ is the jth dimensional settings for the tree number 'r', which is considered randomly and produced in the boundary [-1, 1]. The best-fitted seed is found throughout the search process by making spontaneous choices among the seeds of each tree. The TSA has progressed until the termination criterion is achieved. Figure 7.6 depicts the TSA flowchart in this case.

7.5 Result and Analysis

MATLAB is used to simulate a two-area interconnected hybrid PS (picturized in Figure 7.1) having one NCES and one non-renewable type unit in each area. A little change in demand in the first and second areas requires the use of ΔP_{D1} and ΔP_{D2}, respectively. In the relevant areas, frequency errors ΔF_1 and ΔF_2 are observable, as well as a tie-bar power deviation of ΔP_{tie} between the areas. Thermal and solar power plants are considered in Area-1, whereas nuclear power plants and ocean TPPs are located in Area-2. In this study, the same system is used with three different combinations of CUs, such as TSA:MPC, TSA:PID, and PSO:PID controllers. Comparisons of the various system responses are shown in Figure 7.7, and the outcomes, such as overshoot (OS), undershoot (US), and settling time (ST), are tabulated in Table 7.1. The TSA-optimized parameters of MPCs along with the traditional controllers such as TSA:PID and PSO:PID are listed below in Table 7.2.

A critical remark of Table 7.1 reveals that the solution of the TSA:MPC achieves the smallest value of different parameters as well as the value of J (0.00361) in comparison to the conventional approaches. In terms of OS comparison, TSA:MPC performs better with 96%, 61%, and 100% than PSO:PID and 87%, 12%, and 100% than TSA:PID for respective frequency deviations and tie-bar power error, respectively. In terms of US comparison, TSA:MPC also performs better with 74%, 99%, and 100% than PSO:PID and 44%, 98%, and 100% than TSA:PID for all three different parameters, respectively. For the comparison of ST, similar betterment can also be observed as per Table 7.1. Overall, the proposed TSA-tuned model predictive controller (TSA:MPC) can give the fastest (minimum TS), most

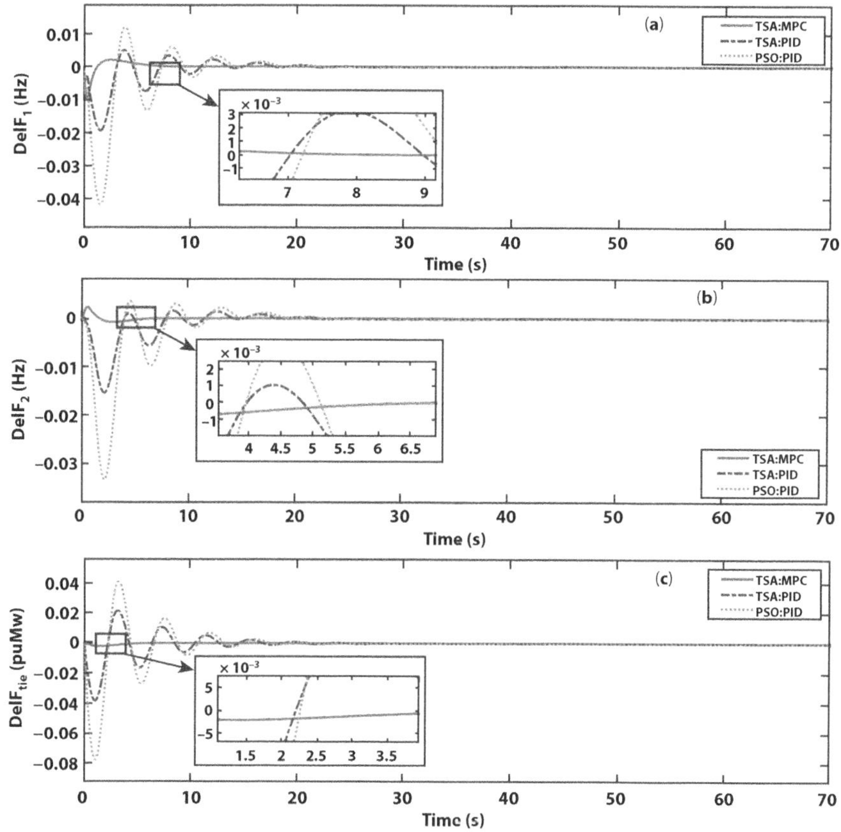

Figure 7.7 Frequency deviation and tie-bar power error in different areas of the hybrid power system controlled by TSA:MPC, TSA:PID, and PSO:PID controllers.

stable (minimum US and OS), and most accurate (oscillation-free) step response for the power system under investigation.

To check the resilience of the suggested TSA:MPC, a random type load (RLP) disturbance is introduced to Area-2 of the 2-AIHPS (Figure 7.1), which includes the ocean thermal power system as a power plant. The disturbance's magnitude is calculated to be ±0.02 puMw. A comparison between resilient behavior and both conventional controlling approaches, such as a PSO-adjusted PID and TSA-adjusted PID, is shown in Figure 7.8. The red line in Figure 7.8 denotes the RLP disturbance.

Table 7.1 Performance analysis of 2-AIHPS using PSO:PID, TSA:PID, and TSA:MPC.

Functions	Parameters	PSO:PID	TSA:PID	TSA:MPC
ΔF_1 (in Hz)	OS	0.0132	0.0038	0.000476
	US	-0.0409	-0.0192	-0.01073
	ST	23.55	21.76	6.94
ΔF_2 (in Hz)	OS	0.0043	0.0019	0.001663
	US	-0.0324	-0.0154	-0.00024
	ST	22.49	21.37	6.57
ΔP_{tie} (in puMw)	OS	0.0402	0.0213	0
	US	-0.0794	-0.0395	-0.000193
	ST	22.47	21.52	3.57

Table 7.2 PSO:PID, TSA:PID, and TSA:MPC controller parameters.

Area	Controller	K_{Pi}	K_{Di}	K_{Ii}	T_S	S	P	Q	R
1st	PSO:PID	0.7239	0.5330	0.7742	-	-	-	-	-
	TSA:PID	0.9531	0.3637	0.5002	-	-	-	-	-
	TSA:MPC	-	-	-	0.021	1	4	20	3.579
2nd	PSO:PID	0.6998	0.1434	0.9724	-	-	-	-	-
	TSA:PID	0.7331	0.0637	0.5553	-	-	-	-	-
	TSA:MPC	-	-	-	0.015	3	3	19.997	3.425

7.6 Conclusion

The work presented here is an integrated MPC-LFC strategy to improve frequency regulation control performance in 2-AIHPS. At first, only PID controllers were utilized to control the power system network. The findings, as shown in Figure 7.7, reveal that the response graph has substantial oscillations when the 2-AIHPS works by being controlled by a PID

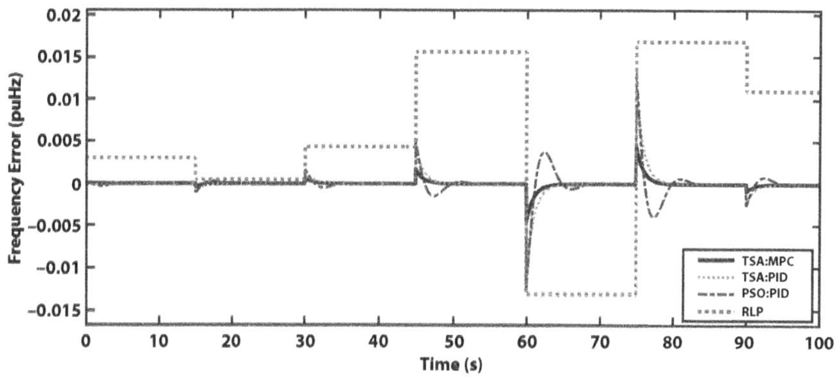

Figure 7.8 Under ±0.02-puMw RLP scenario load frequency fluctuation in ocean thermal unit.

controller. With the efficient insertion of an MPC unit into the power system network, oscillations are drastically reduced, and the system becomes highly stable quickly. As shown in Table 7.1, the suggested controller operates effectively under a variety of settings and, in certain circumstances, outperforms the performance of the traditional controlling strategy (PID). The robustness of the constructed controller is investigated, and as shown in Figure 7.8, this examination demonstrates its efficacy. As a consequence, the control strategy described in this study that employs the TSA optimization technique and controller structure may be an effective alternative to the existing method in LFC analyses of interconnected power system networks comprising renewable energy sources.

Appendix

- For TPP unit:

$$T_{G1} = 0.08s, T_{T1} = 0.3s, K_{G1} = 1, K_{T1} = 1$$

- For solar unit:

$$A = 0.08, B = 0.8, G = 1, F = 0.1$$

- For ocean thermal power plant:

$$K_{SG} = 1, T_{SG} = 0.08s, T_t = 1.25s$$

- For nuclear power plant:

$$T_{G2} = 0.08s, T_{T2} = 0.3s, K_{G2} = 1, K_{T2} = 1$$

- For other parameters:

$$K_{PSi} = 120\frac{Hz}{puMW}, T_{PSi} = 20s, R_1 = R_2 = \frac{2.4Hz}{pu}, \beta_1 = \beta_2 = \frac{0.425\ puMW}{Hz}$$

References

1. Shayeghi, H.A.S.H., Shayanfar, H.A., Jalili, A., Load frequency control strategies: A state-of-the-art survey for the researcher. *Energy Convers. Manage.*, 50.2, 344–353, 2009.
2. Liu, F., *et al.*, Optimal load-frequency control in restructured power systems. *IEE Proceedings-Generation Transm. Distrib.*, 150.1, 87–95, 2003.
3. Vachirasricirikul, S. and Ngamroo, I., Robust LFC in a smart grid with wind power penetration by coordinated V2G control and frequency controller. *IEEE Trans. Smart Grid*, 5.1, 371–380, 2014.
4. Shankar, G. and Lakshmi, S., Frequency control of hybrid renewable energy system with PSO optimized controller. *2015 International Conference on Recent Developments in Control, Automation and Power Engineering (RDCAPE)*, IEEE, 2015.
5. Srinivasa, R,.C., Adaptive Neuro Fuzzy based Load Frequency Control of multi area system under open market scenario. *IEEE-International Conference on Advances in Engineering, Science and Management (ICAESM-2012)*, IEEE, 2012.
6. Sathya, M.R. and Mohamed Thameem Ansari, M., Load frequency control using Bat inspired algorithm based dual mode gain scheduling of PI controllers for interconnected power system. *Int. J. Electric. Power Energy Syst., 64,* 365–374, 2015.
7. Ahmadi, A. and Aldeen, M., Robust overlapping load frequency output feedback control of multi-area interconnected power systems. *Int. J. Electr. Power Energy Syst.*, 89, 156–172, 2017.

8. Lu, K., *et al.*, Constrained population extremal optimization-based robust load frequency control of multi-area interconnected power system. *Int. J. Electr. Power Energy Syst.*, 105, 249–271, 2019.
9. Naidu, K., Mokhlis, H., Terzija, V., Performance investigation of ABC algorithm in multi-area power system with multiple interconnected generators. *Appl. Soft Comput.*, 57, 436–451, 2017.
10. Rodríguez-Abreo, O., *et al.*, Self-tuning neural network PID with dynamic response control. *IEEE Access*, 9, 65206–65215, 2021.
11. Vrdoljak, K., Perić, N., Petrović, I., Sliding mode based load-frequency control in power systems. *Electr. Power Syst. Res.*, 80.5, 514–527, 2010.
12. Dahiya, P., Sharma, V., Naresh, R., Optimal sliding mode control for frequency regulation in deregulated power systems with DFIG-based wind turbine and TCSC–SMES. *Neural Comput. Appl.*, 31.7, 3039–3056, 2019.
13. Rosaline, A.D. and Somarajan, U., Structured H-Infinity controller for an uncertain deregulated power system. *IEEE Trans. Ind. Appl.*, 55.1, 892–906, 2018.
14. Sondhi, S. and Hote, Y.V., Fractional order PID controller for load frequency control. *Energy Convers. Manage.*, 85, 343–353, 2014.
15. Arya, Y. and Kumar, N., BFOA-scaled fractional order fuzzy PID controller applied to AGC of multi-area multi-source electric power generating systems. *Swarm Evol. Comput.*, 32, 202–218, 2017.
16. Saikia, L.C., *et al.*, Automatic generation control of a multi area hydrothermal system using reinforced learning neural network controller. *Int. J. Electr. Power Energy Syst.*, 33.4, 1101–1108, 2011.
17. Khuntia, S.R. and Panda, S., Simulation study for automatic generation control of a multi-area power system by ANFIS approach. *Appl. Soft Comp.*, 12.1, 333–341, 2012.
18. Fathy, A. and Kassem, A.M., Antlion optimizer-ANFIS load frequency control for multi-interconnected plants comprising photovoltaic and wind turbine. *ISA Trans.*, 87, 282–296, 2019.
19. Sharma, D. and Mishra, S., Non-linear disturbance observer-based improved frequency and tie-line power control of modern interconnected power systems. *IET Generation Transm. Distrib.*, 13.16, 3564–3573, 2019.
20. Guha, D., Roy, P.K., Banerjee, S., Disturbance observer aided optimised fractional-order three-degree-of-freedom tilt-integral-derivative controller for load frequency control of power systems. *IET Generation Transm. Distrib.*, 15.4, 716–736, 2021.
21. Sahu, R.K., Panda, S., Sekhar, GT.C., A novel hybrid PSO-PS optimized fuzzy PI controller for AGC in multi area interconnected power systems. *Int. J. Electr. Power Energy Syst.*, 64, 880–893, 2015.
22. Sahu, B.K., *et al.*, A novel hybrid LUS–TLBO optimized fuzzy-PID controller for load frequency control of multi-source power system. *Int. J. Electr. Power Energy Syst.*, 74, 58–69, 2016.

23. Nayak, J.R., Shaw, B., Sahu, B.K., Application of adaptive-SOS (ASOS) algorithm based interval type-2 fuzzy-PID controller with derivative filter for automatic generation control of an interconnected power system. *Eng. Sci. Technol. Int. J.*, 21.3, 465–485, 2018.
24. Afaneh, T., Mohamed, O., Elhaija, W.A., Load Frequency Model Predictive Control of a Large-Scale Multi-Source Power System. *Energies*, 15.23, 9210, 2022.
25. Camacho, E. and Bordons, C., *Model Predictive Control*, 2nd, Springer, Berlin, Germany, 2004.
26. Bemporad, A., Morari, M., Ricker, N.L., Tech. Rep. AUT01-08, Autom. Control Lab, in: *The MPC Simulink library*, ETH, Zurich, Switzerland, 2000.
27. Rerkpreedapong, D., Atic, N., Feliachi, A., Economy oriented model predictive load frequency control. *Proc. Power Eng. Conf. Large Eng. Syst.*, 12–16, 2003, DOI: 10.1109/LESCPE.2003.1204672.
28. Kong, L. and Xiao, L., A new model predictive control schemebased load frequency control, in: *Proc. 2007 IEEE Int. Conf. Cntrl. Autom*, pp. 2514–2518.
29. Venkat, A.N., Hiskens, I.A., Rawlings, J.B., Wright, S.J., Distributed MPC strategies with application to power system automatic generation control. *IEEE Trans. Control Syst. Technol.*, 16, 6, 1192–1206, 2008, DOI: 10.1109/TCST.2008.919414.
30. Shiroei, M., Ranjbar, A.M., Amraee, T., A functional model predictive control approach for power system load frequency control considering generation rate constraint. *Int. Trans. Electr. Energy Syst.*, 23, 2, 214–229, 2013, DOI: 10.1002/etep.653.
31. Mohamed, T.H., Bevrani, H., Hassan, A.A., Hiyama, T., Decentralized model predictive based load frequency control in an interconnected power system. *Energy Convers. Manage.*, 52, 2, 1208–1214, 2011, DOI: 10.1016/j.enconman.2010.09.016.
32. Ogata, K. *Modern Control Engineering*, Prentice Hall, New Jersey, 2002.
33. Kiran, M.S., TSA: Tree-seed algorithm for continuous optimization. *Expert Syst. Appl.*, 42.19, 6686–6698, 2015.
34. Hakimuddin, N., Nasiruddin, I., Singh Bhatti, T., Generation-based automatic generation control with multisources power system using bacterial foraging algorithm. *Eng. Rep.*, 2.8, e12191, 2020.
35. Biswas, S., Roy, P.K., Chatterjee, K., Renewable energy-based multi-source system under deregulated environment using cokha algorithm. *IETE J. Res.*, 63, 9, 1–19, 2021.
36. Hernández-Romero, I.M., *et al.*, Optimal design of the ocean thermal energy conversion systems involving weather and energy demand variations. *Chem. Eng. Processing-Process Intensif.*, 157, 108114, 2020.
37. Hasan, N. and Farooq, S., Real time simulation of automatic generation control for interconnected power system. *Int. J. Electr. Eng. Inf.*, 4.1, 40, 2012.
38. Maciejowski, J.M., *Predictive Control with Constraints*, Prentice Hall, London, 2002.

8

Wide-Area Monitoring, Protection, Automation and Control (WAMPAC) System

Sanchita Kumari[1] and Amrita Sinha[2*]

[1]*Electrical & Instrumentation Department, Thapar Institute of Engineering & Technology, Patiala, Punjab, India*
[2]*Electrical & Electronics Department, Bakhtiyarpur College of Engineering, Bakhtiyarpur, Patna, Bihar, India*

Abstract
The penetration of renewable energy sources in power networks and catastrophic power outages in grids has significant effects on designing wide-area monitoring, protection, automation, and control (WAMPAC) systems. Remote terminal units (RTUs) and SCADA systems communicate with WAMPAC functionalities through established communication protocols, enabling seamless data exchange, control commands, and event reporting. Phasor measurement unit (PMU) and intelligent electronic device (IED) integration with the WAMPAC system in power networks enable enhanced grid visibility, situational awareness, and control, by providing synchronized real-time measurements, local protection, control capabilities, and data acquisition functions. PMUs provide precise time-synchronized measurements of voltage and current phasors by applying different phasor estimation techniques (discrete Fourier and Hartley transform, Taylor weighted least square, and phase-locked loop technique), after preprocessing the raw data. IEDs, a multifunction device, deliver protection, control, monitoring, and automation functionalities and use standardized communication protocols to transmit measurement data, status information, and control commands over communication networks. Communication networks facilitate the exchange of data and control commands among several components (PMU, IED, RTU, and SCADA) of the WAMPAC system. The data collected from different components of WAMPAC are transmitted using communication protocols (DNP3, IEC61850, MODBUS,

**Corresponding author*: amrita.itbhu@gmail.com

Arindam Mondal and Souvik Ganguli (eds.) Controller Design for Industrial Applications, (141–166)

and IEC60870) to central control centers for analysis and automating decisions in wide-area monitoring, protection, and control.

Keywords: Blackouts, WAMPAC, SCADA, PMU, IED

8.1 Introduction

The electric power commerce has gone through substantial changes owing to continuous inflammation in demand, technological growth, and improved quality of life. The addition of new generating units, conventional or nonconventional, interconnection of grids, and expansion of distribution/transmission system requires operational enhancements in terms of efficiency, reliability, and security. It is important to maintain reliable and resilient power infrastructure by means of implementing effective strategies for outage prevention and response.

Figure 8.1 shows the structure of wide-area monitoring, protection, automation, and control systems (WAMPACs), which use system-wide information in power networks. It can check large disturbances and reduce the frequency, duration, and affected area of catastrophic events by incorporating adaptive protection and control strategies aimed at increasing network capacity and minimizing wide-area disturbances. WAMPAC systems have progressed as vital parts of monitoring, protecting, and controlling a complex power system with the advancement in computer and communication techniques which include supervisory control and data acquisition (SCADA), phasor measurement units (PMUs), global positioning system (GPS), remote terminal units (RTUs), intelligent electronic device (IED), etc.

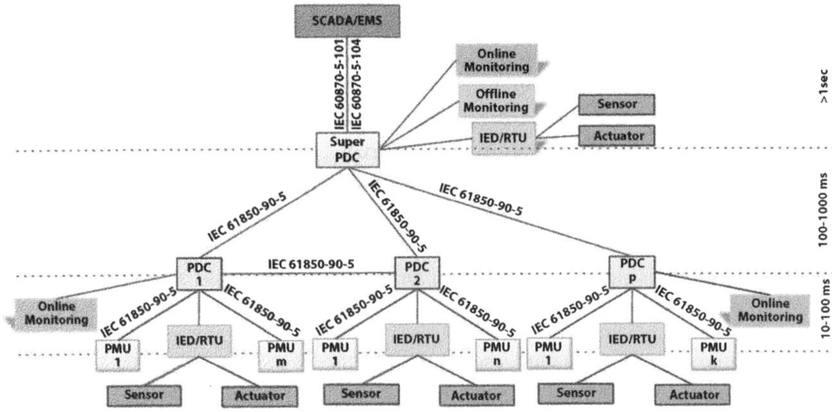

Figure 8.1 Wide-area monitoring, protection, and control (WAMPAC) structure.

The SCADA system collects real-time data from remote locations for monitoring and controlling power distribution as well as generation from a central place. PMUs impart precise GPS-synchronized phasors of the acquired electrical quantities by using appropriate phasor estimation techniques. IED is a microprocessor-controlled, multifunction device used in power networks for protection, automation, and bidirectional control. The communication between SCADA, PMU, and IED has to follow different protocols for the exchange of data and control commands.

8.2 Blackouts

Power system blackout arises due to an event such as mal-operation of the relay after a fault occurrence resulting in cascading outages followed by partial or full system failure. Several reasons behind blackouts can be mal-operation of protection and control systems, transmission line (TL) overloading/tripping, lightning surges, failure of equipment, system frequency variations, loss of synchronism of generators, etc. [1]. Some of the blackouts that occurred across the globe have been listed in Table 8.1 for reference. In India, in the summer of 2012, the tripping of the circuit breaker (CB) of a 400-kV Gwalior-Binar line caused a series of events resulting in system collapse. Restoration of the system took around 14 hours. On consecutive days, the malfunctioning of the relay near the Taj Mahal leads to the shutdown of many power systems. Many million people suffered from power loss, and many infrastructure facilities (e.g., railways, vehicular traffic movement, and metro rail system.) got paralyzed [2, 3]. About 8 million people in 14 provinces were affected in Thailand in 2013 due to lightning strikes [4]. On 25 May 2013, the falling of a tree on the Di-Linh-Tan Dinh 500-kV line in South Vietnam led to the disintegration of the transmission and distribution system. A total loss of 9400-MW capacity from 43 generators has been encountered [5]. Due to a sudden HVDC station outage, a country-wide blackout occurred in Bangladesh, and restoration took a long time. The system frequency rapidly declined to 48.646 Hz within 6.87 seconds. Under frequency, schemes available tried to cope with system frequency but failed to maintain the rated value [6]. In 2015, a blackout took place in Turkey due to the tripping of a 400-kV line due to overloading. Initially, the Osmanca-Kursunlu line carrying 1127 MW/1237 MVA tripped as a consequence of overloading, which further commenced tripping off line distance protective relays of all parallel lines resulting in the loss of synchronism among the eastern and western subsystem of Turkey causing the separation of the both subsystems [7]. A blackout occurred in

Ukraine in December 2015 because of a cyberattack [8]. As per the company personnel, the attacks occurred within a 30-minute time interval of each other at each company and impact was visible on multiple regional and central facilities. A severe storm damaged various transmission towers in South Australia in 2016 resulting in a 52% loss of wind generation in a short duration [9]. Due to cascading faults, the US power grid faced a blackout in 2017 [10]. Due to grid failure, blackouts were experienced in 2019 and 2022 in Argentina, Uruguay, Paraguay [11], and Bangladesh [12], respectively. Recently, blackouts due to unusual voltage fluctuations have been encountered in Pakistan [13]. Practically, blackouts cannot be eliminated completely because of power system complexity, dispersion in a wide geographical area, and synchronous operations of a number of machines.

Table 8.1 Major blackouts occurred across the world in the past decade.

S. no.	Country/ region	Date	Affected people (in millions)	Duration (in hours)	Cause of blackout
1.	India [2]	30 July 2012	620	14	Circuit breaker tripping
2.	India [3]	31 July 2012	700	2–8	Relay mal-operation
3.	Thailand [4]	21 May 2013	8	10	Lightning strike
4	Vietnam [5]	25 May 2013	1.8	1–8	Falling of tree on 500-kV TL
5.	Bangladesh [6]	1 November 2014	150	24	Outage of HVDC system
6.	Turkey [7]	31 March 2015	70	4–7	Power system failure
8.	Ukraine [8]	23 December 2015	230	6	Cyberattack
9	South Australia [9]	28 September 2016	1.7	6.1	Heavy storm damaging transmission infrastructure, cascading fault

(Continued)

Table 8.1 Major blackouts occurred across the world in the past decade. (*Continued*)

S. no.	Country/ region	Date	Affected people (in millions)	Duration (in hours)	Cause of blackout
10.	United States [10]	10 September 2017	7.6	5	Transmission tripping, cascading events
11.	Argentina, Uruguay, Paraguay [11]	16 June 2019	48	14	Grid failure
12.	Bangladesh [12]	4 October 2022	130	7	Grid failure
13.	Pakistan [13]	23 January 2023	230	12	Unusual voltage fluctuation

8.3 Supervisory Control and Data Acquisition System

SCADA systems exhibit an evolving area that is used for the automation of power systems by incorporating communication channels between components and adequate control and monitoring. It comprises components that provide sufficient information to remote-located operators for determining the real-time status of ongoing processes and the condition of equipment helping to take necessary action irrespective of physical presence at that particular location. SCADA has evolved from a simple telemetry system to an advanced integrated control and supervising platform by incorporating communication and data processing technologies. SCADA systems provide real-time data acquisition, visualization, control functionalities, and supervise various components of the electrical grid from a central location.

The following steps are involved in the SCADA system:

i. Collection of data from the location
ii. Conversion of data
iii. Bundling of data into packets
iv. Transmitting the packet of data through communication channel

v. Receiving data at control center
vi. Data decoding
vii. Information received at the operator's end

The commands issued by the operator are received by the devices at the location by similar steps as mentioned above. The set of measuring and controlling elements aids in acquiring data from the field sensors, and implementing the control commands by the actuators. The hierarchy of functioning of SCADA in several layers has been depicted in Figure 8.2.

There are several system architecture types for SCADA, i.e., monolithic, distributed, and networked. Monolithic is the most basic standalone type consisting of a single SCADA system that communicates with RTUs. The mixing of another type of data is not possible in this structure. Distributed architecture is an extension of the aforementioned type in which several SCADA units are connected to a local area network (LAN), thus providing data-sharing options. It also provides redundancy to the system. However, the connection of different networks is based on LAN protocols that limit communication to the local environment. Networked architecture is the current version of advanced networking using standards and open protocols which incorporate wide-area network (WAN) and optic fiber for interfacing with the LAN system. With this architecture, the supervision and data can be accessed from different locations and plants as needed.

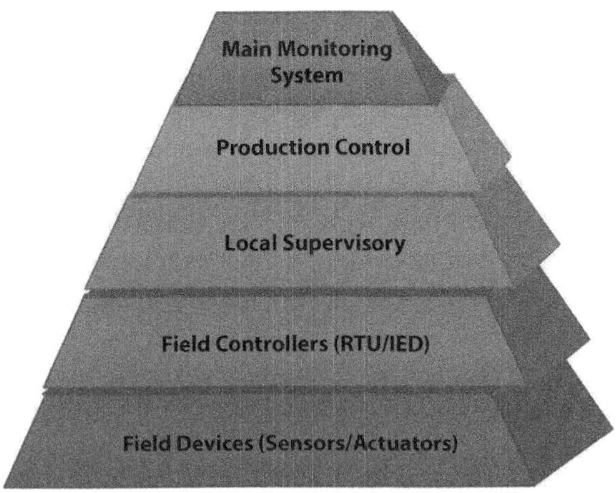

Figure 8.2 Hierarchical functioning of SCADA system.

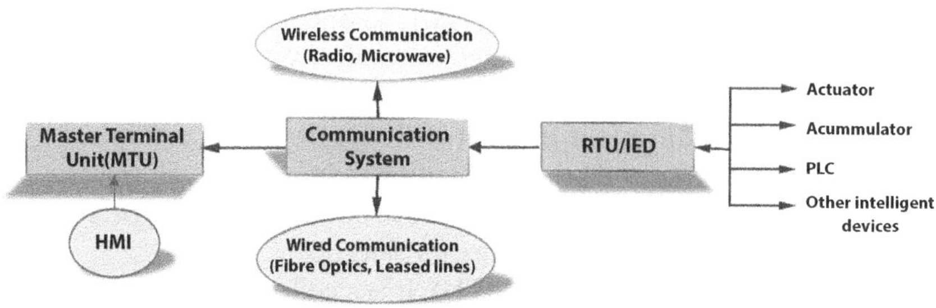

Figure 8.3 Block diagram of SCADA system.

8.3.1 Components of SCADA in Power System

The block diagram of SCADA applicable for power systems has been illustrated in Figure 8.3 and its main components are explained as follows:

1. **RTU/IED:** They are deployed at several key points and substations of the power system for collecting data from different sensors, meters, and relays. The parameters measured by RTUs/IEDs in the power network are magnitude of voltage, active and reactive power, tap changer position, and status of CB. Binary variables are used for mentioning the status of CB, while, for other cases, it is a continuous value that is transformed into DC currents in the range of 4 to 20 mA by using transducers. It builds interfacing between sensors and actuators in the field and master control stations. Data collected from multiple remote locations create extensive monitoring and controlling capabilities. Programmable logic controllers (PLCs) provide additional flexibility in real-time measurement, control, and data acquisition capabilities.
2. **Communication System:** This states communication channels implemented between field equipment and the master station. Speed of communication is governed by the bandwidth of the channel, done by means of wireless (radio and microwave) or wired (fiber optics and leased lines) channels. Open protocols, such as Modbus and distributed network protocol (DNP3), came into existence with the development of digital communication systems in the 1980s [14]. Open protocol advantages include support by multiple manufacturers, software vendors, and install/service organizations. Furthermore, incorporating powerful computers for real-time managing, processing, and visualizing data provides

human-machine interface (HMI) to operators for monitoring and controlling processes more effectively. Nowadays, by combining information technologies (ITs) such as big data analysis [15] and the Internet of Things (IoT) [16], optimization capabilities aid the functioning of smart grids. However, with more exposure to the Internet/online systems, cybersecurity comes into the picture for safety concerns.

3. **Master Terminal Unit (MTU):** It is a collection of computers, peripherals, and appropriate input and output (I/O) systems, enabling the operators to monitor the state of the wide power network and control it remotely. It receives data from several RTUs/IEDs located across the grid, acting as a central gateway; it processes this information and sends control commands accordingly to the devices.

Combining SCADA with a wide area monitoring system gives a broader picture for monitoring and controlling power systems [17]. The application of microprocessor-based RTUs /IEDs with protection capability added further flexibility.

SCADA system is applicable for the overall monitoring, control, and management of power networks that comprise generation, transmission, and distribution. It provides a comprehensive status of power networks to operators for making decisions and taking suitable actions for reliable and efficient power flow.

RTU plays an important role in the functioning of the SCADA system. Its applications can be visualized in substation automation, fault detection and localization, load management, switching operations, voltage regulations and optimization, event recording, serving communication gateway, and security and access control. RTUs deployed at key points within the grid that are geographically dispersed gather real-time data and execute control commands.

RTUs and PMUs are both used for data acquisition monitoring and control, having distinct purposes and functionality. PMUs are mainly applied for accurate and synchronized measurement of phasors required for dynamic system analysis, whereas RTUs are dedicated to data collection from various equipment and sensors located remotely in the network. Thus, amalgamating PMU makes the wide area measurement system different from the SCADA system on the basis of time synchronization along with working time. The difference between PMU and SCADA on the basis of different attributes has been listed in Table 8.2 [18].

Table 8.2 Comparison of SCADA and PMU.

Attribute	SCADA	PMU
Resolution	1 sample/2–4 seconds	10-60 samples/second
Observability	Steady state	Dynamic state
Measurements	IVI, I	IVI, I, frequency, ROCOF, angle
Synchronization	No	Yes
Phase angle	No	Yes
Focus	Local monitoring and control	Wide-area monitoring and control

Measurements of different parameters such as active/reactive power, voltage magnitudes, and system frequency acquired using SCADA are utilized for static state estimators. It provides a state of the system every few minutes. However, tracking system dynamics is not possible using such estimation techniques and has been addressed in [19]. Implementation of PMUs speeds up the estimation process and can update the system state at a sub-second rate [20]. High-level monitoring of the grid is carried out by the implementation of SCADA systems. However, the enormous integration of renewables makes SCADA systems inappropriate for capturing fast system transients. Integration of SCADA along with PMU in the modern power system gives enhanced monitoring, control, protection, and state estimation methods [21].

8.4 Phasor Measurement Units (PMU)

PMUs are an evolving technology that amended the state monitoring of the system. These devices yield precise time-synchronized voltage and current phasor measurements and communicate through the grid at higher time resolution compared to the RTUs. PMUs are widely implemented and have various utilizations in the power system [22]. They can be comprehensively categorized as monitoring, control, and protection and collectively called wide-area measurement system–based applications [23]. The synchronized phasor measured parameters are described in the C37.118.1-2011 and its modification in IEEE Standard C37.118.1a-2014 [24]. The parameters that are emphasized for the accuracy of phasor measurement in steady, dynamic, and transient conditions include frequency error (FE), rate of change of FE (ROCOFE), total vector error (TVE), delay time, response time, etc. Furthermore, PMUs are split into two classes as per the standard, i.e., Protection (P) and Measurement (M) classes on the basis of their performance [25]. The functioning of M-type PMUs needs wider ranges of operating conditions and higher steady-state accuracy than P-type.

Several phasor estimation methods have been suggested contemplating the significance of phasor units [26]. Most of them can be categorized in time and frequency domain techniques. Some of them are the Fourier transform (FT), wavelet transform, Hartley transform, Taylor weighted least square, Kalman filter–based methods, etc. [27–29]. FT is a powerful tool for data analysis but it does not represent sudden variations efficiently. This is due to the fact that it characterizes data as a sum of sine waves that are not localized in time or space, thus having fixed resolution and localized frequency. Furthermore, fault occurrences in power systems always contain exponentially decaying DC offsets which further degrade the accuracy of FT. Many methods have been proposed for the modification of discrete FT such as interpolated technique (IPDFT), nonrecursive, and recursive DFT.

The protection schemes become challenging due to the wide-area network and its complexity. The overhead TLs are most vulnerable to faults as it is open to the environment. Precise fault location is of utmost importance for restoring the network and reducing outage time.

Using phasor data from multiple PMUs located across the power system, fault location can be triangulated. By analyzing the phase angle differences between the phasors measured at different locations, the distance to the fault location can be estimated. Also, various algorithms are applied to the synchronized phasor data for estimating the location of the fault accurately, considering factors that include fault impedance, network topology, and line parameters.

8.4.1 Block Diagram of PMU

Synchronization of time is necessary for phasor measurement at different locations on common time reference and is provided by a GPS pulse signal. Synchrophasor measurement is widely used in state estimation, adaptive protection, fault and disturbance recording, and stability of the system. Figure 8.4 depicts the block diagram of PMU. The input to PMU is the instantaneous values of voltage and current taken from the potential and current transformer respectively. The voltage and current signals are

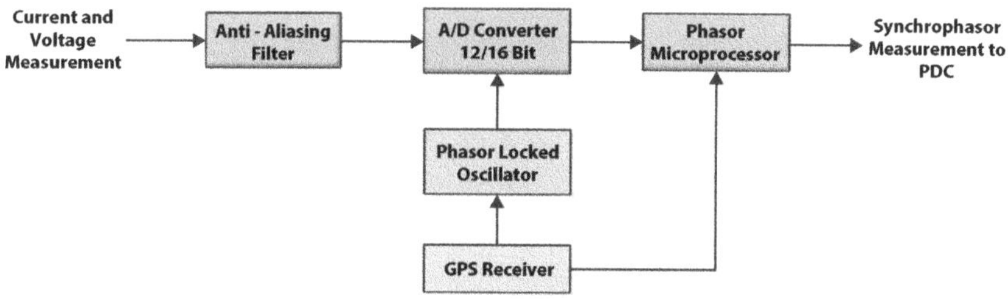

Figure 8.4 Block diagram of PMU.

converted to voltage using appropriate shunt or instrument transformers within the range of ±5 or 10 V to match analog-to-digital (A/D) converter requirements.

The signals are processed using an anti-aliasing filter to eliminate high frequency before feeding to the A/D converter. The sampling clock is phase-locked with a GPS clock pulse and the rate of sampling considered is 12 samples/cycle to 96 or 128 samples/cycle. A higher sampling rate yields improved estimation accuracy. However, it requires more computation. GPS satellites keep accurate clocks providing one pulse per second (1PPS) signal which is utilized by all PMUs. GPS time corrections are made in the GPS receiver to provide a coordinated universal time clock (UTC). The final values evaluated are used to estimate parameters such as frequency, power, and rate of change of frequency (ROCOF) for monitoring, control, and protection aspects.

8.4.2 Synchrophasor Technology

Any sinusoidal signal given by Equation 8.1 can be represented in the form of phasor as in Equation 8.2

$$V(t) = V_m cos(wt + \theta) \tag{8.1}$$

$$V = \frac{V_m}{\sqrt{2}} e^{-j\theta} = \frac{V_m}{\sqrt{2}} (\cos\theta + j\sin\theta) \tag{8.2}$$

where $\frac{V_m}{\sqrt{2}}$ represents the input signal's rms value. Graphical representation of a sine input signal and its related phasor has been depicted in Figure 8.5. The length of the phasor is equivalent to the RMS value of the signal,

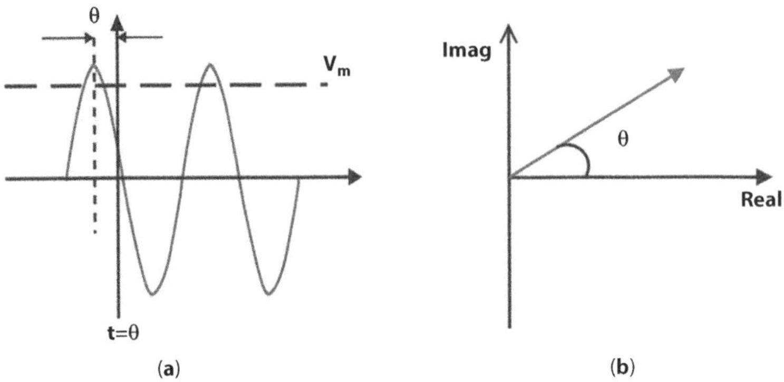

Figure 8.5 (a) Representation of sinusoidal signal. (b) Phasor representation.

and the phase angle depends on a timescale. In a real-time scenario, the input signal is admixed with noise and harmonics. As phasor representation is applicable for single frequency components, it is required to extract the fundamental frequency component for phasor form. The phasor phase angle is the angle between the instant of measurement with respect to the peak of the signal.

8.4.3 Phasor Estimation Techniques

8.4.3.1 Discrete Fourier Transform (DFT)

DFT is one of the most extensively used tools for digital signal processing based on a steady-state signal model. It is analogous to continuous FT (CFT) taking N instants segregated by sample times, i.e., a finite sequence of data. Considering a signal waveform at a nominal frequency as Equation 8.3

$$s(t) = S_m \cos\left(2\pi f_n t + \theta\right) \tag{8.3}$$

where S_m represents signal magnitude, θ represents phase angle, and f_n represents nominal fundamental frequency.

If the signal is sampled at the rate of N samples per cycle, i.e., $f_n = \frac{F_s}{N}$, the above equation can be written as Equation 8.4

$$s(n) = S_m \cos(2\pi n / N + \theta) \tag{8.4}$$

The fundamental phasor of the signal can be specified as Equation 8.5

$$S = \frac{s_m}{\sqrt{2}} e^{j\theta} \tag{8.5}$$

Let s(t) be a continuous signal act as a data source and its N samples be depicted as Equation 8.6

$$s = \left(s[0], s[1], \ldots, s[N-1]\right) \tag{8.6}$$

DFT of s is N-tuple as shown in Equation 8.7 can be defined by Equation 8.8

$$S = \left(S[0], S[1], \ldots, S[N-1]\right) \tag{8.7}$$

$$S[m] = \sum_{n=0}^{N-1} s[n] e^{-\frac{2\pi jmn}{N}} \tag{8.8}$$

S[m] is the DFT of the sequence signal s[n].

Hence, the fundamental phasor of the mentioned signal according to Equation 8.3 can be expressed as Equation 8.9

$$S = \frac{\sqrt{2}}{N} \sum_{n=0}^{n=N-1} s(n) e^{-2\pi jn/N} \tag{8.9}$$

Time and frequency resolution improvement requires capturing followed by processing the block data at a higher rate, or else applying recursive FT (RFT) techniques. Implementation of RDFT techniques requires sample-by-sample upgradation along with flexible computation of sheer time-frequency resolution. For updating of phasor with every new sample data acquired, the window has to be shifted by each sample. It deduces the computation burden by performing a recursive update of the past phasor data acquired from the preceding window. This algorithm for the N sample window can be depicted by Equation 8.10

$$S^{x+1} = S^{x} + \frac{\sqrt{2}}{N}\left(s(N+x) - s(x)\right)e^{-j(x)\theta} \tag{8.10}$$

8.4.3.2 *Taylor Weighted Least Square Technique*

The power system transients and oscillations can be modeled using dynamic electrical waveform and can be represented by Equations 8.11 and 8.12 [27]

$$y(t) = x(t)\cos\left(2\pi f_r t + \varnothing(t)\right) \tag{8.11}$$

$$= \sqrt{2}Re\left\{P(t)e^{2\pi j f_r t}\right\} \tag{8.12}$$

where x(t) and Ø(t) are time-varying amplitude and phase angle at fundamental frequency f_r. P(t) is termed complex dynamic phasor of the oscillation synchronized to f_r and is expressed as Equation 8.13

$$P(t) = \frac{x_n(t)}{\sqrt{2}}e^{j\varnothing(t)} \tag{8.13}$$

As per complex valued approach, P(t) can be approximated in the form of its complex Taylor's series expansion as Equation 8.14

$$P_K(t) = P(t_s) + P'(t_s)(t - t_s) + \frac{P''(t_s)}{2!}(t - t_s)^2 + \cdots\ldots\ldots + \frac{P^K(t_s)}{K!}(t - t_s)^K$$
$$for\ \frac{-T}{2} < (t - t_s) < \frac{T}{2} \tag{8.14}$$

where s=0, 1, 2, …, $P^{(K)}(t_s)$ is the kth-order derivative of P(t) at t_s time

For this work, the approximation is done by considering the second-order Taylor polynomial, i.e., K=2. The abovementioned equation can be modified as Equation 8.15 [27]

$$y_2(t) = \sqrt{2}Re\left(P_2(t)e^{2\pi jf_r t}\right) = \sqrt{2}\left(\frac{1}{2}\left(P_2(t)e^{2\pi jf_r t} + \underline{P_2}(t)e^{-2\pi jf_r t}\right)\right) \tag{8.15}$$

where

$$P_2(t) \cong P(t_s) + P'(t_s)(t - t_s) + \frac{P''(t_s)}{2!}(t - t_s)^2 \tag{8.16}$$

Sampling signal for N samples/cycle $T_1=1/f_r$, every interval of T size having a center at s^{th} interval can be expressed by M linear equations and are depicted in the form matrix.

The matrix equation can be expressed as Equation 8.17

$$y_s = Z_2 p_{2,s} \ for \ s = 1,2,3\ldots \tag{8.17}$$

$p_{2,s}$ vector coefficients are associated with phasor derivation as per Equations 8.18 and 8.19

$$P_{0s} = P(t_s) \tag{8.18}$$

$$P_{Ks} = \frac{p^{(K)}(t_s)}{K!(Nf_0)^K} \ for \ K = 1,2 \tag{8.19}$$

8.4.3.3 *Phase-Locked Loop (PLL) Technique*

The PLL technique has been utilized for extracting frequency and phase angle information in electrical systems. A schematic diagram of the PLL-based phasor estimation has been depicted in Figure 8.6.

It consists of abc/dq transformation in the first step for conversion using the transformation matrix. After conversion, the output is passed through a low-pass filter to eliminate high-frequency components. The imaginary (q) component after the transformation is fed to the first-order PI controller for obtaining the value of theta, which is further passed through the integrator to obtain frequency data of the input signal. The automatic gain controller

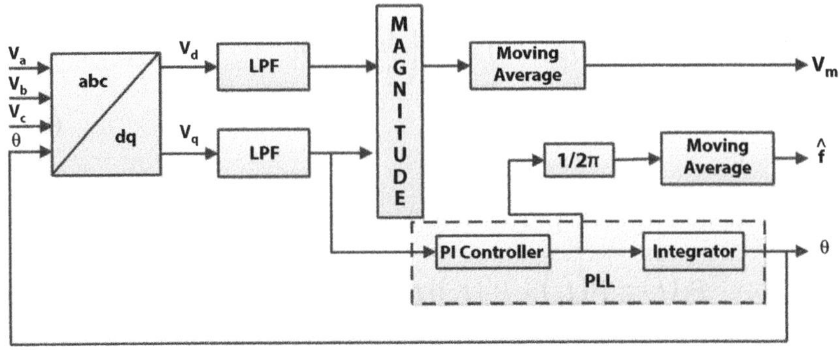

Figure 8.6 PLL-based phasor estimator.

is required if the parameters of the input signal vary over a wide range. The incorporation of a moving average block is for smoothening waveforms of magnitude and frequency by suppressing ripples present in it.

8.4.3.4 *Kalman Filter (KM)–Based Technique*

It is a mathematical algorithm employed for state estimation based on linear dynamical systems in state space format and has many applications, such as signal processing and control systems. Any discrete-time linear system can be expressed as Equation 8.20

$$x_k = A_k x_{k-1} + B_k u_k + w_k \tag{8.20}$$

where x represents the system state, A represents the system matrix (define system dynamics), B represents the control input matrix applied to input vector 'u', w represents process noise (corresponding to lack of information on system description), and k represents discrete-time variability.

The relationship between the available measurement and state for time step k is given by Equation 8.21

$$z_k = h_k(x_k) + e_k \tag{8.21}$$

where z_k is the measurement vector referred to n^{th} instant. h_k is a measurement function matrix that defines the relationship between state and measured quantities, and e_k is the corresponding zero mean measurement error.

The Kalman filter's function is to proffer an estimate of x_k at time k, provided initial estimate (x_0), the series of measurements, $z_{1...}z_k$, and the details of the system given by variables mentioned in the above equations. This approach to estimating the system state is done in two steps, i.e., prediction and update using Equations 8.22 to 8.27

i) Prediction

$$\text{Predicted state estimate} \quad \widetilde{x_k} = A_k \widetilde{x_{k-1}} + B_k u_k \tag{8.22}$$

$$\text{Predicted error covariance} \quad \widetilde{P_k} = A_k P_{k-1} A_k^{\,T} + Q_k \tag{8.23}$$

ii) Update

$$\text{Measurement residual} \quad \widetilde{y_k} = z - H_k \widetilde{x_k} \tag{8.24}$$

$$\text{Kalman gain} \quad K_k = \widetilde{P_k} H_k^{\,T} \left(H_k \widetilde{P_k} H_k^{\,T} \right) + R_k \tag{8.25}$$

$$\text{Updated state estimation} \quad \widehat{x_k} = \widetilde{x_k} + K_k \widetilde{r_k} \tag{8.26}$$

$$\text{Updated error covariance} \quad \widehat{P_k} = \left(I - K_k H_k \right) \widetilde{P_k} \tag{8.27}$$

where I is the identity matrix. '^' and '~' denote estimated and predicted quantities, respectively. P, Q, and R are the state, process noise, and measurement error vector covariance matrix, respectively.

8.5 Intelligent Electronic Devices (IEDs)

IEDs play a crucial role in modern power systems by performing a variety of functions related to protection and control. Figure 8.7 shows the block diagram of an IED.

IEDs consist of advanced microprocessors, communication interfaces, and software that enable them to perform intelligent and automated tasks to ensure the safety, reliability, and efficient operation of the power grid. IEDs can perform intelligent functions of relays, protection devices,

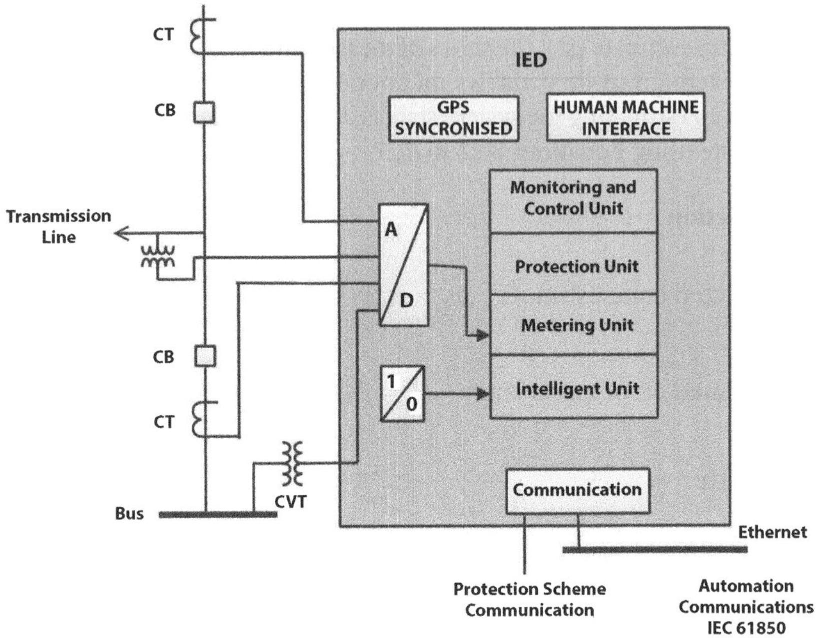

Figure 8.7 IED block diagram.

meters, PMUs, controllers, and more. These devices can communicate with each other within a substation or control centers through messaging following communication protocols (IEC 61850, DNP3, etc.) to monitor, control, protect, and manage [30]. This permits real-time data sharing as well as decision-making across a wide area. The data collected by IEDs can be analyzed to identify trends, patterns, and anomalies, aiding in predictive maintenance and system optimization. IEDs can make informed decisions and acknowledge disturbances quickly, ensuring stability as well as reliability of the network.

IEDs are multi-purpose intelligent devices with multiple protection and control functions. IEDs can perform protection functions such as overcurrent, over/under voltage, over/under frequency, and auto-reclosers. IEDs can implement schemes like distance, differential, directional, thermal overload, and unit protection of devices. Control of remote units, load shedding, maintaining frequency and voltage levels, and synchronization of different sources connected to the grid can also be executed by IEDs. It can avert overloading, exchange data, record events, and monitor dynamic line ratings.

Time-synchronized acquired signals of voltage and current can be processed using different algorithms and represented in phasor form [31]. The phasor data can be communicated in real time to a central location, SCADA system, or phasor data concentrator (PDC) using standard protocols. The received data can be applied for real-time monitoring, disturbance analysis, oscillation detection, and stability assessment. The phasor data received from multiple IEDs are used to evaluate and sustain wide-area grid dynamics, identify abnormalities, and optimize real-time grid operation. The integration of IEDs, PMUs, and SCADA systems gives a comprehensive view of the grid's health for faster decision-making with greater accuracy.

Voltage stability across a wide area can be monitored by IEDs and initiates control actions to maintain voltage within permissible limits, prevent voltage collapse, and avert cascaded failures.

IEDs help in maintaining voltage stability and minimizing voltage fluctuations. It can implement coordinated voltage control in retort to deviations by adjusting tap changers and reactive power sources in the network.

IED observes system frequency and its deviation to coordinate and control generation output, load shedding, and energy storage units. Wide-area phasor data of IEDs detects power oscillations and also calculates damping coefficients. On the basis of collected information, generator outputs can be adjusted to dampen oscillations and improve system stability. Active and reactive power generation can be redistributed among generators to relieve TL congestion, minimize losses, and improve the utilization of the available resources.

IEDs enable faster fault clearance and system restoration by locating and isolating faults accurately through analysis of fault-generated transients in a wide-area network. Backup protection can be initiated using data of distant points, in case of primary protection failure. During an emergency, load shedding, and islanding strategies can be implemented using IEDs, based on real-time data. This helps in maintaining supply to critical loads and averts widespread blackouts.

IEDs can enhance system controllability, voltage support, and power flow by controlling FACTs devices. Overloads and voltage stability can be averted using IEDs based on real-time data by implementing load shedding strategies. After the stabilization of the system, it can help in system restoration by reconnecting the loads intelligently. IEDs can contribute to the integration of renewable energy resources by controlling their outputs and optimizing their participation in grid stability and control.

IEDs can calculate the dynamic thermal rating of TLs using real-time weather data and line temperature measurements, maximizing line utilization for power transfer within defined safety limits.

8.6 Communication Protocols

Secure communication between IEDs is crucial to avert unauthorized access or manipulation of protection schemes. Wide-area protection requires planning, integration, and coordination between several devices. Therefore, data exchange requires compliance with standard communication protocols such as IEC 61850 [32]. The aim is to boost the ability of utility operators to use real-time synchronized data to maintain the reliability and stability of the grid, specifically during abnormal conditions. IEDs can exchange data and collaborate in real time with other IEDs, SCADA systems, and central control centers through well-defined control algorithms. This data exchange supports coordinated decision-making and control strategies. Furthermore, proper cybersecurity measures are crucial to protect communication and control channels from potential threats [33]. The integration of advanced control functionalities into IEDs enables more effective management of the grid and responds to dynamic conditions in a coordinated manner. A few common communication protocols used in WAMPAC systems are discussed in the subtopics below.

8.6.1 IEC 61850

This is a widely adopted international standard for communication and integration of IEDs in power substations and across power systems [32]. Generic Object-Oriented Substation Event (GOOSE) is a communication mechanism defined by IEC 61850 for high-speed, peer-to-peer communication between IEDs within a substation, used for fast-tripping and interlocking applications. It enhances the interoperability, flexibility, and efficiency of substation automation and control systems. Devices in an IEC 61850 network are organized into logical nodes that represent specific functions or pieces of equipment. Data flows between these logical nodes following a standardized structure. It has transformed the way substations are designed, operated, and maintained, resulting in more reliable and responsive power systems.

8.6.2 Distributed Network Protocol 3 (DNP3)

DNP3 was developed to address the specific needs of remote telemetry and control applications, making it well-suited for communication between IEDs, RTUs, and supervisory control systems. It can facilitate real-time monitoring of grid apparatus, transformers, meters, breakers, etc. DNP3 enables remote control of devices, such as opening and closing CBs, tap changers, and load shedding [34]. DNP3 has three layers of the Open Systems Interconnection model as follows:

1. Application Layer - Handles data structures, command supervision, and data authentication.
2. Data Link Layer - Manages error detection, correction, and addressing and framing of messages.
3. Physical Layer - Deals with the physical transmission of data as electrical signals and radio waves.

DNP3 follows master-slave architecture. The master device initiates communication with remote (slave) devices to gather data or issue control commands. Slaves can spontaneously report to the master to provide real-time updates about events/conditions that require attention. This helps in fast reporting and monitoring. It is also suitable for communication over long distances and in noisy environments. DNP3 is designated as robust and reliable.

8.6.3 MODBUS

MODBUS protocol is used for communication between devices and control systems. Power system protection and control uses the serial (RTU) variant for binary-coded data transmission over RS232/RS485 serial interfacing. It also uses master-slave architecture as DNP3 but lacks built-in security features.

8.6.4 Inter-Control Center Communications Protocol (ICCP)

Inter-Control Center Communication Protocol known as IEEE 60870-6 is used in power systems for the exchange of information between control centers. It aids coordination and management of power systems over wide areas and regional centers across large geographical areas. It manages secure information transfer between utilities, independent system operators, and regional transmission organizations. The protocol can be implemented using different transport protocols, including TCP/IP and dedicated networks.

8.6.5 C37.118

It is part of the IEEE C37 series of standards, covering various aspects of communication, protection, and control in power systems. C37.118 is a communication protocol for synchrophasor data exchange across wide-area networks. The data provide significant information for monitoring and controlling the dynamic behavior of a wide-area network.

8.6.6 IEC 60870-5-101 and IEC 60870-5-104

The IEC 60870-5-101 and IEC 60870-5-104 are communication protocols used in power system automation and control for telecontrol and remote applications. The choice of the different subsections of the protocol depends on specific requirements and constraints of the power system application [35]. It improves and enhances data transmission efficiency and supports modern communication networks.

8.6.7 IEC 61400-25

This communication protocol has been defined for wind power plant control and monitoring. The standard provides a structured approach to ensure interoperability between different wind turbine manufacturers and SCADA system vendors. This model ensures a common understanding of data across different systems that represent various aspects of wind turbine operation, such as operational states, power generation, wind conditions, alarms, and diagnostics.

8.7 Conclusions

The logical nodes and devices at the substation level collect the data of current/voltage and the status of CBs through sensors and actuators connected to RTUs/IEDs and send them to local data concentrators. The PMUs/IEDs from key locations also send the GPS-synchronized phasors of voltage/current signals through communication protocols to data centers. IEDs can perform protection/control functions and take decisions as per events occurring in the power system network. The collected data by PMUs/IEDs at a faster rate than the SCADA system can make decisions and consider dynamic conditions to avoid cascaded failure. These devices work in synchronization to prevent blackouts in geographically widespread areas. Many researchers are working on the dynamic stability of grids considering the changes in voltage and frequency of the buses, fault identification and location in TLs, and device status installed in remote locations to prevent future power outages.

References

1. Kundur, P., P., Power system stability. *Power System Stability Control, 10*, 7–1, 2007.
2. Tang, Y., Bu, G., Yi, J., Analysis and lessons of the blackout in Indian power grid on July 30 and 31, 2012. *Proc. Chin. Soc. Electr. Eng.*, 32, 167–174, 2012.
3. Rampurkar, V., Pentayya, P., Mangalvedekar, H.A., Kazi, F., Cascading failure analysis for Indian power grid. *IEEE Trans. Smart Grid*, 7, 1951–1960, 2016.
4. Phuangpornpitak, N. and Tia, S., Opportunities and challenges of integrating renewable energy in smart grid system. *Energy Procedia*, 34, 282–290, 2013.
5. Son, P.T. and Voropai, N., The major outage in South Vietnam in 2013: the nature of blackout, security measures and strategy of national power system modernization, in: *International Conference on Problem of Critical Structures*, pp. 177–182, 2015, June.
6. Kabir, M.A., Sajeeb, M.M.H., Islam, M.N., Chowdhury, A.H., December. Frequency transient analysis of countrywide blackout of Bangladesh Power System on 1st November, 2014, In: *2015 International Conference on Advances in Electrical Engineering (ICAEE)*, pp. 267–270, IEEE, 2015.
7. Saribulut, L., Ok, G., Ameen, A., A Case Study on National Electricity Blackout of Turkey. *Energies, 16*, 11, 4419, 2023.
8. Liang, J., Weller, S.R., Zhao, J., Luo, F., Dong, Z.Y., The 2015 Ukraine blackout: Implications for false data injection attacks. *IEEE Trans. Power Syst., 32*, 4, 3317–3318, 2016.
9. Yan, R., Saha, T.K., Bai, F., Gu, H., The anatomy of the 2016 South Australia blackout: A catastrophic event in a high renewable network. *IEEE Trans. Power Syst., 33*, 5, 5374–5388, 2018.
10. Mukherjee, S., Nateghi, R., Hastak, M., A multi-hazard approach to assess severe weather-induced major power outage risks in the us. *Reliab. Eng. Syst. Saf., 175*, 283–305, 2018.
11. Argentine president Macri faces criticism after massive 14-hour power failure, 24, 18 June, 2019.

12. Hindustan Times, Power back in Bangladesh after grid failure caused massive blackout, 5 Oct, 2022, https://www.hindustantimes.com/world-news.
13. Masood, S., Power outage sweeps Pakistan drop-ping millions into darkness, in: *The New York Times*.
14. Si, Y., Korada, N., Ayyanar, R., Lei, Q., A high performance communication architecture for a smart micro-grid testbed using customized edge intelligent devices (eids) with spi and modbus tcp/ip communication protocols. *IEEE Open J. Power Electron., 2*, 2–17, 2021.
15. Zhang, Y., Huang, T., Bompard, E.F., Big data analytics in smart grids: a review. *Energy Inform.*, 1, 8, 2018.
16. Radhakrishnan, G. and Gopalakrishnan, V., Applications of internet of things (IOT) to improve the stability of a grid connected power system using interline power flow controller. *Microprocess. Microsyst., 76*, 103038, 2020.
17. Cheng, G., Lin, Y., Abur, A., Gómez-Expósito, A., Wu, W., A Survey of Power System State Estimation Using Multiple Data Sources: PMUs, SCADA, AMI, and Beyond. *IEEE Trans. Smart Grid*, 15, 1, 1129–1151, 2023.
18. Usman, M.U. and Faruque, M.O., Applications of synchrophasor technologies in power systems. *J. Mod. Power Syst. Clean Energy, 7*, 2, 211–226, 2019.
19. Dubey, A., Chakrabarti, S., Terzija, V., SCADA and PMU measurement based methods for robust hybrid state estimation. *Electr. Power Compon. Syst., 47*, 9–10, 849–860, 2019.
20. Costa, A.S., Albuquerque, A., Bez, D., An estimation fusion method for including phasor measurements into power system real-time modeling. *IEEE Trans. Power Syst., 28*, 2, 1910–1920, 2013.
21. Karvelis, G.I., Korres, G.N., Darmis, O.A., State estimation using scada and pmu measurements for networks containing classic hvdc links. *Electr. Power Syst. Res., 212*, 108544, 2022.
22. De La Ree, J., Centeno, V., Thorp, J.S., Phadke, A.G., Synchronized Phasor Measurement Applications in Power Systems. *IEEE Trans. Smart Grid*, 1, 1, 20–27, June 2010, doi: 10.1109/TSG.2010.2044815.
23. James, J.Q., Lam, A.Y., Hill, D.J., Li, V.O., A unified framework for wide area measurement system planning. *Int. J. Electr. Power Energy Syst., 96*, 43–51, 2018.
24. Synchrophasor Measurements for Power Systems— Amendment 1: Modification of Selected Performance Requirements, vol. C37.118.1a, IEEE Standard, Apr. 2014.
25. Roscoe, A.J., Abdulhadi, I.F., Burt, G.M., P and M class phasor measurement unit algorithms using adaptive cascaded filters. *IEEE Trans. Power Deliv., 28*, 3, 1447–1459, 2013.
26. Nanda, P., Panigrahi, C.K., Dasgupta, A., Phasor estimation and modelling techniques of PMU-a review. *Energy Procedia, 109*, 64–77, 2017.
27. Vats, M. and Kamboj, S., Application of phasor & frequency estimation techniques in phasor measurement unit, in: *2016 7th India International Conference on Power Electronics (IICPE)*, pp. 1–6, IEEE, 2016.

28. de Oliveira Nascimento, Francisco A., Hartley transform signal compression and fast power quality measurements for smart grid application. *IEEE Trans. Power Deliv.*, 38, 6, 4134–4144, 2023.
29. Muscas, C., Pegoraro, P.A., Sulis, S., Pau, M., Ponci, F., Monti, A., New Kalman filter approach exploiting frequency knowledge for accurate PMU-based power system state estimation. *IEEE Trans. Instrum. Meas.*, *69*, 9, 6713–6722, 2020.
30. Das, R., Looking into the future protection, automation and control systems, *14*, 2016.
31. Castello, P., Ferrari, P., Flammini, A., Muscas, C., Rinaldi, S., A new IED with PMU functionalities for electrical substations. *IEEE Trans. Instrum. Meas.*, *62*, 12, 3209–3217, 2013.
32. Saadi, K. and Abbou, R., On IEC 61850 communication networks in smart grids system: Methodology of implementation and performances analysis on an experimental platform. *Int. J. Energy Res.*, *46*, 1, 89–103, 2022.
33. Hahn, A., Ashok, A., Sridhar, S., Govindarasu, M., Cyber-physical security testbeds: Architecture, application, and evaluation for smart grid. *IEEE Trans. Smart Grid*, *4*, 2, 847–855, 2013.
34. Lu, X., Wang, W., Ma, J., An empirical study of communication infrastructures towards the smart grid: Design, implementation, and evaluation. *IEEE Trans. Smart Grid*, *4*, 1, 170–183, 2013.
35. Mai, K., Qin, X., Silva, N.O., Cardenas, A.A., IEC 60870-5-104 network characterization of a large-scale operational power grid. In: *2019 IEEE Security and Privacy Workshops (SPW)*, pp. 236–241, IEEE Computer Society, 2019.

9

An Efficient Smart Prepaid Interface Design for Power Industries

Antara Kundu[1], Harsh Kumar Shaw[2] and Maitrayee Chakrabarty[1]*

[1]Department of Electrical Engineering, JIS College of Engineering, Kalyani, West Bengal, India
[2]Department of Computer Science Engineering, JIS College of Engineering, Kalyani, West Bengal, India

Abstract
This chapter introduces electricity distribution by seamlessly integrating an industrial power system control application with a prepaid system. Empowering consumers to purchase electricity efficiently and cost-effectively from the closest available power station to their location, the smart prepaid interface design can identify and optimize overall electricity flow using the shortest path. The research explores the intricacy of this innovative approach related to the hybrid model power distribution in the electricity markets. Additionally, the proposed design highlights its dynamic fusion of centralized and decentralized features. The proposed system streamlines the user experiences by categorizing the users based on location and priority for efficient power distribution in the area. Users can conveniently purchase and store chosen power in a digital wallet which is created at the time of registration to enhance accessibility and usability. This abstract encapsulates the transformative essence of the research, promising a resilient and consumer-centric future in energy distribution.

Keywords: Smart prepaid interface, hybrid model, microgrid

**Corresponding author*: moitry29@gmail.com

Arindam Mondal and Souvik Ganguli (eds.) Controller Design for Industrial Applications, (167–188)

9.1 Introduction

This chapter represents a paradigm shift in electricity distribution, seamlessly combining microgrid technology with a prepaid system. This integration empowers consumers to purchase electricity at favorable rates from the closest microgrid nodes, ensuring efficiency and cost-effectiveness. Basically, this system uses the shortest path for connecting the consumer to their nearest microgrid generation point, for optimal flow of the electricity. This approach not only provides users with better control over their energy consumption but also aligns with the objective of providing reliable, cost-effective, and sustainable power solutions. Also, the prepaid model will allow the user to manage their expenses efficiently, and after its integration with the power grid, our target is to provide electricity services, with the commitment to accessible and affordable power. This research discusses this innovative approach, highlighting its potential to change the energy distribution system to a more resilient and consumer-centric.

Prepaid electricity is a system in which the customer pays for the electricity in advance in the form of credit or tokens. This approach defies the traditional distribution system, where consumers are billed after the usage of electricity within a specific period of time. It provides several advantages over the existing models making it a more efficient and consumer-centric system. Firstly, prepaid systems will eliminate the problem of payment delays, account posting, and the need for an additional billing system. It will help in improving customer service and providing them with control over their energy usage. A prepaid electricity system is a better, cost-effective, and consumer-empowered system as compared to conventional models.

The hybrid model is a fusion of both the PoolCo and the bilateral model. In this approach, flexibility allows the blend between the two providing a middle ground between the two models. Figure 9.1 shows the PoolCo model of the restructured power system.

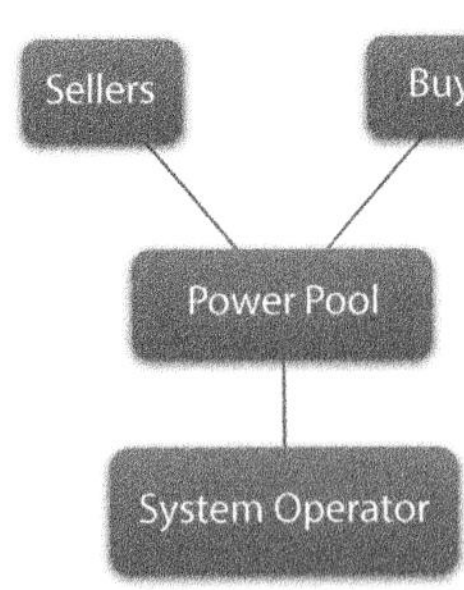

Figure 9.1 PoolCo model.

The hybrid model is way different than the PoolCo model because, in this approach, the utilization of power exchange is not mandatory. In the hybrid model, consumers get the option to choose bilateral contracts instead of participating in a centralized pool. It empowers the consumer to select the supplier directly by providing the flexibility to engage in individual agreements. The bilateral model has been shown in Figure 9.2.

Figure 9.2 Bilateral model.

The California model is an example of this hybrid model which incorporates the features of both PoolCo and bilateral contract models. This model is very useful in providing maximum flexibility to end users. It allows them to make the purchase of electricity based on their preference. Figure 9.3 shows the hybrid model of the restructured power system.

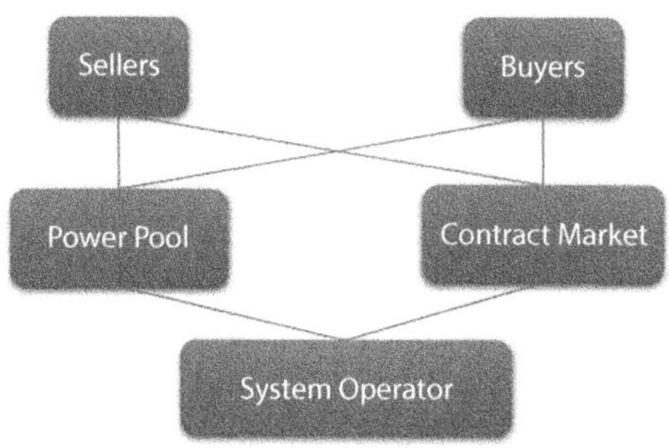

Figure 9.3 Hybrid model.

This approach provides the consumer with the option to actively track their energy consumption by providing real-time usage data and budget through a prepaid metering system. The ability of hybrid models to accommodate different procurement strategies is directly beneficial for the implementation of prepaid electricity systems. The combination of these two not only helps to reduce theft but also empowers the consumer to make informed decisions.

9.2 Literature Survey

The study [1] is an analysis of the economic logistical and technological challenges of prepaid electricity experiments in South Africa. This experiment was very beneficial for consumers, but it resulted in some drawbacks and challenges. This experiment can be termed as a success because of its efficient marketing unique and customer-centric tariffs and a strong and efficient administration. We can learn a lot from this experiment conducted by South Africa for designing the policies of prepaid electricity in the emerging markets of developing countries.

One of the major benefits of this experiment was that it helped the customer to improve their energy consumption. As they were aware of the energy consumption, they were more effectively controlling their electricity spending by putting off lights, geysers, fans, and various other appliances when not in use. It also helped them to make electricity a part of their budget, which also helped a lot in controlling excessive electricity consumption by controlling the budget.

The weaknesses in the traditional electricity metering system resulted in significant financial losses for consumers as well as distribution companies due to cases of power theft. There are many challenges faced by traditional metering systems such as the malfunctioning of meters, miscalculations in the computation of power use, manual errors in monitoring the units, and some other inefficiencies. To deal with all these kinds of problems, a proposed solution called "Smart Electric Meter" is operated on a prepaid system. The implementation of this system is described in [2]. A system is designed integrated with a Node MCU 8266 Wi-Fi module. The implementation of this system helped the consumer to manage their electricity consumption in a more efficient manner and also provided features such as alert settings and email notifications. This system has the potential to solve the problems in the traditional meters such as unpaid bills and electricity theft and helps the consumer to reduce the cost of electricity.

The daily electricity sector in India faces lots of challenges for its operations such as nonpayments of customers, increasing fiscal losses at the Regional Electricity Board (reb), power theft overstaffing, and lack of financial resources. To solve these problems, a large-scale reform of the overall structure and operations of the electricity sector was required. Research completed by Srivastava and Shahidehpour [3] examines the prevailing situation in India's power sector. Basically, this study investigates the alternative to the current electricity market by introducing some reforms such as setting up a wholesale power market, introducing competition, and predetermining the generation tariffs in a competitive market environment. The goal of this study is to provide useful insights into the reforms that should be taken into effect in India's electricity market by referring to the learning from the worldwide restructuring process of electricity distribution.

In this study [4], the impact of the use of prepaid metering systems in the households of Bongor, a city in Indonesia, was examined. The main aim of the study was to check the change in energy-related behavior due to the switching from a standard postpaid system to a prepaid system. A total of 600 families were interviewed in October and November 2015. Using some common variables, dashed thoroughly investigated the factors influencing the decision in household energy usage.

The study highlighted the fact that the financial growth and the number of families had a significant impact on the choice to switch. The families with the lower number of members were more interested in converting due to their regular usage. The education level played an important role in the realization and the decision to switch. The awareness campaign should be created by the city officials for Energy Conservation by targeting certain demography.

A study conducted in [5] discusses the elements that influence customer acceptance of prepaid metering systems in India and also explains the reason for their adoption. In India, the transmission and distribution losses are as high as 31%, which is one of the highest in the region, and this study identifies critical problems related to these losses. The losses are attributed to the following factors such as outdated infrastructure, insufficient investment, low meter efficiency, problems with the collection of bills, and power theft. When compared with the regional data on electricity losses, India stands on top of the list just after Pakistan and Nepal, and China has the least number of losses; it is evident from the fact that this problem needs a solution. The report from the above studies suggests that by acceptance of a prepaid metering system, the revenue management and driving utility can benefit a lot. It can significantly improve the efficiency and effectiveness of India's electricity distribution system.

This study is dedicated to evaluating different methods for service restoration with the help of the Floyd Warshall algorithm, which is generally known for determining the shortest path in a graph. The IEEE 14 bus system was used for applying this technique which is a standard system used by various testing. The main objective [6] of this study was to efficiently recover loads after a disturbance.

The most important part of the research was the study of different clustering strategies for optimal service restoration. Clustering is a process of grouping several components or subsystems on the basis of specific criteria, and it can be used for many practical applications. This technique was used to find the best sites for Starting the recovery process. This study presented a noble method for finding a problem unit that will help minimize the switching operations which will directly affect the active and reactive power losses which will be reduced by a significant amount. It will also help a lot in improving disaster recovery planning. This study is focused on improving the dependability and resilience of the power system by optimizing the restoration process, particularly at the time of blackout or disturbance.

The proposed method stated in [7] suggests the use of microgrids as an effective mechanism at the time of catastrophic disaster for restoring important loads on the distribution network. This approach considers the dynamic performance of distributed generators.

This study highlights the importance of dynamic restriction of the microgrid-based restoration system while assessing its feasibility. Microgrids are self-sufficient energy system that plays a key role in the case of critical load during service restoration. The proposed method provides a more realistic and practical solution to service restoration after a disaster by changing the dynamic performance of distributed generators and adding some restrictions to the generating resources.

The study in [8] provides an overall view of the decision-making process in the distribution network operators which focuses on the deployment based on a priority list which is made from the data from various units. Making a priority list simplifies the calculation and increases the computation efficiency. It is also extremely useful for the companies which work on renewable energy production. In this study, the dependency on traditional generation is reduced by introducing some emission limits in the optimization process. This decision helped the generators by encouraging them to invest in renewable energy generators. After simulating the above conditions, the active power losses in the multi-microgrid system were minimized when it was included in the energy management processes objective function. This system helped a lot in improving the operating efficiency of the microgrid system and also contributed to reducing greenhouse emissions by prioritizing renewable energy sources in the energy management strategy.

A study was conducted in [9] to analyze the operational dynamics of a microgrid in both autonomous and nonautonomous modes in a distribution network for utilizing energy from both renewable and nonrenewable resources. To operate autonomously, this distribution system requires meticulous design and strategic thinking. It provides a comprehensive pathway for creating an independent and sustainable microgrid.

The proposed technique introduced methods for structural modifications as well as the sizing and siting of distributed generators specifically designed for autonomous micro-grades. Optimizing the selection of sites and size of renewable resources was a significant challenge for optimizing the resources to ensure effective autonomous operation. This process utilizes advanced approaches, such as particle swarm optimization and genetic algorithm–based optimization.

The strategy proposed in [10] was important for protecting the interests of both seller and buyer microgrids by providing them the infinitive to choose their strategy for maximizing their utilities. The most important part of this technique was to encourage local energy trading, which was achieved by providing various incentives to the seller microgrids. They were actively encouraged to participate in the energy exchanges which allowed them to have priority access to the energy from the local market.

It also allowed the seller microgrids to determine the amount of energy they wanted to distribute on the basis of their surplus. This step enabled them to make decisions that were important and with respect to their operational requirements. After the study, a thorough, numerical analysis was done to compare the advantages of a disproportioned system to both baseline and multi-leader–multi-follower systems. The results showed that the efficiency and benefits of this strategy are very helpful in improving the local energy market environment.

In [11], the implementation of smart grid technology efficiently addressed the challenges posed by today's complicated electricity systems. It was discovered that synchronized technology was helpful for grid operation and monitoring. The distribution industry faced several issues in minimizing AT & C losses, peak load management, outage management, energy storage, etc. Advanced metering infrastructure was highly useful in meeting these challenges. Large-scale distributed generation using renewable energy sources (solar and wind) depended heavily on advancements in the smart grid industry.

According to an implementation in [12], to summarize, a prepaid energy meter was more user-friendly due to its numerous advantages, which included a decrease in labor costs. It was similar to the mechanization of the first recharge and service prototype. The created meter could control the load in the sector, charge energy in advance, and keep track of its present status. It was capable of being used for household, industrial, and small business purposes. Additionally, it had excellent two-way communication capabilities between the smart meter and the user via the GSM module for sending and receiving instant incentives. On the other hand, the Bluetooth module was helpful for suppliers as it allowed them to adjust rates as needed.

When environmental factors were taken into account, using fossil fuels was also not a very good choice [13]. When the primary resource, like solar and wind power, was available, using local renewable sources was the most practical answer for a large number of circumstances. Depending on the availability of resources, this method could be economically attractive while minimizing environmental concerns.

If someone wanted to compare their household's consumption to that of other households with similar demographics and housing factors, the current study in [14] could also offer intriguing data and findings. As a result, they would be able to recognize how much they differed from the typical or average household with similar characteristics, and this would motivate them to look for ways to cut down on the amount of energy they use by implementing energy-saving practices like updating their outdated appliances and lightbulbs and adding insulation to their homes.

In [15], the entire world experienced electricity losses of over $25 billion, a staggering sum. Governments needed to take the lead in taking the necessary actions to deal with this scenario. Subsidies for electricity prices had to be offered. If the government offered financial incentives to users, they could work on modest projects with the money. In these projects, power was generated by various area-specifically advantageous methods and thereafter repaid to grid stations and utilities. Educating people about the positive and bad uses of media, such as radio, newspapers, and television, was a crucial first step.

At the distribution level in Ghana, Otchere-Appiah *et al.* [16] investigated the issue of electricity theft, meter tampering, and other nontechnical energy losses. They examined information from a pilot program involving digital prepaid meters, in which 46% of the 1666 residential customers included switched to tamper-proof prepaid meters from older, more prone-to-tampering meters. They were able to conduct a causal inference analysis of the relationship between digital prepayment meters and measured electricity usage, utility revenues, and electricity theft thanks to the project structure. The group with prepaid meters saw a 13.2% rise in power consumption, and all income quintiles saw an increase in measured consumption with the exception of the lowest quintile. Their findings implied that digital electric technologies could enhance the distribution grid's efficient operation, notwithstanding data restrictions.

The usage of wireless gateways and intelligent mobile apps for home automation was covered in this study [17], which is easy to install and configure. This method also included tracking energy usage and applied image processing for identifying human activities. It includes the deployment of a smart grid–based Internet-of-Things system which will lead to home automation for improved energy saving. The benefits of using these kinds of servers and sensors with the Internet of Things can make a better home automation solution by taking the factor of temperature, light intensity, and humidity for a proper energy management system.

According to Sharma and Saini (2015) [18], smart meter standards were changing due to market pressures and government policies. A number of topics related to smart meters were covered such as integrated circuit standards, the effect of harmonic factors, privacy and security measures, and the infrastructure for supporting smart meters. This book also highlighted the difficulties that are faced by the smart meter industry. It also highlighted the importance of smart grids in the distribution system to have secure communication and data security.

Accessibility to modern energy services is very important for both economic growth as well as human well-being. A study conducted under the direction of Jamasb *et al.* (2018) [19] discusses the business model, distribution network, and application in underdeveloped nations. The distribution industry in the developing nation was suffering from inefficiency and losses. Based on the theory of entrepreneurship and the current developing technologies, a better business model was proposed for separating the companies into different sectors, such as communication, IT, metering, and infrastructure. It also recommended the participation of financial institutions, independent regulators, and insurance firms. This model can be very beneficial for decreasing losses and increasing customer satisfaction.

The study in [20] discussed that smart meters are generally used to track and measure energy usage as their ability to interact with other gadgets by the use of the Internet of Things makes them a very important part of monitoring and managing electricity consumption in smart cities. Information on pricing offers and safety alerts was provided to the customer. Many different factors are measured by the smart meter such as voltage level, total current drawn, power factor, and energy usage. Establishing a smart environment is very important for the smart metering infrastructure. There are many benefits that are related to smart meters such as they can provide data on energy usage, there is no reading cost, and it also has theft and tamper detection. In India, smart city and smart grid projects are being implemented by the government at a large level, and establishing value-added services and support for smart metering infrastructure will be very helpful for the consumer in the future.

A study was conducted by a research team led by Vijay Suryad from the Indian Institute of Technology Bombay (2017) [21] on microgrids in India. Microgrids in India are controllable systems that can operate on both isolated and grid-connected networks. They are made up of distributed sources, loads, and energy storage devices. Microgrids are very helpful for providing electricity in places where the main grid is inaccessible or unreliable. In rural areas, the government has been supporting the use of these microgrids by using locally accessible resources and renewable energy resources to supply electricity in these areas. In the past 10 years, there has been a rise in the installation of ac microgrids which has encouraged businessmen to establish isolated power systems. The electricity generated by these renewable energy sources is then distributed through planned microgrids and mini-grids. In rural areas, the remote communities are suffering from an economic perspective due to the lack of access to power. Decentralized renewable energy and solar household systems can be very beneficial to serve a wide range of consumers throughout the nation. Various microgrid initiatives have been initiated in many states of India including Orissa Bihar and Sikkim. Sikkim utilizes its locally available hydro resources for many micro-hydro projects. The electricity produced by microgrids is cheaper than that of the main grid. This survey included 18,452 communities that were not connected to the electricity network.

In order to provide electricity in under-served places, the combination of diesel generators and storage was discussed in [22]. It made the case for running diesel generators at their highest efficiency while examining the engineering and economics of this technological combination. The usage of prepaid metering systems to pay for the initial expenses of storage was also covered in the book. It provides insight into the best operating policies and a proper model for the generation and storage of electricity. The advantage of using prepaid energy and the possible savings of combining diesel generation with storage is covered in this text.

The study [23] proposed a method for enhancing the features of household energy meters in India by implementing some technical improvements. It suggested the switch from the static meter system to a dynamic metering system that could monitor the electricity usage in real time and store the data in the cloud for easy accessibility. The Internet of Things is used in this system to provide facilities such as power cut notifications, tampering alerts, and power disconnection and reconnection alerts which are provided to the user for a better experience. The software and hardware that were used for making this system are an ARM Cortex M4 processor, a display unit, a keypad, a Wi-Fi module, and sensors for alerts about power outages and tampering. Disproportion gave the customer the option of combining the features of both prepared and postpaid metering systems which was able to give a higher accuracy, speed, and dependability when processing and transmitting data. The experimental findings and the conclusion highlight the benefit of using the system and the Internet of Things for energy metering services. It provided a very effective approach for the development of energy metering technology with the possibility of using it to streamline the billing procedure and decrease electricity theft.

Shahzad and associates (2023) [24] investigated the potential difficulties and situations of microgrids in the future. Microgrids are self-sufficient energy systems that can produce, store, and distribute electricity to supply local communities. Establishing microgrids can solve the problem of the energy crisis in Pakistan as well as the sustainability issues worldwide. There are many obstacles to establishing a microgrid such as lack of technological advancements, legal concerns, and funding. It is very useful for providing energy independence and offers a reliable and resilient energy solution in isolated locations that are vulnerable to natural calamities. Microgrids help to reduce the dependency on fossil fuels and are very efficient for recovery and disaster response as well. Microgrids are the most practical and affordable solution for supplying electricity once the obstacles are removed.

According to Alahakoon and Yu [25], "Smart Electricity Meter Data Intelligence for Future Energy Systems" highlighted the significance of smart metering and smart energy for future energy consumption. Intelligent meters were here to stay, in spite of resistance. A framework for connecting smart meter data to stakeholders and applications was presented in this study. Building blocks for smart meter analytics were also identified. In the future, the smart grid and smart meters will be a part of the larger Internet of Things. Accurate information and real-time decision-making were essential for smart grid management. Monitoring energy use and reconstructing network topology could both benefit from shorter time interval observations. Valuable data for price intelligence and forecasting were provided by smart meters. Smart metering required cloud platforms and expanded analytic capabilities.

Energy information display systems for intelligent homes had the potential to influence user behavior, as noted by Wood and Newborough (2006) [26]. Cutting-edge energy consumption displays could encourage eco-friendly behaviors and energy awareness. It would be best to use a hybrid display strategy that combines central and appliance-specific displays. While there was not enough study on mobility for energy-consuming displays, usability was essential for large energy savings. If and where energy information should be presented depended on how many people used an item and what kinds of behaviors, they engaged in. Low automation and complex appliances presented chances for energy-saving practices. Practical micro-behaviors, the relationship between appliance control and end-use, and the energy consumption of specific appliances should all be taken into account. Activity-based or central displays ought to go hand in hand with appliance-specific displays.

9.3 Proposed System

The proposed system brings substantial benefits to consumers by revolutionizing their electricity access experience. Through a secure and user-friendly platform, individuals can articulate their specific power needs, leading to a tailored and efficient selection process. Categorization based on location, demand, and priority ensures a swift connection to the nearest available substation, optimizing power distribution.

The algorithm prevents overload and ensures a proper supply of electricity. The user is provided with the option to choose from various power sources near them and the purchased electricity is stored in their credit wallet. This system is equipped with the technology to enhance the user experience and provide accessible and sustainable energy solutions.

To use this proposed system, at first, they will have to go through a simple sign-up process in which they are asked to fill out the basic details. After the successful sign-up, they are asked to log-in to their account. After successfully logging into the account, the customer has the choice to choose from the available energy resources near them. The system then examines the availability of power at the nearest stations and figures out the optimal part for supplying the electricity. To properly implement this system, the user is categorized on the basis of location, demand, priority, and type to ensure optimal distribution.

To ensure the efficient distribution of electricity, the capacity of the nearest substations is calculated on a regular basis to get the available slots in the substations. After that, on the basis of the demand by the consumer,

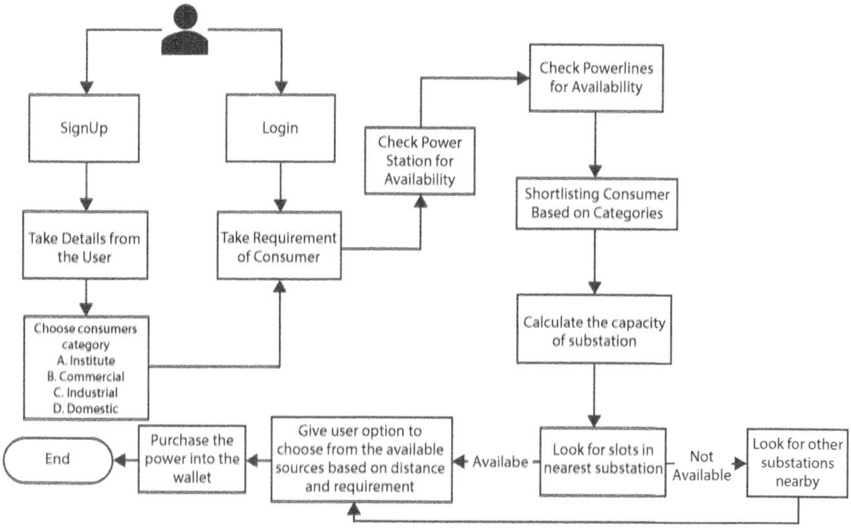

Figure 9.4 Flow chart of the proposed system.

they are connected to the nearest substation with available slots for the supply of electricity. In case the nearest substitution does not have the slots for the supply, the user can seek electricity from a different source with available slots. An algorithm is implemented for regularly checking every distributing substation for available slots and providing data to the consumer for an efficient and optimal supply of electricity as per the request of the consumer. This system optimizes the utilization of existing infrastructure, prevents overloading, and ensures sustainable and effective operations.

Initially choosing a wide range of available power sources for electricity purchases. It is then stored in a secured wallet of the consumer which can be used by them at any point in time which provides them with the flexibility and accessibility of their energy usage. Figure 9.4 shows the overall flowchart of the proposed system.

9.3.1 Methodology

In the proposed method, the consumers are categorized into institutional, industrial, commercial, and residential factors, which help in making strategies and prioritizing the needs of every segment. This approach ensures the requirements and priorities of different types of consumers by ensuring special sectors, such as institutional and industrial, have uninterrupted supply. The proposed system enhances the quality and electricity supply by creating unique solutions by analyzing the specific demands of each category of consumer. For example, industries require a large amount of electricity, so we can focus on providing consistent supply in this sector, whereas in the case of household consumers or residential consumers, we may prioritize the affordability and reliability of supply. This system helped to provide a more personalized and responsive service for optimal energy distribution and improving customer satisfaction.

A web interface was designed by using the facility of Wix website builder to create an online platform for implementing the proposed method. The process of building the website included a comprehensive assessment of the requirements and the objective of our system, which helped us f to determine the key features and functionalities to include in the interface.

In the initial phase of planning the website, we discussed the structure, arrangement of pages, and navigation. It was a seamless task to build this website with the help of wits because of its wide variety of templates and the features of dragging and dropping the objects for creating the website aligned with our goal of making an aesthetic and functional website.

In the next phase of customization, the specific branding and design requirements were fulfilled. It also included the designing part of the website which allowed us to easily modify the colors, fonts, animations, etc.

The entire process of building this website included regular testing and optimization of the various functionalities. Finally, the system was deployed on the web interface, making it accessible for the users with the help of Wix's hosting service, which easily deployed the website, and it was live and ready to use.

9.3.2 Implementation

The user interaction is initiated with a structured sign-up process.

Step 1

In the initial step, the user is asked to provide essential personal information, such as their name, email, password, address, and contact information. After the collection of these data, the user is systematically categorized into a distinct sector, which helps to provide efficient electricity service. These factors are residential, commercial, industrial, and institutional. Figure 9.5 shows the evidence of sign up process.

After the successful completion of the sign-up process, the details provided by the users are verified by the authorities. This verification is conducted to ensure the accuracy and legitimacy of the information which is provided by the user. It also helps to ensure and maintain the integrity and security of the platform. A one-step verification process is completed, and

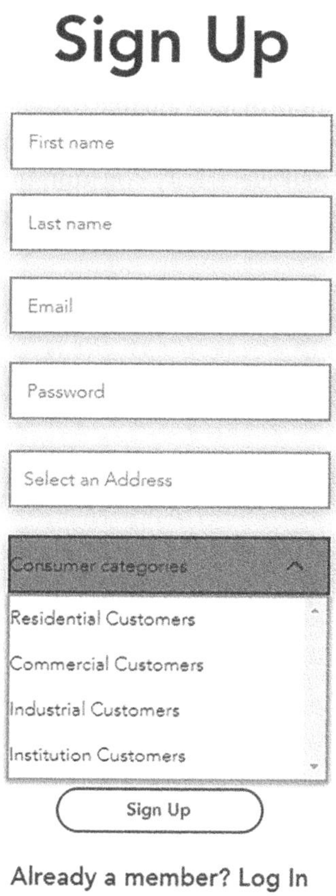

Figure 9.5 Sign-up process.

the user account is created. After the creation of an account, the user can seamlessly use the features and services of the platform.

Step 2

In the second step, the user is introduced to the home page of the website, which is the central hub for all the facilities that are provided via the platform. The user can freely navigate through each and every part of the website, which is designed for a comprehensive experience and engaging the user in purchasing electricity.

On the home page, the user can initiate the purchase of electricity by clicking on the purchase option after which they will be navigated to the purchase section where they can select their proper requirement and make transactions for their purchase.

The platform also provides a feature to monitor the electricity consumption of the user, which can provide valuable insight into their energy usage pattern. It can be extremely useful for making informed decisions and making them aware of their electricity consumption. Figure 9.6 shows the evidence of the proposed home page interface system.

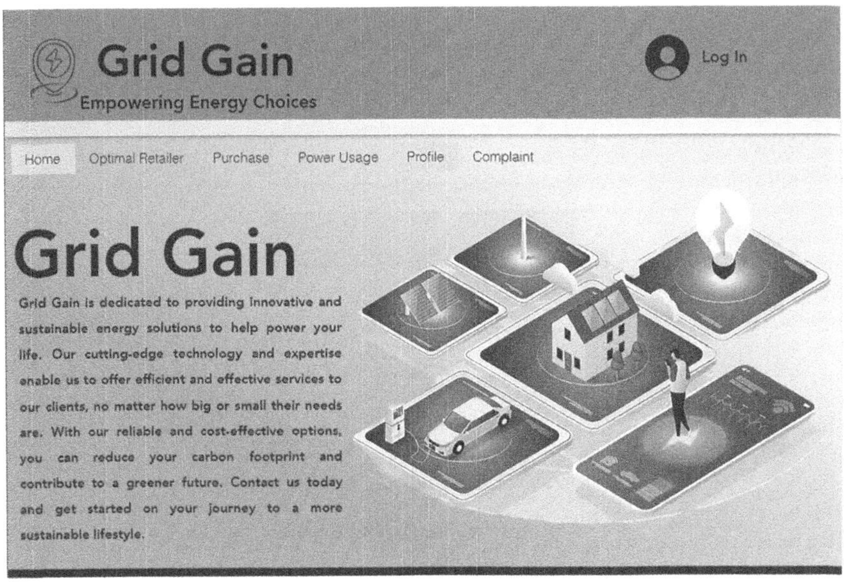

Figure 9.6 Home page.

Step 3

On the website, the user is provided with the option to choose from the nearest electricity-generating source with the help of an interactive map. The map aligns with the geographical location of the user, offering them a personalized way to select their preferred electricity source. It enables the user to explore and identify the nearest electricity-generating sources in their area. By connecting the nearest generating station to the user, the platform aims to provide an uninterrupted supply of electricity which enhances the overall user experience and contributes to providing a more resilient and stable distribution network. Figure 9.7 shows the evidence of the overview of the location of different electricity-generating sources of the proposed home page interface system.

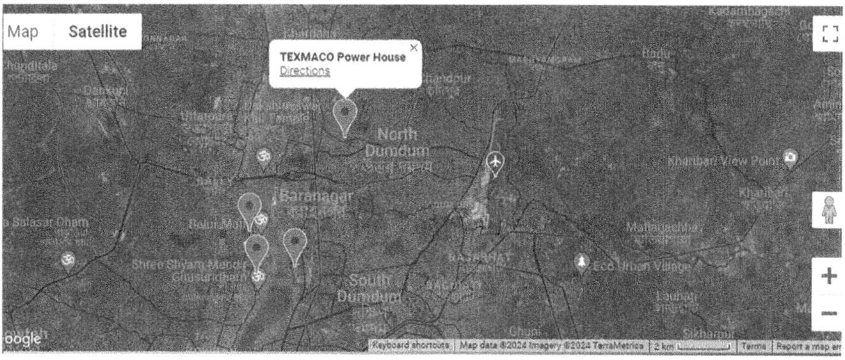

Figure 9.7 The overview of the location of different electricity-generating sources.

Step 4

In step 4, the user is guided to specify their electricity needs by providing the desired quantity of electricity they want to purchase. The implementation of the load forecasting technique at this stage helps the user to determine and suggest an optimal quantity of electricity that they may require by studying the historical usage pattern of the user by utilizing analytical insights. These data empower the consumer with valuable information that enables them to make informed decisions for their electricity consumption. Figure 9.8 shows the evidence of the prices of different power-generating sources of the proposed home page interface system. Figure 9.9 shows the evidence of the power purchase form of the proposed home page interface system.

Figure 9.8 Outlook of different power-generating sources with prize.

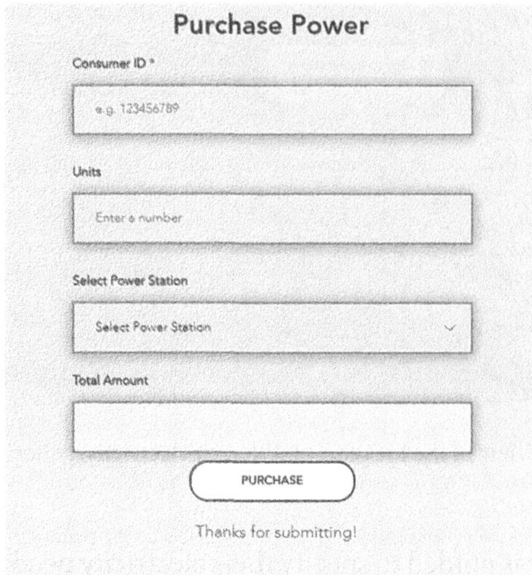

Figure 9.9 The power purchase form.

Step 5

In step 5, the platform allows the user to visualize their energy usage pattern with the help of an interactive graph. This feature allows the user to monitor their energy consumption habit on a daily and monthly basis by visualizing the consumption trends. Figure 9.10 shows the evidence of the load curve of different time frame of the proposed home page interface system.

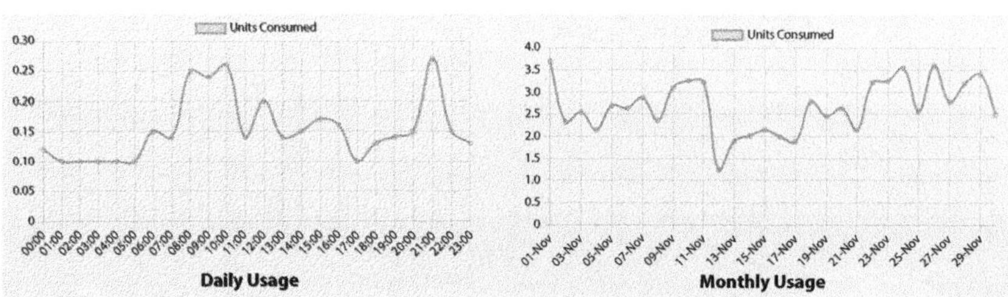

Figure 9.10 The electricity usage of the user.

It helps the user to understand their usage pattern and makes them aware and responsible for informed decision-making in electricity consumption. User can easily identify the trends, such as peak usage periods to optimize their consumption and create a sustainable usage behavior.

9.4 Industrial Designing Application for Power Systems Control

A smart prepayment meter can be used in an industrial context to track energy use and keep an eye on how much electricity different industrial equipment uses. Industrial companies can save production costs by identifying equipment energy consumption problems and implementing targeted energy-saving solutions by evaluating meter data. Another advantage of this proposed system is that the landlord may monitor each tenant's electricity consumption at any time which is based on the prepaid electricity meter system's real-time data collection feature. Reminders can be sent out appropriately in case of excessive load power to guarantee safe electricity use, even for a few days without consuming any electricity.

9.5 Conclusion

In conclusion, by using the hybrid model in this project, the overall energy distribution landscape can be significantly revolutionized. The advantage of incorporating a hybrid model with the benefits of this project is a transformative shift toward a more sustainable and consumer-centric approach to the electricity distribution system. The flexibility of the hybrid model allows it to integrate with various energy sources which ensures a diversified and reliable energy supply. This approach optimizes the price of electricity based on different sources and also provides the freedom to choose them based on their preferences. The advantages of implementing a prepaid electricity system outweigh the disadvantages. The shifting to a prepaid system helps to improve the cash flow in the businesses and also efficiently recover bad debt. The elimination of fixed charges and reduced administrative interference turns out to be a consumer-friendly nature of the prepared system. The system is easy to install, and it has an effective fraud control system which enhances the overall benefits of this approach. Most importantly, it empowers the consumer by helping them to properly understand their energy consumption and usage with the facility to conveniently purchase electricity in the wallet which has a positive impact on the end user. There are some concerns, such as the large initial investment in installation costs and the limitation of handling large amounts of current, but the overall advantages, such as enhanced safety, improved revenue generation, and increased efficiency, make the prepaid electricity distribution system a better solution for modern energy distribution scenario.

This project attempts to solve the problem of electricity theft and ensures fair and legal consumption patterns. It also makes the consumer aware of and empowers the consumer with the information of their usage pattern for making informed decisions and taking a step forward toward sustainable energy consumption and conservation. Lastly, this chapter contributes to the development of efficient energy distribution systems, optimizing resources, minimizing losses, and using renewable resources for a sustainable society.

References

1. Tewari, D.D. and Shah, T., An assessment of South African prepaid electricity experiment, lessons learned, and their policy implications for developing countries. *Energy Policy*, *31*, 9, 911–927, 2003.
2. Misri, S., Navghare, S., Sawant, S., Chauhan, J., IoT enabled prepaid electricity meter, in: *2nd International conference on advances in science & technology (ICAST)*, 2019.
3. Srivastava, A. and Shahidehpour, M., Restructuring choices for the Indian power sector. *IEEE Power Eng. Rev.*, *22*, 11, 25–29, 2002.
4. Nugroho, S.B., Zusman, E., Nakano, R., Takahashi, K., Koakutsu, K., Kaswanto, R.L., Fujita, T., The effect of prepaid electricity system on household energy consumption–the Case of Bogor, Indonesia. *Procedia Eng.*, *198*, 642–653, 2017.
5. Bandyopadhyay, K., User acceptance of prepayment metering systems in India. *Int. J. Indian Cult. Bus. Manage.*, *1*, 4, 450–465, 2008.
6. Sarkar, D., Chakrabarty, M., De, A., Goswami, S., Emergency restoration based on priority of load importance using Floyd–Warshall shortest path algorithm, in: *Computational Advancement in Communication Circuits and Systems: Proceedings of ICCACCS 2018*, pp. 59–72, Springer, Singapore, 2020.
7. Xu, Y., Liu, C.C., Schneider, K.P., Tuffner, F.K., Ton, D.T., Microgrids for service restoration to critical load in a resilient distribution system. *IEEE Trans. Smart Grid*, *9*, 1, 426–437, 2016.
8. Aghdam, F.H., Ghaemi, S., Kalantari, N.T., Evaluation of loss minimization on the energy management of multi-microgrid based smart distribution network in the presence of emission constraints and clean productions. *J. Clean. Prod.*, *196*, 185–201, 2018.
9. Kirthiga, M.V., Daniel, S.A., Gurunathan, S., A methodology for transforming an existing distribution network into a sustainable autonomous microgrid. *IEEE Trans. Sustain. Energy*, *4*, 1, 31–41, 2012.

10. Jadhav, A.M., Patne, N.R., Guerrero, J.M., A novel approach to neighborhood fair energy trading in a distribution network of multiple microgrid clusters. *IEEE Trans. Ind. Electron.*, *66*, 2, 1520–1531, 2018.
11. Jha, I.S., Sehgal, Y.K., Sen, S., Kumar, R., Smart grid in indian power system. *NPSC 2012*, 1–6, 2012.
12. Rashid, F., Rasheed, S., Farooq, A., Ali, M., Ruben, Y., Jehanzeb, M., Wahab, A., Design and implementation of prepaid power billing system in smart grid environment. *Int. J. Electr. Comput. Eng.*, *12*, 4, 3365–3374, 2022.
13. Blasques, L.C.M. and Pinho, J.T., Metering systems and demand-side management models applied to hybrid renewable energy systems in micro-grid configuration. *Energy Policy*, *45*, 721–729, 2012.
14. Desipri, K., Legaki, N.Z., Assimakopoulos, V., Determinants of domestic electricity consumption and energy behavior: A Greek case study. In IISA 2014. *The 5th International Conference on Information, Intelligence, Systems and Applications*, pp. 144–149, IEEE, 2014, July.
15. Anas, A.M., Avoiding Electricity Theft using Smart Meters in Smart Grid. *Proc. COMSATS Inst. Inf. Technol. Islamabad 2012*, 1–59, 2012.
16. Otchere-Appiah, G., Takahashi, S., Yeboah, M.S., Yoshida, Y., The impact of smart prepaid metering on non-technical losses in ghana. *Energies*, *14*, 7, 1852, 2021.
17. Prakashraj, K., Vijayakumar, G., Saravanan, S., Saranraj, S., IoT Based Energy Monitoring and Management System for Smart Home Using Renewable Energy Resources. *Int. Res. J. Eng. Technol.*, *7*, 2, 1790–1797, 2020.
18. Sharma, K. and Saini, L.M., Performance analysis of smart metering for smart grid: An overview. *Renew. Sustain. Energy Rev.*, *49*, 720–735, 2015.
19. Jamasb, T., Thakur, T., Bag, B., Smart electricity distribution networks, business models, and application for developing countries. *Energy policy*, *114*, 22–29, 2018.
20. Janardhana, S. and Shashikala, M.S.D., Challenges of smart meter systems, in: *2016 International Conference on Electrical, Electronics, Communication, Computer and Optimization Techniques (ICEECCOT)*, pp. 78–82, IEEE, 2016, December.
21. Suryad, V.A., Doolla, S., Chandorkar, M., Microgrids in India: Possibilities and challenges. *IEEE Electrif. Mag.*, *5*, 2, 47–55, 2017.
22. Porter, J., Pedrasa, M., Woo, A., Poolla, K., Combining storage and generation for prepaid electricity service, in: *2015 IEEE International Conference on Smart Grid Communications (SmartGridComm)*, pp. 422–427, IEEE, 2015, November.
23. Rajesh, B.R., Kumar, S.M., Sharief, N.Z., Kothari, S., Ezhilarasan, K., IoT based automatic energy metering system with prepaid/postpaid configurability, in: *2017 International Conference On Smart Technologies For Smart Nation (SmartTechCon)*, pp. 592–595, IEEE, 2017, August.

24. Shahzad, S., Abbasi, M.A., Ali, H., Iqbal, M., Munir, R., Kilic, H., Possibilities, Challenges, and Future Opportunities of Microgrids: A Review. *Sustainability*, 15, 8, 6366, 2023.
25. Alahakoon, D. and Yu, X., Smart electricity meter data intelligence for future energy systems: A survey. *IEEE Trans. Ind. Inf.*, 12, 1, 425–436, 2015.
26. Wood, G. and Newborough, M., Influencing user behaviour with energy information display systems for intelligent homes. *Int. J. Energy Res.*, 31, 1, 56–78, 2007.

10

PV System Maximum Power Point Tracking Under Partial Shadowing Using Gray Wolf Optimization Algorithm

Snehashis Ghoshal and Arindam Mondal*

Dr. B. C. Roy Engineering College, Durgapur, West Bengal, India

Abstract
Solar energy is utilized efficiently with photovoltaic (PV) installation when operated at or near the point of maximum or peak power. In order to implement such an operation, proper tracking becomes essential, which is undoubtedly a challenging objective as PV systems possess a multitude of peaks on their power-voltage characteristics due to varying weather conditions, particularly during partially shaded conditions. The main drawback of conventional MPP is the proper step size selection. Otherwise, the system may be operated at a local MPP instead of the global MPP. In the present study, gray wolf optimization was adopted in the MPPT controller and a comparative analysis was carried out between GWO and P&O-based results on the same system simultaneously. The outcomes reflected the efficacy of the proposed methodology during partially shaded conditions.

Keywords: Photovoltaic system (PV), tracking of maximum power, partially shaded condition, perturb and observe technique, gray wolf optimization

10.1 Introduction

The ever-increasing demand for energy and the detrimental impact of pollution make it necessary to put significant efforts into environment-friendly renewable resources. Among different renewable resources, solar energy possesses a good potential for power generation due to its abundance throughout the globe [1]. A major part of solar energy is extracted through

**Corresponding author*: armsvist@gmail.com

Arindam Mondal and Souvik Ganguli (eds.) Controller Design for Industrial Applications, (189–210)

photovoltaic (PV) systems which work on the photoelectric effect. The main drawback of the PV mode of energy conversion is lower output which is generally addressed by connecting several panels of suitable ratings in a series-parallel combination. Solar PV–based DC systems often employ power conditioning devices to supply the load reliably [2]. The efficient operation of PV systems requires their operation at the point where maximum power can be extracted. Numerous algorithms have been adopted to form controllers in order to address the issues related to tracking of maximum power [3–7]. During consistent insolation, the behavior of power-voltage characteristics may exhibit a singular peak, allowing accurate tracking. Nevertheless, real-world scenarios introduce variations from this pattern like shading in a portion of the module due to the clouds, large buildings, dust accumulation, or the presence of large trees. Such an obstruction between the module and sunlight leads to operating the module in multimodal mode. Employing conventional MPPT techniques, the pursuit of MPP can inadvertently get trapped within a local peak, leading to a decline in the effective utilization of solar energy. The major challenge is to operate the system at the global maxima adhering to their power-voltage curve irrespective of local maxima [8]. In order to achieve this, several optimization techniques were combined in certain cases [9, 10]. Particle swarm optimization (PSO) was utilized to form an MPPT controller in a PV system connected to the grid [11]. A boost converter with an MPPT mechanism and a grid-tied three-phase inverter were inserted between the PV system and the grid. According to the literature neural networks (NN) [12], fuzzy logic control (FLC) [13], ant colony optimization (ACO) [14], artificial bee colony (ABC) [15], etc. have been employed to achieve the global maxima during operation of similar PV system.

In this chapter, a controller working on gray wolf optimization (GWO) logic is presented to operate a PV system at MPP. A well-designed iterative termination strategy is introduced to mitigate power oscillations once the system has achieved stability. Simulation outcomes indicate that the proposed MPPT approach can swiftly and accurately trace the global MPP even amidst variations in irradiance and partial shading conditions.

10.2 Analysis and Modeling of Solar PV Systems

Mathematical models may be useful to investigate the performance of PV systems. They enable the simulation and forecasting of PV system behavior under various operating conditions. In order to model a solar PV system, power converters, loads, and a regulating mechanism for maximizing

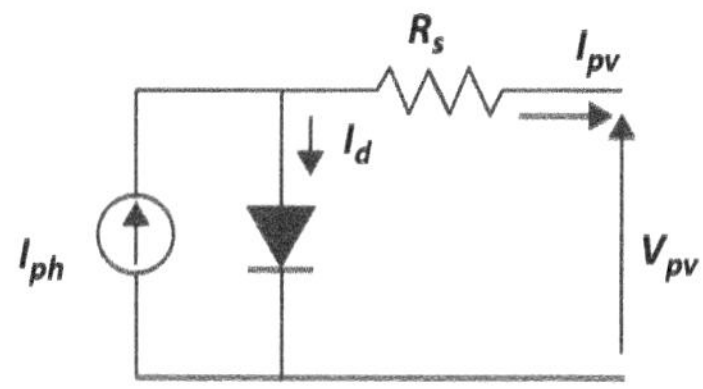

Figure 10.1 One diode model representation of solar cell [2].

power are often interfaced. A PV system may be modeled in a variety of ways using the simulation platforms that are available, including mathematical block modeling, embedded programming, and physical block modeling. Simulink's physical modeling is used in the current work. One diode model serves as the foundation for the Simulink model of a PV system. Figure 10.1 shows one diode model depiction of a solar cell.

In this assumption, a solar cell is treated as a constant magnitude I_{ph}. The bypass diode is linked in R_S, which is the circuit's series resistance and represents the equivalent internal resistance of the system that is open-circuited. The following equation is produced when KCL is applied in the circuit, as shown in Figure 10.1.

$$I_{pv} = I_{ph} - I_d \tag{10.1}$$

Shockley diode equation describes that

$$I_d = I_s\left[\exp\left(\frac{qV_{OC}}{NKAT_o}\right) - 1\right] \tag{10.2}$$

Combining Equations 10.1 and 10.2 yields Equation 10.3 as below.

$$I_{pv} = I_{ph} - I_s\left[\exp\left(\frac{qV_{OC}}{NKAT_o}\right) - 1\right] \tag{10.3}$$

The following equation explains how the system is affected by internal resistance:

$$I_d = I_s\left[\exp\left(\frac{q(V + IR_S)}{NKAT_o}\right) - 1\right] \tag{10.4}$$

Equation 10.4 can be rewritten with the help of Equation 10.3 as below.

$$I_{pv} = I_{ph} - I_s\left[\exp\left(\frac{q(V + IR_S)}{NKAT_o}\right) - 1\right] \tag{10.5}$$

In order to produce significant output, PV panels are linked in a suitable series-parallel arrangement, as can be seen in Equation 10.6.

$$I_{pv} = N_p I_{ph} - N_p I_s\left[\exp\left(\frac{qV_{OC}}{NKAT_o}\right) - 1\right] \tag{10.6}$$

The output of the cell is independent of the internal resistance in the circuit connected and is reliant on the amount of incident light on its surface as well as the ambient temperature. Equation 10.7 may be used to explain this.

$$I_{ph} = \left[I_{sc} + K_i\left(T_o - T_i\right)\right] \times \frac{G}{G_{ref}} \tag{10.7}$$

However, if the impact of the solar cell's leakage current on the system is taken into account, this study may be made more practical. In mathematical modeling, the effect of the leakage current is taken into account as a parallel-connected resistor R_{sh}, as depicted in Figure 10.2.

From a mathematical perspective, the output equation in relation to the current for the configuration represented in Figure 10.1 can be formulated as the following.

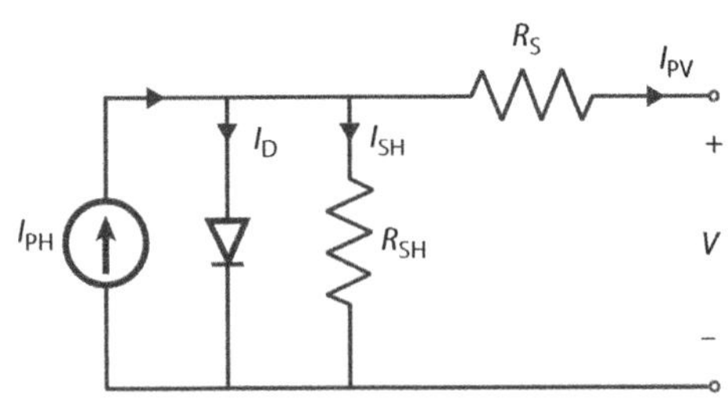

Figure 10.2 Representation of a solar cell using one diode model and leakage current [1].

With the above-mentioned approach, Equation 10.1 is modified as

$$I_{pv} = I_{ph} - I_d - I_{sh} \tag{10.8}$$

The voltage across the diode and current in the shunt branch is represented by Equations 10.9 and 10.10, respectively as below.

$$V_d = V + I_{PV} R_S \tag{10.9}$$

$$I_{sh} = \frac{V + I_{PV} R_S}{R_{sh}} = \frac{V_d}{R_{sh}} \tag{10.10}$$

Considering the impact of leakage current, Equation 10.5 can be rewritten as

$$I_{pv} = I_{ph} - I_s \left[\exp\left(\frac{q(V + IR_S}{NKAT_o} \right) - 1 \right] - \frac{V_d}{R_{sh}} \tag{10.11}$$

10.3 Impact of Partial Shading in a PV System

When referring to PV modules, the term "partial shading" describes conditions in which just a section of a solar panel is exposed to sunlight while the remainder is shaded. The generation of a PV module may be significantly impacted by this partial shading. The effect of partial shading on a PV system is outlined below.

Reduction in Output
The output of a PV module reduces significantly even when a very small portion of the PV system is shaded partially. The main reason behind this is the series of connections of the panels where the current disrupts significantly due to the partial shading.
Voltage Loss
Besides, partial shading leads to a reduction in output voltage. Voltage drops drastically across cells that are shaded reducing the overall output voltage of the whole module.

Mismatch Losses

During partial shading conditions, shaded and unshaded cells produce different amounts of power causing a mismatch, and hence, the overall output of the system is reduced.

Formation of Hotspot

Shaded cells develop a small amount of power but cannot effectively dissipate heat, thereby forming hotspots. Continuous formation of hotspots overheats the panel, ultimately leading to the potential damage to the shaded cells and the lifespan of the module.

Impact of Partial Shading on Overall System Performance

The partially shaded condition reduces the output from a PV module. In the case of multi-module strings, the overall performance of the system degrades due to partial shading. Besides, it causes potential damage to the PV modules, and hence, system engineers should adopt suitable mitigation strategies.

10.4 Performance Optimization of Solar Panel During Partial Shading Condition

It is evident that partially shaded conditions significantly affect the performance of a PV system. During partial shading, certain areas of panels are covered by the shades of trees, buildings, etc. To overcome the detrimental effects of the partially shaded condition, the implementation of proper shading mitigation techniques becomes essential. A few such mitigation techniques are discussed below.

Optimal Array Design

Designing an array possesses a crucial role in mitigating the impact of partial shading. Optimal designing of the array involves panel positioning, string arrangement, monitoring system, etc. An optimally designed array system ensures stable and efficient operation of the system, extracting maximum power out of it. The selective design method enhances system performance on a long-term basis and maximizes return on investment.

Bypass Diodes

The bypass diode acts as a protective measure during partially shaded conditions. The application of such diodes bypasses the circulating current produced during the partially shaded condition. As a result, an unshaded portion of the module remains unaffected, and the module operates without overheating. Eventually, losses in the system are reduced, and the overall efficiency is improved.

Power Conditioning at Module

The introduction of micro-inverters or power optimizers in the module enhances the system performance. Such converters are placed alongside individual panels in order to mitigate the impacts of partially shaded conditions.

Solar Trackers

A solar tracker effectively minimizes the effect of a partially shaded situation in a PV system. The main purpose of such a tracker is to change the position of the panel along the movement of the Sun throughout the day to extract maximum power from it. Both single-axis trackers and dual-axis trackers are used in solar MPPT applications. A brief discussion of the same is carried out in the following. Trackers that maintain a single axis of tracking are known as single-axis trackers, whereas trackers working along two axes are known as dual-axis trackers.

a. Single-Axis Tracker

Single-axis trackers are classified into horizontal- and tilted-axis trackers based on the tracking mechanism adopted.

Horizontal-Axis Trackers (Single-Axis, East-West Tracking)

Trackers working on a single axis in the horizontal direction which rotates the panels along the movement of the Sun (East-West movement) are known as horizontal-axis trackers. This kind of tracking mechanism is economical and very simple to implement. This kind of tracking mechanism finds usefulness, particularly in low-cost applications where the trajectory of the Sun remains more or less consistent throughout the year.

Tilted-Axis Trackers (Single-Axis, Tilt, or Polar Tracking)

In this mode of tracking, panels are rotated along an inclined axis in order to capture more energy. In this case, there is a provision for the selection of the tracking along the North-South or East-West direction. Such flexibility catches up with the trajectory of the Sun in a better way and modifies the tilt angle of the system in a more effective way. In higher latitudes, the angle of the Sun varies throughout the season, and hence, in these applications, such a tracking arrangement becomes very promising.

b. Dual-Axis Tracker

Trackers maintaining tracking mechanisms along two different axes, which have a provision of rotating the panels along two different axes, are known as dual-axis trackers. In practice, dual-axis trackers consist of higher cost and complex implementation compared to single-axis trackers. In this mode of tracking, the trajectory of the Sun is most accurately followed across all the seasons. This kind of tracking mechanism is generally employed in precise tracking, such as in concentrated solar power (CSP) installations.

Depending on the specific requirement, a more particular and application-specific tracking mechanism may be developed depending on the applicability and cost of the same.

10.5 Significance of MPPT in Partial Shading Conditions

Solar PV systems provide a satisfactory operation while they are operated at the peak of the power-voltage characteristic. Continuous adjustment of the panel is necessary to achieve this. In this case, MPPT algorithms play a vital role. It is evident that panels possess a specific functional limit of voltage and current to extract maximum power depending on the irradiation, surrounding temperature, and shading effects. The controllers employing the MPPT algorithm keep track of panel output in every instant and regulate the power conditioning equipment for proper impedance matching with the system. Eventually, the system is shifted to the close proximity of its MPP. As a result, energy extraction from the system is maximized.

In the case of partial shading, solar irradiation is nonuniformly distributed among different cells or panels as the case may be. Due to this, there is a mismatch in the output of the system leading to overheating. MPPT-based controllers play a significant role in this scenario as they help maintain the unshaded part of the system operating at its MPP, whereas power from the shaded portion depends on the degree of shading. During its operation, the MPPT controller can identify and isolate the part of the system, performing unusually due to shading. Likewise, the impacts of partial shading are minimized significantly. Besides, the MPPT mechanism consistently monitors and updates the output from the unshaded portion of solar panels to ensure that they work properly and safely. As a result, the MPPT controller efficiently utilizes available irradiation by bypassing the shaded panels, thereby possessing the capability of dynamic adjustment of the healthy parts of the system. To operate the system at or near MPP, efforts are made to reduce mismatching losses, thereby minimizing overheating.

10.6 MPPT Techniques for Partially Shaded Scenario

In this work, the performance of a PV system was analyzed with two separate MPPT algorithm–based controllers (P&O and GWO) for similar environmental conditions. A brief discussion of these algorithms is presented in the following.

10.6.1 Perturb and Observe (P&O) Method

This technique has a widespread application for MPPT due to its simplicity. In this method, a comparison regarding the power output of the module is carried out at every instant to its previous instant. To maintain the peak output from the module, the voltage reference is promptly adjusted at all instants to alter the duty cycle. Figure 10.3 depicts the flowchart of this logic [16]. This

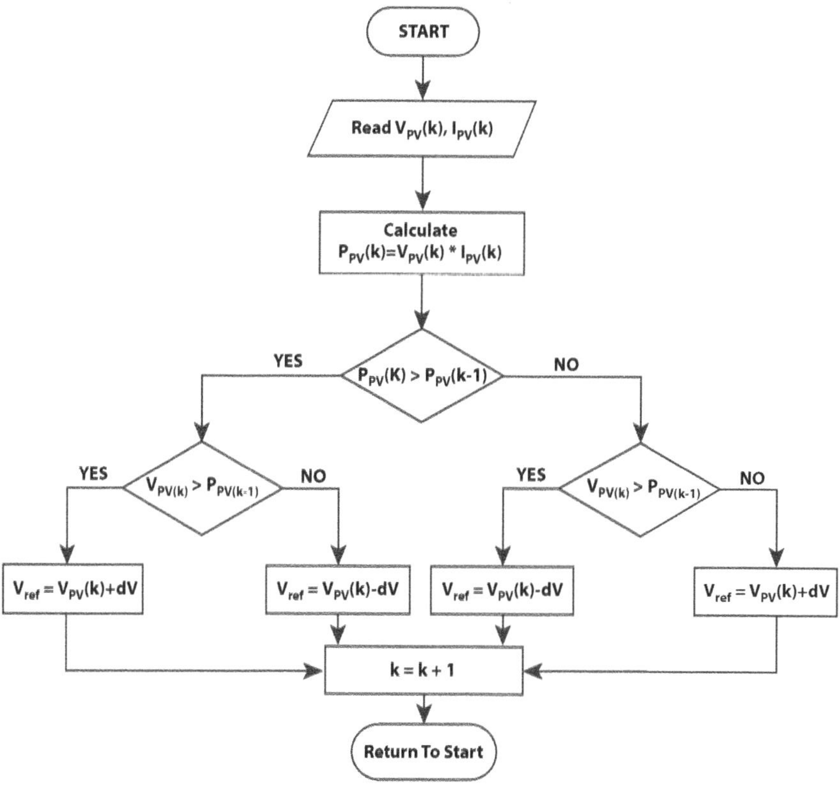

Figure 10.3 Flowchart of the P&O algorithm [16].

method provides reliable performance and is easy to implement. However, there are certain limitations of this method. In the case of P&O logic applications, oscillation of the operating point near MPPT requires quicker adjustments, which may lead to potential wear and tear in the system in the long term. Besides, during dynamically changing scenarios, this method exhibits sluggish response affecting the accuracy of the MPPT. This slow response leads the system to operate at a local maximum point rather than the global maximum.

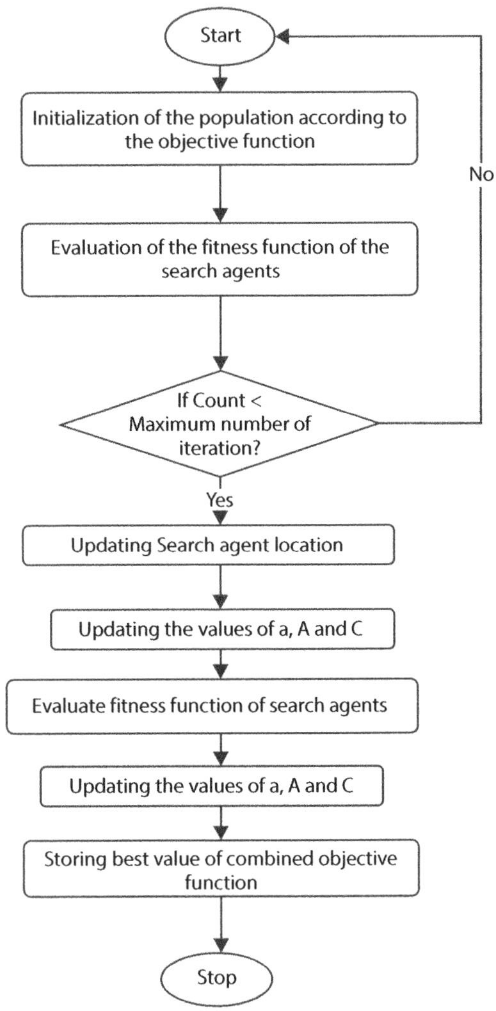

Figure 10.4 Flowchart of the GWO algorithm [17].

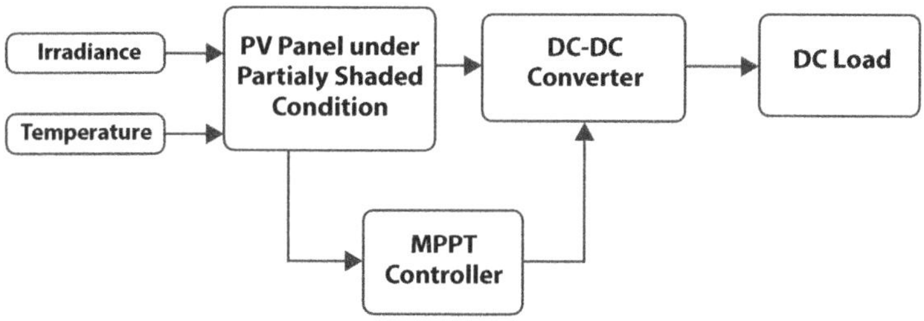

Figure 10.5 Block diagram representation of the proposed system.

10.6.2 Gray Wolf Optimization (GWO) Method

To address the drawbacks of a P&O-based controller, especially during partially shaded conditions, this study proposes a controller working on the GWO technique. This metaheuristic iterative algorithm was developed on the basis of the nature of hunting of gray wolves. Proper tuning of the system parameter is essential to yield optimal results applying the GWO technique. Figures 10.4 and 10.5 depicts the flowchart of the algorithm and block diagram of the system respectively [17]. In the present study, a comparative analysis was carried out where two different algorithms (P&O and GWO) were applied to the controllers simultaneously.

10.7 Simulation Result and Analysis

To conduct this study, a model was developed in MATLAB/Simulink environment and the output adopted by different controllers was analyzed. Table 10.1 below shows the specifications of the PV system under consideration. Figure 10.6 represents the Simulink diagram of the proposed system.

Case I (Constant Temp. and Insolation)

The power-voltage and current-voltage characteristics of the array under consideration due to uniform irradiance (1000 W/m^2 in each panel at temperatures 25°C and 45°C) are shown in Figure 10.7.

However, during partial shading conditions, each module is subjected to a nonuniform irradiance level (1000 W/m^2 in one and 500 W/m^2 in the other panel). Consequently, the output of the panel is of a fluctuating nature showing a multitude of its operating characteristics, as given in Figure 10.8.

Table 10.1 Specification of the PV module under consideration.

Parameters	Variable	Value
Current at short-circuited condition	I_{sc}	7.84 A
Voltage during the open-circuited condition	V_{oc}	36.3 V
Voltage at MPP	V_{MPP}	29 V
Current at MPP	I_{MPP}	7.35 A
Maximum Power	P_{MPP}	213.15 W
Temp. coef. for voltage	K_v	-0.36099
Temp. coef. for current	K_i	0.102
Cells linked in series	N_s	60

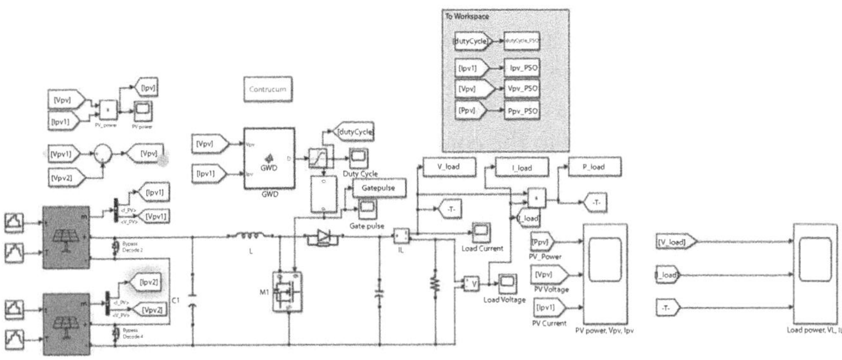

Figure 10.6 Simulink model of the proposed scheme.

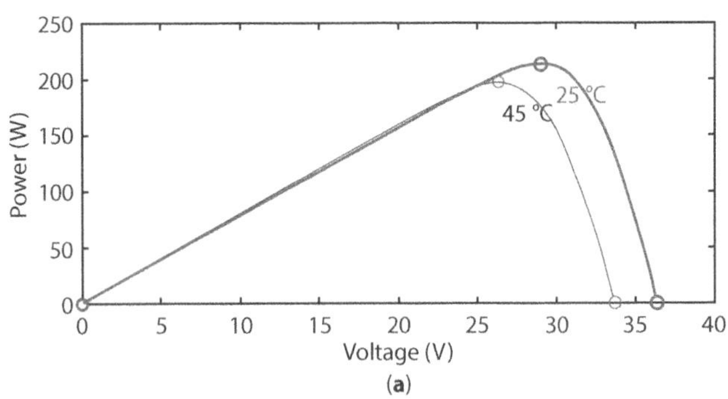

Figure 10.7(a) Power-voltage characteristics of the at 25°C and 45°C.

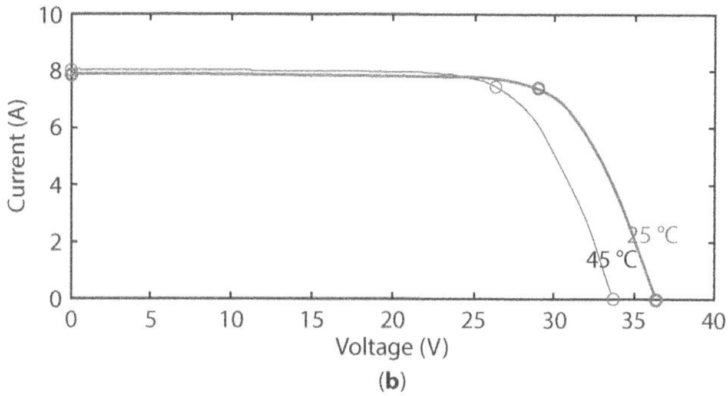

Figure 10.7(b) Current-voltage characteristics of the array at 25°C and 45°C.

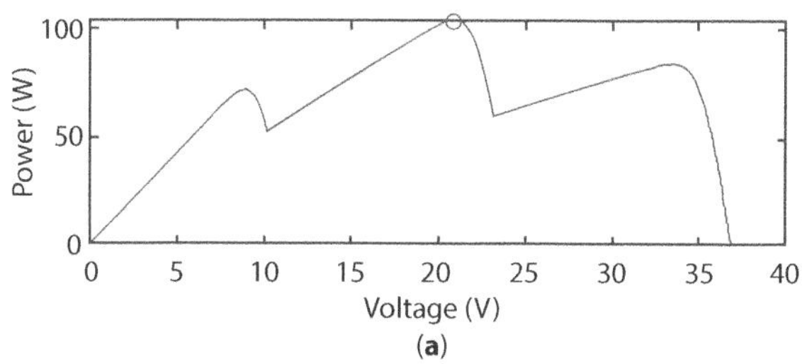

Figure 10.8(a) Power-voltage characteristics of the array during partially shaded conditions.

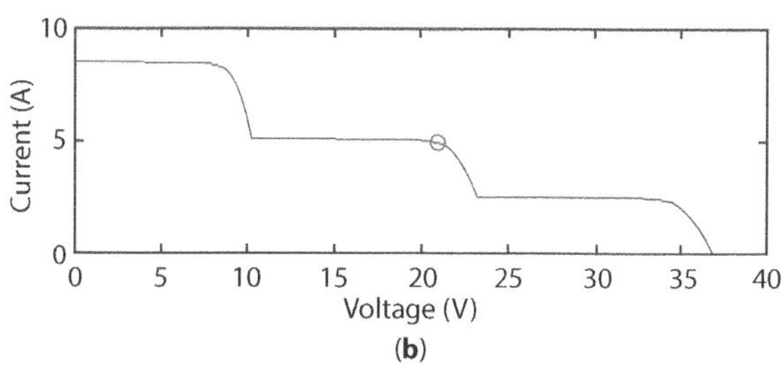

Figure 10.8(b) Current-voltage characteristics of the array during partially shaded conditions.

The variable output from the panel was fed into a boost converter with its duty cycle controlled by an MPPT controller. As previously mentioned, two distinct control algorithms (P&O and GWO) were implemented in the design of the MPPT controller, and a comparative analysis was conducted. This study primarily focused on three profiles: the power profile, voltage profile, and current profile. These profiles were examined both at the panel output and on the load side. Similar operating conditions were adopted during the design of the MPPT controller using both algorithms. Due to partial shading condition, panel parameters were fluctuating as evident in Figure 10.9.

The main goal of this study was to mitigate the undesirable fluctuations, especially since the load was exclusively DC. To achieve this objective, two different controller logic, namely, P&O and GWO, were implemented under two distinct scenarios. Table 10.2 represents the parameter specification used for the implementation of the GWO algorithm.

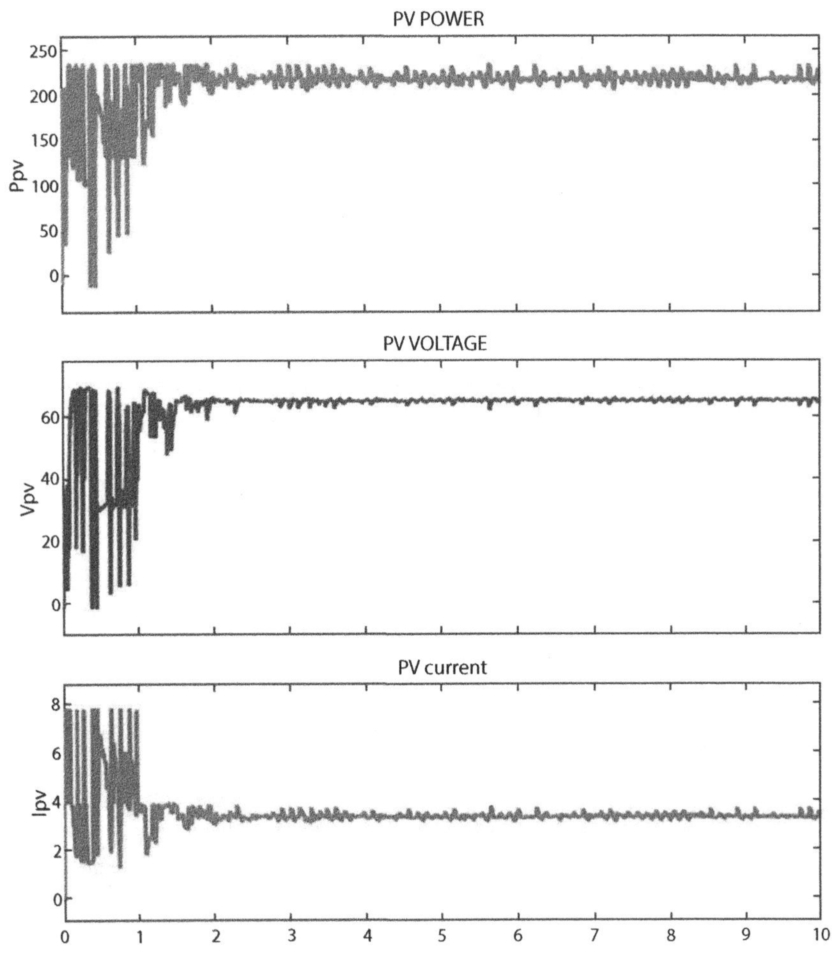

Figure 10.9 Power, voltage, and current profile of the panel due to partial shading.

Table 10.2 Parameters of GWO algorithms.

Parameters	**GWO**
Number of dimensions	3
Lower bound	-10
Upper bound	10
Number of gray wolves	50
Maximum number of iterations	100

Simulation results with P&O controller logic are shown in Figures 10.10 and 10.11, where Figure 10.10 represents the duty cycle profile of the converter and Figure 10.11 depicts the load profile of the proposed system.

Figures 10.12 and 10.13 illustrate the results of the proposed scheme with GWO logic. Figure 10.12 displays the duty cycle profile of the power converter, while Figure 10.13 portrays the load profile of the proposed scheme.

As evident from Figures 10.9, 10.11, and 10.13, output at the load was boosted by minimizing the undesired fluctuation of the panel parameters showing the effectiveness of both algorithms. It is also evident from Figures 10.11 and 10.13 that a relatively better result was obtained with GWO over P&O in terms of total power obtained at load. Besides, load voltage and load current were also more while adopting GWO.

Case II (Variable Temp. and Insolation)

In this case, one module gets irradiance between 400 and 800 W/m^2, whereas the other module gets an irradiance level between 200 and 400 W/m^2 as it was shaded partially. However, the temperature level was the same for both panels, and it varied between 25°C and 45°C.

Figure 10.14 represents the duty ratio of the converter and Figure 10.15 shows the load side power, voltage, and current profile.

The simulation result of the response of the proposed system under similar dynamic weather conditions but with GWO MPPT logic is shown in Figures 10.16 and 10.17 where Figure 10.16 shows the duty cycle of the system under consideration, whereas Figure 10.17 shows the power, voltage, and current profile at the load.

It is evident from Figures 10.15 and 10.17 that the GWO-based controller produced results that were somewhat superior to its P&O-based counterpart in view of total power delivery at the load. Additionally, load voltage and load current were higher when the GWO-based controller was used.

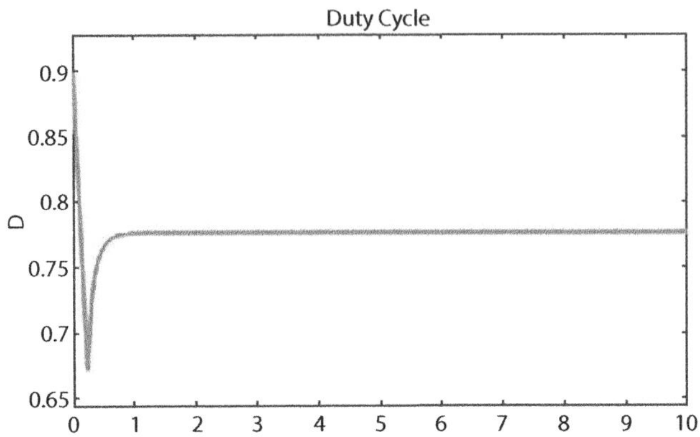

Figure 10.10 Duty cycle of the converter adopting the P&O logic controller.

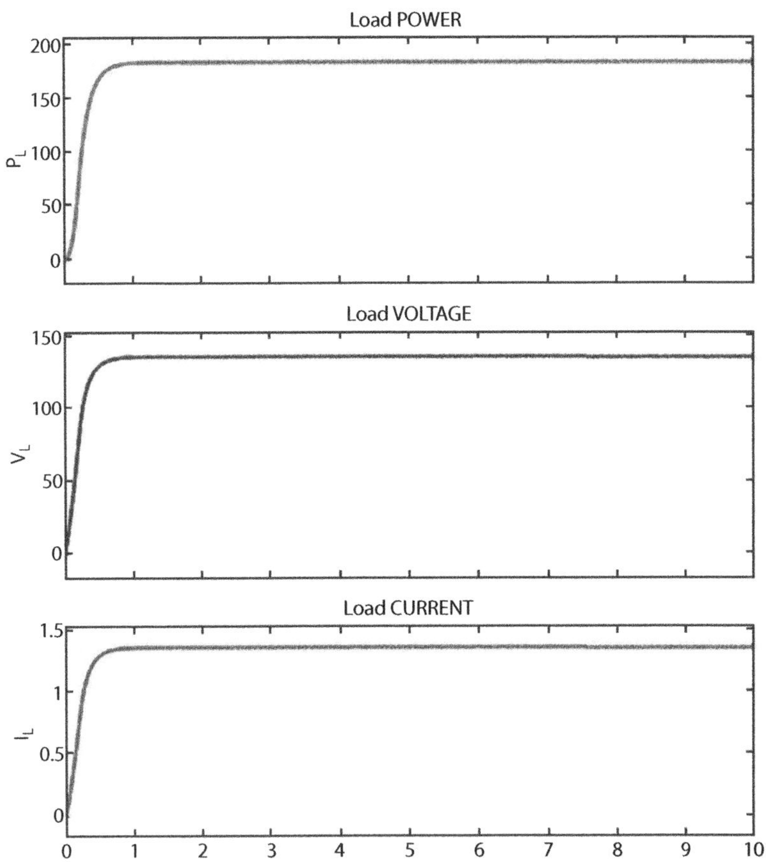

Figure 10.11 Power, voltage, and current profile at the load using the P&O logic controller.

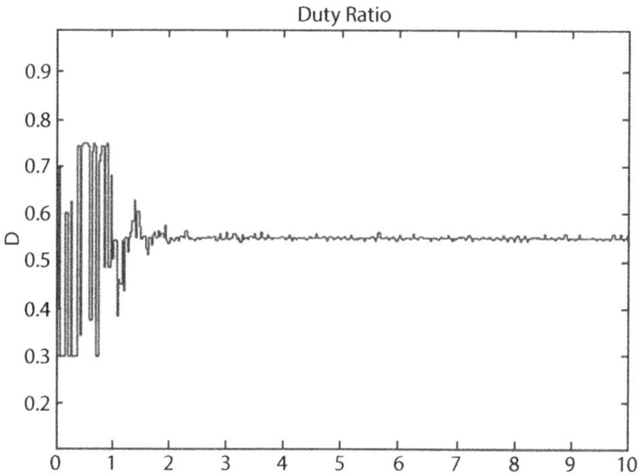

Figure 10.12 Duty cycle of the converter adopting the GWO logic controller.

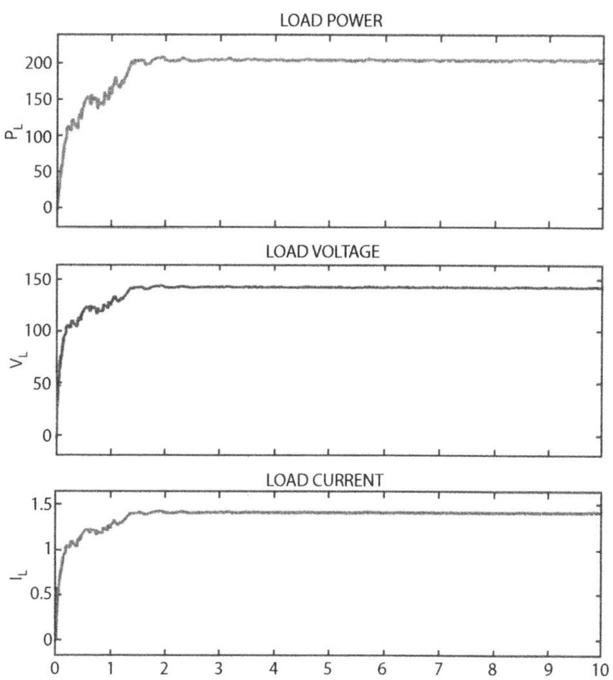

Figure 10.13 Power, voltage, and current profile at the load using the GWO logic controller.

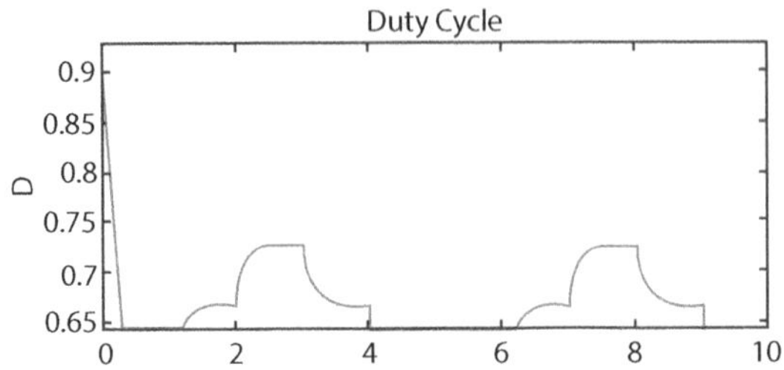

Figure 10.14 Duty cycle of the power converter adopting the P&O logic controller.

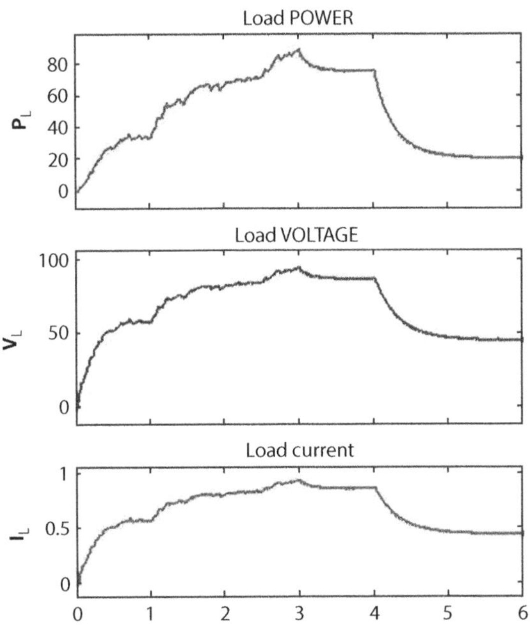

Figure 10.15 Power, voltage, and current profile at the load using the P&O logic controller.

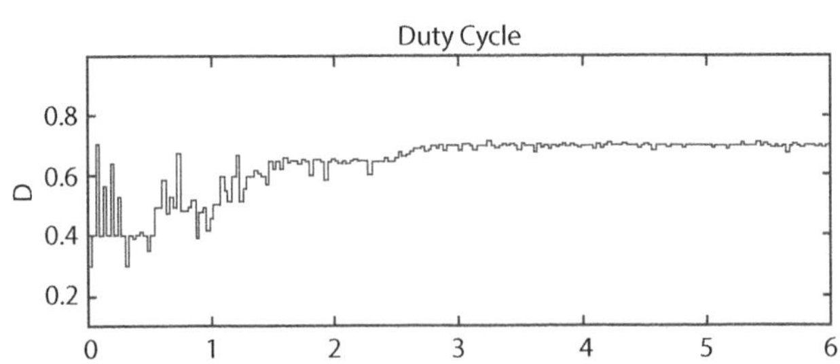

Figure 10.16 Duty cycle of the power converter adopting GWO algorithm.

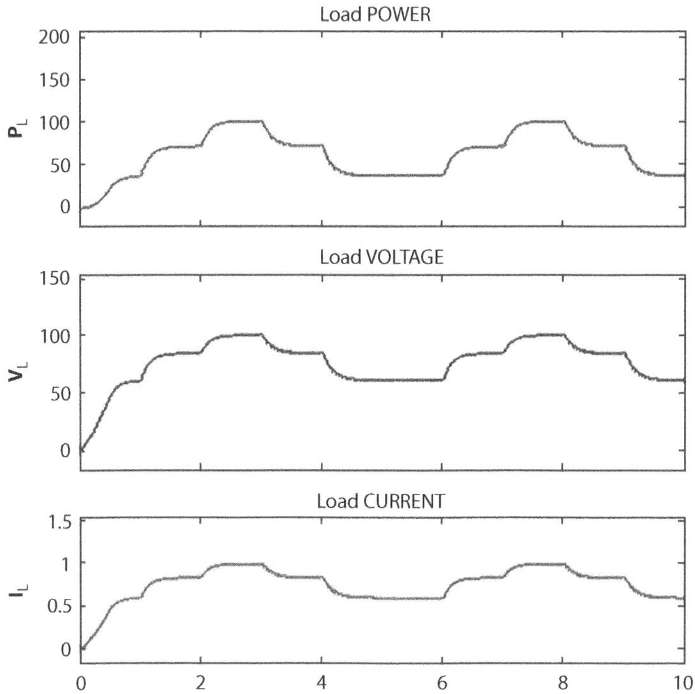

Figure 10.17 Power, voltage, and current profile at the load using the GWO algorithm.

10.8 Conclusion

In solar energy extraction, the implementation of MPPT requires proper system optimization. Tracking of maximum power becomes challenging when a PV array encounters operations involving multiple maxima and minima on its power-voltage curve due to varying external conditions, such as partial shading. In this study, a controlling mechanism was developed working on the GWO technique, and its results were compared with the results obtained from a controller working on a traditional P&O technique operating under similar environmental conditions. The PV system with an MPPT controller was modeled in MATLAB/Simulink to validate the effectiveness of the proposed scheme. Simulation of this system was performed for two distinct case studies: fixed temperature and fixed irradiance levels of partially shaded arrays and variable temperature and variable irradiance levels of partially shaded arrays. Outcomes from the simulation clearly illustrate that the proposed scheme provides better results, as the system in this case adaptively adapts to changes in external environmental conditions and effectively manages situations involving partial shading, showing the efficacy of the GWO approach over the traditional P&O technique.

References

1. Kordestani, M., Mirzaee, A., Safavi, A.A., Saif, M., Maximum Power Point Tracker (MPPT) for Photovoltaic Power Systems-A Systematic Literature Review. *2018 European Control Conference (ECC)*, pp. 40–45, 2018.
2. Chowdhury, S.R. and Saha, H., Maximum power point tracking of partially shaded solar photovoltaic arrays. *Sol. Energy Mater. Sol. Cells*, 94, 9, 1441–1447, 2010.
3. Villalva, M.G. and E. R. F., Analysis and simulation of the P&O MPPT algorithm using a linearized PV array model. *2009 35th Annual Conference of IEEE Industrial Electronics*, pp. 231–236, 2009.
4. Gaga, A., Errahimi, F., Es-Sbai, N., Design and implementation of MPPT solar system based on the enhanced P&O algorithm using Labview. *2014 International Renewable and Sustainable Energy Conference (IRSEC)*, pp. 203–208, 2014.
5. Ahmed, A., Ran, L., Bumby, J., Perturbation parameters design for hill climbing MPPT techniques. *2012 IEEE International Symposium on Industrial Electronics*, pp. 1819–1824, 2012.
6. Bahari, M., II, Tarassodi, P., Naeini, Y.M., Khalilabad, A.K., Shirazi, P., Modeling and simulation of hill climbing MPPT algorithm for photovoltaic application. *2016 International Symposium on Power Electronics, Electrical Drives, Automation and Motion (SPEEDAM)*, pp. 1041–1044, 2016.
7. Safari, A. and Mekhilef, S., Incremental conductance MPPT method for PV systems. *2011 24th Canadian Conference on Electrical and Computer Engineering (CCECE)*, pp. 000345-000347, 2011.
8. Shankar, N. and SaravanaKumar, N., Reduced Partial shading effect in Multiple PV Array configuration model using MPPT based Enhanced Particle Swarm Optimization Technique. *Microprocess. Microsyst.*, 103287, 2020.
9. Seyedmahmoudian, M., Rahmani, R., Mekhilef, S., *et al.*, Simulation and hardware implementation of new maximum power point tracking technique for partially shaded PV system using hybrid DEPSO method. *IEEE Trans. Sustain. Energy*, 6, 3, 850–862, 2015.
10. Ishaque, K. and Salam, Z., A deterministic particle swarm optimization maximum power point tracker for photovoltaic system under partial shading condition. *IEEE Trans. Ind. Electron.*, 60, 8, 3195–3206, 2013.
11. Saba, J. and Ishaque, K., A comprehensive analyses with new findings of different PSO variants for MPPT problem under partial shading. *Ain Shams Eng. J.*, 13, 5, 2090–4479, 101680, 2022.

12. Subha, R. and Himavathi, S., Neural network based maximum power point tracking scheme for PV systems operating under partially shaded conditions. *2014 International Conference on Advances in Green Energy (ICAGE)*, pp. 39–43, 2014.
13. Yingquan, Z., Yan, F., Wang, X., Zhang, J., An efficient fuzzy logic control algorithm for photovoltaic maximum power point tracking under partial shading condition. *J. Franklin Inst.*, 357, 6, 3135–3149, 2020.
14. Titri, S., Larbes, C., Toumi, K.Y., Benatchba, K., A new MPPT controller based on the Ant colony optimization algorithm for Photovoltaic systems under partial shading conditions. *Appl. Soft Comput.*, 58, 465–479, 2017.
15. González-Castaño, C., Restrepo, C., Kouro, S., MPPT Algorithm Based on Artificial Bee Colony for PV System. *IEEE Access*, 9, 43121–43133, 2021.
16. Villalva, M.G., Gazoli, J.R., Filho, E.R., Analysis and simulation of the P&O MPPT algorithm using a linearized PV array model. *10th Brazilian Power Electronics Conference (COBEP)*, pp. 231–236, 2009.
17. Chawla, V., Chanda, A., Angra, S., The scheduling of automatic guided vehicles for the workload balancing and travel time minimization in the flexible manufacturing system by the nature-inspired algorithm. *J. Project Manage.*, 4, 1, 19–30, 2019.

11

An Efficient Optimization Approach for Solving the Relay Coordination Problem

Maitrayee Chakrabarty[1]*, Sudipta Chakraborty[1], Suparna Pal[1] and Raju Basak[2]

[1]*Department of Electrical Engineering, JIS College of Engineering, Kalyani, Nadia, West Bengal, India*

[2]*Department of Electrical Engineering, Techno India University, Kolkata, West Bengal, India*

Abstract

An efficient optimization approach has been required to manipulate any form of real-time fault mitigation. Right here, defensive relays need to be cautiously calibrated and synchronized to isolate and hit upon complicated portions inside the energy device. Distribution networks are frequently set up to allow for protection and exclusive load situations. As a result, each over-current relay (OCR) configuration ought to be arranged in any enterprise utility. Furthermore, there is a crucial prerequisite—relay configuration to satisfy each viable network configuration primarily based on the sort and place of troubles. It is not a smooth venture to identify the right OCR to set up with a strength engineer manually or use the traditional way to maintain all the situations stated above. As a result, an optimization approach is required to modify the proper relay configuration for a shorter relay operation time. Moreover, accomplishing the bottom feasible fault clearance time requires the most advantageous coordination. As a result, the most efficient OC relay coordination is created by way of fixing the relay operating time discount trouble while keeping a sure quantity of constraint by way of adopting an optimization method. The over-modern-day relay coordination trouble can be handled via the use of a time dial setting (TDS) that is based on predefined choose-up contemporary settings that have been adjusted through unique limitations.

Keywords: Over-current (OC), relay coordination (RC), particle swarm optimization (PSO), time-dial setting (TDS), coordination time interval (CTI)

**Corresponding author*: moitry29@gmail.com

Arindam Mondal and Souvik Ganguli (eds.) Controller Design for Industrial Applications, (211–232)

11.1 Introduction

Distribution networks are frequently installed to allow for renovation and unique load situations. As a result, every OC relay configuration needs to be arranged. Moreover, there may be a critical prerequisite—relay configuration to satisfy each viable community configuration based totally on the type and region of problems. A crucial issue of electrical energy systems is relay coordination, which ensures the appropriate operation of protective relays in case of malfunctions or uncommon situations. It entails configuring circuit breakers and defensive relays so that their time-contemporary characteristics allow downstream relays to synchronize their response, with the protective relay nearest to the trouble operating first.

Selectivity, or the capacity of the relay nearest to the problem to isolate it as speedy as possible without needlessly triggering close-by relays or interrupting crucial additives of the electric machine, is the primary goal of relay coordination. In doing so, downtime is decreased, the effect of mistakes is lessened, and gadget reliability is preserved. Relay coordination considers fault kind, area, and relay's time-contemporary characteristics.

The primary aim is to reduce disturbances as much as possible even as retaining machine integrity. The dynamic and complicated shape of energy systems means that reaching the greatest coordination is a tough assignment. Within the context of relay coordination, optimization techniques—from state-of-the-art device learning models to classical algorithms—are included. This proposed approach has used mathematical expressions to stability selectivity and dependability in defensive relay features. The objective of this abstract is to provide a radical understanding of the relay coordination trouble (RCP) and the function optimization performed in fixing it by contemplating both its theoretical foundation and realistic consequences.

To deal with the RCP [5–29], several researchers have placed out quite some analytical and adaptive techniques in current years. Among these algorithms are ant colony optimization (ACO), coaching-learning-based totally optimization (TLBO), PSO, and genetic algorithms (GAs), in conjunction with differential evolution (DE). Numerous researchers then subtle the GA strategies to offer a hybrid in addition to continuous GA, which extensively shortened the relay operation time. Furthermore, scholars and researchers have delicate and altered the techniques defined in the literature. Modified PSO, modified adaptive TLBO, and adaptive DE are some of those stepped-forward editions.

The RCP has been solved lately with the use of a ramification of optimization techniques, including the maximum precious player algorithm (MVPA), root tree set of rules (RTA), symbiotic organism search (SOS), firefly set of rules (FA), chaotic firefly algorithm (CFA), harmony search set of rules (HAS), and gray wolf optimization (GWO). The majority of these computational strategies have several disadvantages, which include a sluggish fee of convergence, a high population wants, a steady danger of converging to local minima, and the requirement for additional pc attempts to determine the correct relay configuration. Therefore, the look for the precise relay configuration that might assist reduce working time and, in turn, the time needed to clear faults keeps unabated.

11.2 Literature Survey

Engineers had to spend a lot of time and energy doing calculations by hand to coordinate directional relays. But now, thanks to computers, they do not have to do all that hard work anymore [15–17]. They can use mathematical models of over-current relays (OCRs) on the computer to make things easier. There are aureate algorithms that let the engineer and the computer work together to find the best coordination [18–20]. There are two ways to do this: the traditional way and the optimization way. The traditional way is all about analyzing all the possible problems and situations beforehand [21, 22]. The problem with this method is that it cannot handle situations that were not included in the analysis. It is just too difficult to deal with those.

These days, scientists like to use fancy computer programs called artificial intelligence to make things work better. They do this because the old ways took too much time [23]. One way they use artificial intelligence is by using something called neural networks and fuzzy logic [24]. They use these to figure out the best time for something to happen. Another way they use artificial intelligence is by analyzing things really smartly [25]. They use this to figure out the best settings for protecting something far away. They look at things like how far away it is, how well it works, and how sensitive it is to electricity.

The RCP [5–32] has been the subject of several analyses. Some specifics of the optimization function with limits, restrictions, etc. have been emphasized in this section.

Reference no.	Technology used	Type of optimization technique	Optimization function	Constraints/limit/ boundaries
[6]	Three divisions within a feeder distribution systems	Water cycle algorithm (WCA)	$Minimize\ z = \sum_{i=1}^{n} W_j \times T_{i_k}$	• Operating time constraints • TDS constraints • Coordination constraint
[7]	Swarm intelligence optimization algorithm	Crow search algorithm (CSA)	$Minimize\ z = \sum_{i=1}^{n} W_j \times T_{i_k}$	• Operating time constraints • TDS constraints • Coordination constraint
[3]	Computational method	Particle swarm optimization (PSO)	$Minimize\ z = \sum_{i=1}^{n} W_j \times T_{i_k}$	• Fitness function • Particles movement

Reference no.	Optimization function	Merits	Demerits	Goals/main objective
[6]	Water cycle algorithm (WCA)	Requires less number of insensitive user parameters	It is possible that the WCA has a limited capacity to adjust to changing situations or dynamic surroundings. When the optimization landscape changes, it may be difficult for the algorithm to quickly modify its search plan and identify the best answers.	Based on the published and experimental findings of this study, the WCA provides competitive solutions when compared to other meta-heuristic optimizers.
[7]	Crow search algorithm (CSA)	Quick implementation, minimal control parameters, and a straightforward structure	Gradual convergence and effortless local optimal drop	For a crow to be capable of tracking another crow to its secret meal, the source is the aim of this meta-heuristic.
[3]	Particle swarm optimization (PSO)	Requires fewer parameters tuning, good for multi-objective optimization	Needs memory to update velocity, low-quality solution, early convergence	Improve overall performance, global convergence speed, solution quality, and robustness

In this section has been highlighted some merits and demerits of the optimization function.

This study's primary goal is to determine TDS and reduce equipped relays' total time of operation. A linear equation is used to generate the goal function, and TDS is treated as an optional variable. The PSO method is used in this work to solve the OC-RCP. The PSO method is assessed throughout.

11.3 Overview of the System

Over-Current Relay (OCR): OCR currently surpasses a specified value, and activates or detects changes. Over-current protection shields electrical power systems from excessive currents in cases of short circuits, grounding faults, and other issues. OCRs may be used to protect almost every component of a power system, such as motors, transformers, generators, and transmission lines. Several OCRs would be employed for feeder protection to safeguard various feeder components. These relays have to communicate with each other for the OCR closest to the problem to activate first.

OCR Ratings: The ratings need to be appropriately set for an OC safety device to function as intended. These ratings include interrupting, ampere, and voltage. One may consider current limiting to be an extra rating for over-current protection devices, but only certain over-current protection devices have to have this feature.

Voltage Rating: The over-current safety device's voltage rating needs to match or exceed the circuit voltage. The rating of the over-current safety device may exceed the system voltage, but it will never fall below it.

Ampere Rating: In normal circumstances, the ampere rating of an over-current safety device should not be greater than the conductors' current-carrying capability. An over-current protection device's ampere rating is often set to 125% of the steady load current. Relays are classified according to their operating time as follows:

1. Instantaneous OCR
2. Inverse time OCR
3. IDMT OCR
4. Very inverse relay
5. Extremely inverse relay

Relay time-current characteristics: Relay time-current characteristics, often known as the "trip curve" or the "TCC" (time-current characteristic) curve, are essential to protective relay functionality in electrical systems. These curves are crucial for guaranteeing the dependable and focused operation of safety devices like fuses and circuit breakers.

The formula below can be used to determine the period of operation of any inverse relay, based on the recommendations of the IEC (International Electrotechnical Commission) [2].

Here,

$$k = \frac{\beta}{(\frac{I}{I_o})^{\alpha} - 1} \tag{11.1}$$

where K is the activation time and α and β are constants that vary depending on the kind of relay. I_o is the pick-up current, and I = fault current. Various OCR characteristics and their corresponding values for α and β are known.

11.4 Problem Formulation and Constraint Criterion

An illustration of a two-bus radial distribution system may be found in Figure 11.1 below. There is a systemic flaw k at location F, which is beyond bus B. Relay R2 assumes control of it first since it is the neighboring relay to the issue. If R2 stops operating for whatever reason, R1 takes over after a certain period of time. Relay R1 therefore handles backup protection. Based on the assumption that R2's time of operation has been fixed at 0.1 s, R1's operation time is set plus an additional 0.1-s time delay that is applied to both R1's overshoot time and the operating periods of CB at bus A.

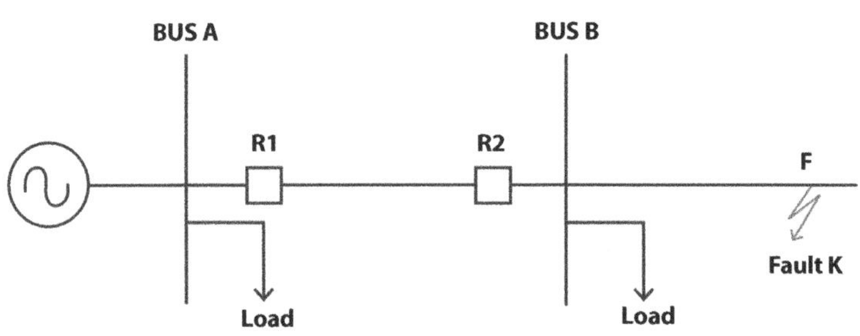

Figure 11.1 A system of radial distribution.

The distribution network's OC relay coordination may be formulated as a challenge involving optimization. It is represented mathematically as "z," the duration of all equipped relays' functioning, for a specific failure point. The following actions should be taken to reduce this to a minimum:

$$Minimize\ z = \Sigma_{i=1}^{n} W_j \times T_{i_k} \tag{11.2}$$

where z stands for the problem statement's objective function in this study. n is the total number of relay. T_{ik} is operational time of the relay consider the fault location at k site. The weighting factor is W_j. The shortest possible individual operating duration would be approached by reducing the weighted total time of operation of all primary relays. This is one possible application of the weighting factor that guarantees that the losses cause the lowest system disruption possible. Because the distribution system's short lines have about equal weights and lengths, an issue might arise at any point along the line. As a result, W_j is regarded as 1 $(W_j = 1)$ for every relay. Additionally, it is a coefficient that indicates the probability of a defect occurring. Consequently, there is an equal chance that a defect will develop on each line since the lines have the same length and weight. W_j is often set to 1 due to this equal likelihood [6–8].

11.4.1 Limitations on Operating Time

The distribution end has a constantly fluctuating load. These little systemic interruptions cause a relay to stall for a few cycles before detecting and documenting a failure condition. The following is a list of restrictions on relay operating time:

$$T_{i_{kmin}} \leq T_{i_k} \leq T_{i_{kmax}} \tag{11.3}$$

Here $T_{i_{kmin}}$ and $T_{i_{kmax}}$ represent the lower and higher boundaries of relay operating time, respectively. $T_{i_{kmin}} = 0.1$ s and $T_{i_{kmax}} = 1.1$ s are typical values [3]. $T_{i_{kmax}}$ occasionally varies depending on network configuration. This limitation shows that as operating time limits are lowered, relay response times will speed up and TDS will therefore decrease.

11.4.2 TDS Constraints

Relay speed is contingent on the appropriate selection of the relay's TDS setting. Its boundaries can be shown as follows:

$$TDS_{imin} \leq TDS \leq TDS_{imax} \tag{11.4}$$

where TDS_{imin} and TDS_{imax} are the ith TDS's lower and upper limits, respectively. Normal values for $TDS_{imin} = 0.025$ and $TDS_{imax} = 1.2$ in this work are taken into account [4].

11.4.3 Coordination Limitation

When a system malfunction is detected, both the primary and backup relays can act concurrently. Therefore, according to a preset period identified as CTI, the backup relay would only activate once the primary relay fails to avoid false tripping.

As a result, the primary and backup relays' coordination requirements may be mathematically stated as follows:

$$T_{bi_k} - T_{pj_k} \geq CTI \tag{11.5}$$

where T_{pj_k} is the operating time of the primary relay at the same fault k, and T_{bi_k} is the operating time of the ith backup relay. CTI stands for coordination time interval. The time difference between the primary and backup relays is known as the coordination time interval or CTI. The circuit breaker speed determines the CTI, which can vary from 0.1 to 0.5s.

11.4.4 Relay Characteristics

There are many relays available in the OC relay family. Given its sophisticated characteristics, the inverse definite minimum time (IDMT) relay is the most favored OC relay. The expression that follows can be used to express this.

Here Tikmin and Tikmax represent the lower and higher boundaries of relay operating time, respectively

$$T_{op} = \frac{\beta \times TDS}{(PSM)^{\alpha} - 1} = \frac{0.14 \times TDS}{(PSM)^{0.02} - 1} \tag{11.6}$$

where T_{op} is the relay's operating time and β and α are typical parameters for IDMT relay characteristics. The ratio of the relay's real fault current to its set relay current, or the plug setting multiplier (PSM), is what makes the relay operate. Every time the fault current approaches or exceeds the relay current setting, the TDS modifies the amount of time before the relay activates. The behavior of Equation 11.6 is nonlinear, even if sustaining PSM remains constant with a preset pick-up current value. It is feasible to convert Equation 11.6 into a linear equation.

$$T_{op} = a_i \times (TDS) \tag{11.7}$$

where

$$a_i = \frac{0.14}{(PSM)^{0.02} - 1} \tag{11.8}$$

To get the relay coordination issue's objective function, replace Equation 11.2 with Equation 11.8.

$$Minimize\ z = \Sigma_{i=1}^{n} a_{ik} \times TDS \tag{11.9}$$

in which the constant term representing the ith relay at point k is denoted by a_{ik} [14].

11.5 Proposed PSO Algorithm

One popular optimization technique is the particle swarm optimization (PSO). This is a heuristic search algorithm that draws inspiration from fish and bird social behavior. PSO is often utilized to find optimal solutions in complex, multidimensional search domains.

PSO is incredibly good at solving optimization problems, particularly when the search space is a complex and nonlinear one. It can be used in a variety of tasks, function optimization, and neural network training and is not difficult to implement.

Nevertheless, similar to many other optimization methods, it requires careful tuning of several parameters. It is important to say that PSO is a randomness-based technique, which means that even if it usually works well by converging to good solutions in real-world scenarios, it cannot guarantee that the best solution is found.

11.5.1 What Makes the Particle Swarm Optimization (PSO) Algorithm So Helpful?

PSO (particle swarm optimization) has benefits in addition to disadvantages. The algorithm has adaptability. To effectively analyze its efficiency, numerous factors to consider have to be made such as toughness, flexibility, merging quickness, and the top quality of the option.

PSO establishes itself apart from its strategy, which is affected by crowd knowledge and also imitates the team actions of fish or birds. The option area might be effectively discovered thanks to this decentralized method. Its performance is extremely based on specification change; therefore, the criteria chosen will certainly affect how well it executes. PSO often reveals faster merging than present techniques such as water cycle algorithm (WCA), genetic approaches (GA), or simulated annealing (SA), especially in high-dimensional areas. This makes it a great suitable for concerns where fast convergence is necessary. Nonetheless, PSO can discover it hard to leave substandard options in complex settings with a number of local optima.

In conclusion, while PSO offers notable advantages such as rapid convergence and simplicity, its performance should be carefully evaluated based on the specific optimization problem at hand. In [1], the pseudocode depicting the general algorithm of PSO has been discussed.

11.6 Simulation Outcomes and Discussions

11.6.1 Case I

Seven OCRs have been used to form a single-end loop distribution system, as seen in Figure 11.2 [6].

Unlike relays 1 and 5, which are nondirectional OCR, the other relays are directional OCR. CT ratios, relay plug settings, and the relationship between primary, backup, and fault current via relay have been taken from [6].

This case study looks at 12 constraints in all. The coordination criteria account for five of the seven restrictions that are taken into account while calculating the minimal operational time. There is a 0.2-second CTI.

In Equation 11.17, s_1 through s_7 display the seven OC relays' TDS. Subscripts $s_1,\ldots, s_7$ denote the relay numbers.

The required function can be represented by the following computations for case I. The various fault sites, current sensing relays, and a_i constant have been taken from [6], which are used to determine the value of the a_i constant in the subsequent Equations 11.18 to 11.24.

Figure 11.2 shows the fault sites M and N detected by relay 1. In mathematical notation, it may be expressed as

$$a_1 = 3.252 + 6.872 = 10.124 \tag{11.10}$$

Only failure point M is detected by relay 2. In mathematical notation, it may also be expressed as:

$$a_2 = 43.776 \tag{11.11}$$

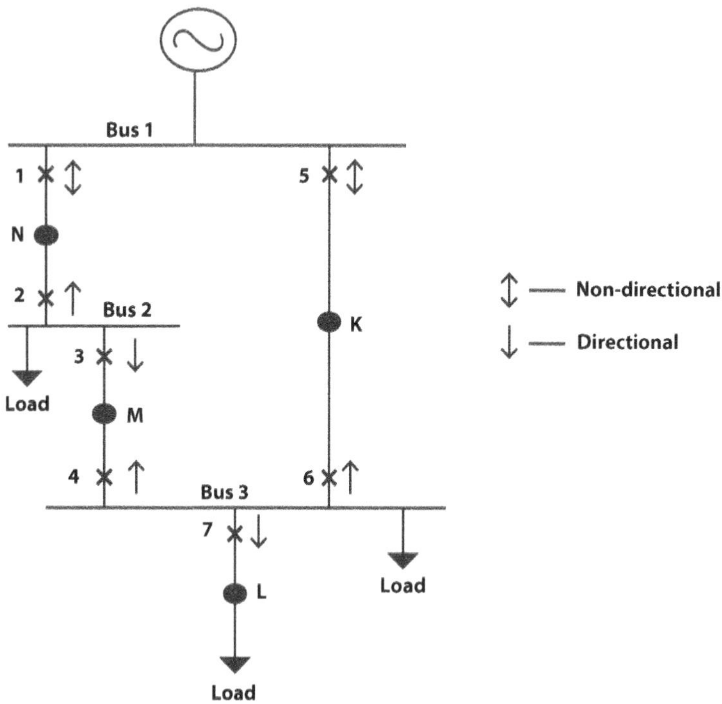

Figure 11.2 A seven-OCR-connected single-end loop.

Relay 3 is mathematically described as follows:

$$a_3 = 6.872 + 5.809 + 4.044 = 16.725 \tag{11.12}$$

Similar to relay 1, relay 4 detects the fault sites M and N. Its mathematical expression is as follows:

$$a_4 = 8.159 + 5.811 = 13.97 \tag{11.13}$$

Relay 5 senses the same fault sites K, L, and N that relay 3 does. Its mathematical expression is as follows:

$$a_5 = 4.811 + 5.809 + 14.01 = 24.63 \tag{11.14}$$

Relay 6 senses a single fault point, K, and its mathematical expression is as follows:

$$a_6 = 22.165 \tag{11.15}$$

Relay 7 senses the single fault point L as well. Its mathematical expression is as follows:

$$a_7 = 4.145 \tag{11.16}$$

Hence, the objective function for case I may be derived from the computations performed in Equations 11.10 to 11.16.

$$Min\,z = 10.24s_1 + 43.77s_2 + 16.725s_3 + 13.97s_4 + 24.63s_5 + 22.16s_6 + 4.145s_7 \tag{11.17}$$

The operating time limitations state that 0.1 s is the minimum operational time. The information required for Equations 11.18 to 11.24 is taken from [6]. Consequently, the minimal operational time has been subjected to the following limits:

$$3.255s_1 \geq 0.1 \tag{11.18}$$

$$43.77s_2 \geq 0.1 \tag{11.19}$$

$$4.044s_3 \geq 0.1 \tag{11.20}$$

$$5.811s_4 \geq 0.1 \tag{11.21}$$

$$4.811s_5 \geq 0.1 \tag{11.22}$$

$$22.165s_6 \geq 0.1 \tag{11.23}$$

$$4.14s_7 \geq 0.1 \tag{11.24}$$

The TDS restriction outlined in TDS limitations is broken by the limitations placed in Equations 11.19 to 11.24. Consequently, these limitations are updated with a minimum TDS of 0.025:

$$s_2 \geq 0.025 \tag{11.25}$$

$$s_3 \geq 0.025 \tag{11.26}$$

$$s_4 \geq 0.025 \tag{11.27}$$

$$s_5 \geq 0.025 \tag{11.28}$$

$$s_6 \geq 0.025 \tag{11.29}$$

$$s_7 \geq 0.025 \tag{11.30}$$

Furthermore, the coordination constraint must be considered, as mentioned in the Coordination Constraint. The CTI in this case is expected to be 0.2 seconds. Therefore, the following is the textual constraint related to the coordination criterion and CTI:

$$43.77s_2 - 8.159s_4 \geq 0.2 \tag{11.31}$$

$$6.872s_1 - 4.044s_3 \geq 0.2 \tag{11.32}$$

$$14.01s_5 - 5.811s_4 \geq 0.2 \tag{11.33}$$

$$6.872s_3 - 22.165s_6 \geq 0.2 \tag{11.34}$$

$$5.809s_3 - 13.988s_7 \geq 0.2 \tag{11.35}$$

The output response of the suggested technique for coordinating a single-end loop distribution network is displayed in Figure 11.3.

11.6.2 Case II

Eight OCRs have been used to form this multi-loop distribution system. The next five relays are classified as directional OCRs; OCR-1, OCR-2, and OCR-3 are considered nondirectional relays. The primary/backup connection between all eight relays and the maximum fault current and the details of the a_i constant and relay current are taken from [6]. The CT ratio for all eight relays is 100:1. In addition, eight constraints were taken into account for the coordination requirement and the minimum operational time of the relay. Since this system is more complex than the one in the prior example, the CTI is determined to be 0.6. In Equation 11.36, the TDS of eight OCRs is represented by s_1 through s_8. Figure 11.4 shows the evidence of multi-loop-single-end distribution network.

The objective function for case II is constructed using a mathematical method that is similar to the case I previously discussed. The information that is needed for Equation 11.36 is taken from [6]. The objective function may be expressed as follows:

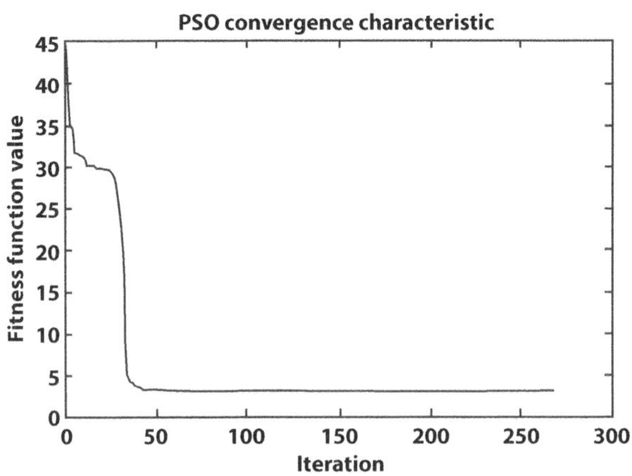

Figure 11.3 The convergence of proposed PSO.

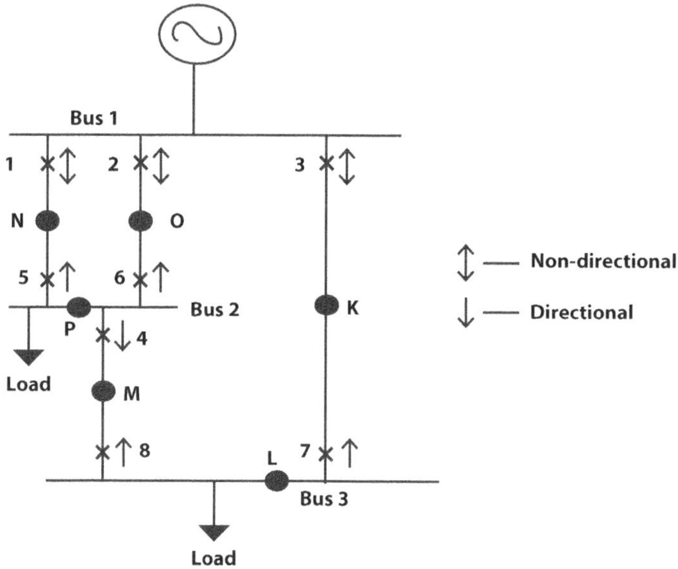

Figure 11.4 Multi-loop-single-end distribution network.

$$Min\,z = 28.975s_1 + 28.975s_2 + 37.736s_3 + 11.502s_4 + 3.297s_5 + 3.297s_6 + 4.9804s_7 + 30.799s_8 \quad (11.36)$$

The minimum operating time is limited to 0.1 seconds, as per the Operating Time Constraints. Information for Equations 11.45 through 11.52 has been gathered from [6]. The following are the restrictions that are subject to the relay's minimum operational time:

$$2.97s_1 \geq 0.1 \quad (11.37)$$

$$2.97s_2 \geq 0.1 \quad (11.38)$$

$$5.584s_3 \geq 0.1 \quad (11.39)$$

$$4.980s_4 \geq 0.1 \quad (11.40)$$

$$3.297s_5 \geq 0.1 \quad (11.41)$$

$$3.297s_6 \geq 0.1 \tag{11.42}$$

$$4.980s_7 \geq 0.1 \tag{11.43}$$

$$10.035s_8 \geq 0.1 \tag{11.44}$$

The constraints presented in Equations 11.39, 11.40, 11.43, and 11.44 defy the boundaries for the lowest relay TDS value. Thus, these constraints are rewritten as follows, with a minimum TDS value of 0.025.

$$s_3 \geq 0.025 \tag{11.45}$$

$$s_4 \geq 0.025 \tag{11.46}$$

$$s_7 \geq 0.025 \tag{11.47}$$

$$s_8 \geq 0.025 \tag{11.48}$$

The primary/backup relay pair data in this case study are derived from [6], with a CTI of 0.6 seconds assumed. As a result, the criteria constraints are stated as follows:

$$5.749s_3 - 5.749s_8 \geq 0.6 \tag{11.49}$$

$$5.584s_1 - 3.551s_4 \geq 0.6 \tag{11.50}$$

$$5.584s_2 - 3.551s_4 \geq 0.6 \tag{11.51}$$

$$4.980s_4 - 4.980s_7 \geq 0.6 \tag{11.52}$$

$$4.281s_1 - 2.971s_4 \geq 0.6 \tag{11.53}$$

$$4.980s_3 - 4.980s_8 \geq 0.6 \tag{11.54}$$

$$10.035s_8 - 3.297s_5 \geq 0.6 \tag{11.55}$$

$$10.035s_8 - 3.297s_6 \geq 0.6 \tag{11.56}$$

The output response of the suggested technique for coordinating a multi-loop–single-end distribution network is displayed in Figure 11.5.

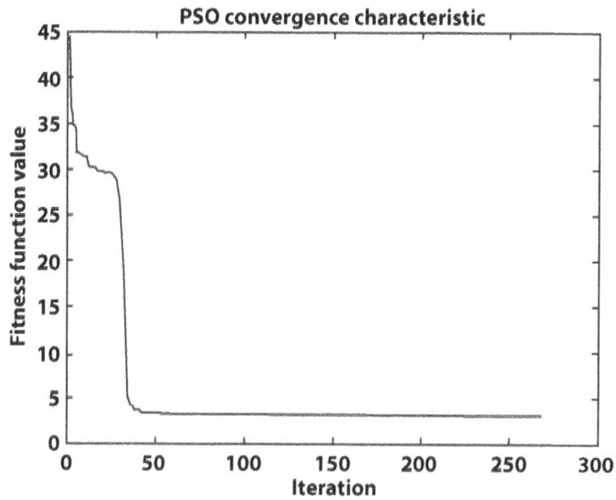

Figure 11.5 The convergence of proposed PSO.

11.7 Industrial Application Design Consideration

The Transmission and Distribution Company has employed the proposed technique to guarantee security. The following components are the main focus of the coordinated setup to regulate the relays and accessories: transformers, generators, line protection, islanding, and H.V. Buses. The primary goal of the protective system is to reduce the duration of any disruptions to the electric system while simultaneously identifying and isolating all defective components as soon as feasible. The security system must be dependable, safe, and quick when needed. The protective system will be ineffective and occasionally become a liability without this fundamental criterion.

11.8 Conclusion

Finally, it is important to emphasize how complex the RCP is inside power system protection before we conclude our research using an optimization method. The challenge of relay coordination is striking a fine balance between avoiding needless disconnections that might interfere with the power supply and guaranteeing prompt and targeted tripping of malfunctioning components.

To maximize the TDS of the OC-RCP with the real pickup current, this study employs the PSO optimization technique. The presented result analysis makes it abundantly evident that the PSO approach is producing better results than the current ones in every way. As a result, the PSO algorithm works well to address the problem of OCRC. As a result, under abnormal conditions, the PSO minimizes the amount of time that the OC relays are operational.

Through the use of an optimization framework, we enter a field where mathematical models and algorithms are essential for optimizing relay settings. This method makes it easier to choose the best coordination scheme to reduce operating times, increase selectivity, and improve system dependability. However, it is important to recognize that creating and resolving these optimization difficulties has inherent complexity. The environment around the power system is dynamic, with variables and uncertainties affecting its properties. This makes it difficult to continue running an efficient relay coordination mechanism that can adjust to shifting circumstances.

References

1. Dai, H.P., Chen, D.D., Zheng, Z.S., Effects of random values for particle swarm optimization algorithm. *Algorithms*, 11, 2, 23, 2018.
2. Acharya, S., Jha, S.K., Shrestha, R., Pokhrel, A., Bohara, B., An analysis of time current characteristics of adaptive inverse definite minimum time (IDMT) over-current relay for symmetrical and un-symmetrical faults, in: *2017 International Conference on Smart grids, Power and Advanced Control Engineering (ICSPACE)*, IEEE, pp. 332–337, 2017, August.
3. Asadi, M.R. and Kouhsari, S.M., Optimal over-current relays coordination using particle-swarm-optimization algorithm, in: *2009 IEEE/PES Power Systems Conference and Exposition*, IEEE, pp. 1–7, 2009, March.

4. Zhu, H., Wang, Y., Wang, K., Chen, Y., Particle Swarm Optimization (PSO) for the constrained portfolio optimization problem. *Expert Syst. Appl.*, 38, 8, 10161–10169, 2011.
5. Bedekar, P.P. and Bhide, S.R., Optimum coordination of over-current relay timing using continuous genetic algorithm. *Expert Syst. Appl.*, 38, 9, 11286–11292, 2011.
6. Kudkelwar, S. and Sarkar, D., Online implementation of time augmentation of over-current relay coordination using water cycle algorithm. *SN Appl. Sci.*, 1, 12, 1628, 2019.
7. Sarkar, D., Kudkelwar, S., Saha, D., Optimal coordination of over-current relay using crow search algorithm. *Smart Sci.*, 7, 4, 282–297, 2019.
8. Bedekar, P.P., Bhide, S.R., Kale, V.S., Optimum coordination of over-current relay timing using simplex method. *Electr. Power Compon. Syst.*, 38, 10, 1175–1193, 2010.
9. Mahari, A. and Seyedi, H., An analytic approach for optimal coordination of over-current relays. *IET Generation Transm. Distrib.*, 7, 7, 674–680, 2013.
10. Noghabi, A.S., Sadeh, J., Mashhadi, H.R., Considering different network topologies in optimal over-current relay coordination using a hybrid GA. *IEEE Trans. Power Deliv.*, 24, 4, 1857–1863, 2009.
11. Razavi, F., Abyaneh, H.A., Al-Dabbagh, M., Mohammadi, R., Torkaman, H., A new comprehensive genetic algorithm method for optimal over-current relays coordination. *Electr. Power Syst. Res.*, 78, 4, 713–720, 2008.
12. Zeineldin, H.H., El-Saadany, E.F., Salama, M.M.A., Optimal coordination of over-current relays using a modified particle swarm optimization. *Electr. Power Syst. Res.*, 76, 11, 988–995, 2006.
13. Eskandar, H., Sadollah, A., Bahreinincjad, A., Hamdi, M., Water cycle algorithm–A novel metaheuristic optimization method for solving constrained engineering optimization problems. *Comput. Struct.*, 110, 151–166, 2012.
14. Paithankar, Y.G. and Bhide, S.R., Fundamentals of power system protection, PHI Learning Pvt. Ltd, 2022. https://d1wqtxts1xzle7.cloudfront.net/58422750/Protection.pdf?1550385033=&response-content-disposition=in-line%3B+filename%3DProtection.pdf&Expires=1736253477&Sig-nature=BdBAanJ3VTNNxiMJ7quUmcUNMe1zN5qlKuGbVVH-qZo6REFVZd0DX5FGJtjVKuLybS0mRfxa-1ImN0j2TWsCW5rDgw-cej3FLZ6M6mmTV0bo8Yky1qt5DnAnU5IrIoUizD6Uv7HgT23H0Hl-b6Aokhj5JO~d8aHDQCLASMhWyTaDGsceu6aOvN4p805myz5rqav1I-8pDDj4AXmqfNyOecJ~YNVpov-olRMuxRE4z~g6lDLI3PlfKcZI1yD-mk-0aNw~PVU2feR8oinzgeBrz2e11ojH4nAUmGSkBp3zrGmADE5R-aWir6A2MtCwb-9RbhiAzpVVKJE8XQ-Xt-hy0gKC-N6Q__&Key-Pair-Id=APKAJLOHF5GGSLRBV4ZA.
15. Tsien, H.Y., An automatic digital computer program for setting transmission line directional over-current relays. *IEEE Trans. Power Apparatus Syst.*, 83, 10, 1048–1053, 1964.

16. Radke, G.E., A method for calculating time-over-current relay settings by digital computer. *IEEE Trans. Power Apparatus Syst.*, 82, 189–205, 1963.
17. Sachdev, M.S., Mathematical Models Representing Time-Current Characteristics of Over-current Relays for Computer Applications, in: *IEEE PES Winter Meeting*, pp. 1–8, 1978.
18. Damborg, M.J., Ramaswami, R., Venkata, S.S., Postforoosh, J.M., Computer aided transmission protection system design part I: Alcorithms. *IEEE Trans. Power Apparatus Syst.*, 1, 51–59, 1984.
19. Damborg, M., Computer Aided Transmission Protection System Design, Part II: Implementation and Result. *IEEE Trans. Power Apparatus Syst.*, 103, 60–65, 1984.
20. Ramaswami, R., Damborg, M.J., Venkata, S.S., Jampala, A.K., Postforoosh, J., Enhanced algorithms for transmission protective relay coordination. *IEEE Trans. Power Deliv.*, 1, 1, 280–287, 1986.
21. Laway, N.A. and Gupta, H.O., Adaptive coordination of directional over-current relays of electrical power systems using network decomposition approach. *J. Inst. Eng. (IEJ) India*, 75, 143–149, 1995.
22. Abdelaziz, A.Y., Talaat, H.E.A., Nosseir, A., II, Hajjar, A.A., An adaptive protection scheme for optimal coordination of over-current relays. *Electr. Power Syst. Res.*, 61, 1, 1–9, 2002.
23. Alam, M.N., Das, B., Pant, V., An interior point method based protection coordination scheme for directional over-current relays in meshed networks. *Int. J. Electr. Power Energy Syst.*, 81, 153–164, 2016.
24. Hussain, M.H., Rahim, S.R.A., Musirin, I.J.P.E., Optimal over-current relay coordination: a review. *Procedia Eng.*, 53, 332–336, 2013.
25. Abyane, H.A., Faez, K., Karegar, H.K., A new method for over-current relay (O/C) using neural network and fuzzy logic. In TENCON'97 Brisbane-Australia. *Proceedings of IEEE TENCON'97. IEEE Region 10 Annual Conference. Speech and Image Technologies for Computing and Telecommunications (Cat. No. 97CH36162)*, vol. 1, IEEE, pp. 407–410, 1997, December.
26. Foqha, T., Alsadi, S., Omari, O., Refaat, S.S., Optimization Techniques for Directional over-current Relay Coordination: A Comprehensive Review. *IEEE Access*, 12, 1–55, 2024.
27. Shih, M.Y., Conde, A., González, J.C., Zárraga, F.L., Directional over-currrent relay coordination optimization, in: *Smart Energy and Electric Power Systems*, pp. 121–146, Elsevier, 2023. https://doi.org/10.1016/B978-0-323-91664-6.00011-5.
28. Bansal, R., *Power system protection in smart grid environment*, CRC Press, 2019. https://www.google.co.in/books/edition/Power_System_Protection_in_Smart_Grid_En/pzX3DwAAQBAJ?hl=en&gbpv=1&dq=Bansal,+R.,+Power+system+protection+in+smart+grid+environment,+CRC+-Press,+2019.&pg=PP1&printsec=frontcover.

29. Gupta, S. and Deep, K., Optimal coordination of over-current relays using improved leadership-based grey wolf optimizer. *Arabian J. Sci. Eng.*, 45, 2081–2091, 2020.
30. Trivedi, I.N., Purani, S.V., Jangir, P.K., Optimized over-current relay coordination using Flower Pollination Algorithm, in: *2015 IEEE International Advance Computing Conference (IACC)*, IEEE, pp. 72–77, 2015, June.
31. Atteya, A., II, El Zonkoly, A.M., Ashour, H.A., Optimal relay coordination of an adaptive protection scheme using modified PSO algorithm, in: *2017 Nineteenth International Middle East Power Systems Conference (MEPCON)*, IEEE, pp. 689–694, 2017, December.
32. Alam, M.N., Das, B., Pant, V., A comparative study of metaheuristic optimization approaches for directional over-current relays coordination. *Electr. Power Syst. Res.*, 128, 39–52, 2015.

12

Intelligent Control for Energy-Efficient HVAC System Modeling and Control

R. Sanjeevi*, J. Anuradha, Sandeep Tripathi and Prashantkumar B. Sathvara

Department of Environmental Science, Nims Institute of Allied Medical Science and Technology, NIET, Nims University Rajasthan, Jaipur, Rajasthan, India

Abstract

The intricate challenges inherent in modeling and controlling HVAC systems stem from their complex dynamics, featuring interconnected components like heat exchangers, sensors, and actuators. The dynamic, nonlinear nature of HVAC systems, compounded by uncertainties from external factors like weather and occupancy, complicates the formulation of effective control strategies. Incorporating renewable energy sources and advanced technologies amplifies complexity, requiring sophisticated algorithms to optimize energy usage. Traditional HVAC systems lack comprehensive sensor networks, emphasizing the need for robust sensing technologies and data analytics. Intelligent control systems, crucial for sustainable building management, employ machine learning algorithms to create adaptive predictive models. Reinforcement learning refines control strategies based on system responses and occupant feedback. Integration with smart grids enhances efficiency during low-demand or high-renewable periods. Intelligent control extends to indoor environmental quality (IEQ), incorporating smart technologies and data analytics for real-time monitoring and adaptive responses. Case studies demonstrate transformative impacts, showcasing significant energy savings and improved comfort in commercial and residential settings. Ultimately, intelligent solutions drive sustainability goals by enhancing energy efficiency, occupant comfort, and overall building management.

***Keywords*:** HVAC modeling, environmental quality, real-time monitoring, control strategies, technologies

**Corresponding author*: r.sanjeevi@nimsuniversity.org

Arindam Mondal and Souvik Ganguli (eds.) Controller Design for Industrial Applications, (233–256)

12.1 Introduction

HVAC (heating, ventilation, and air-conditioning) systems are essential for preserving acceptable interior climates in a variety of contexts, including homes, businesses, and industrial sites. Ensuring occupant comfort, reducing environmental impact, and optimizing energy efficiency all depend on the efficient control and management of HVAC systems. The goal of this research is to improve the performance, sustainability, and dependability of HVAC systems by the application of cutting-edge control strategies and technology [1].

Building environmental control is primarily based on the HVAC system, which is essential for preserving indoor comfort, air quality, and energy efficiency. HVAC systems have developed over time from basic, manually operated devices to complex, computer-controlled systems that can adjust to a variety of ambient situations. The rising complexity of building structures, the increased emphasis on energy efficiency, and technological improvements have all contributed to this transition. The main purposes of HVAC systems in the past have been to control temperature and offer thermal comfort. But the importance of HVAC systems has grown as worries about energy efficiency, indoor air quality, and environmental sustainability have grown in prominence [2]. It is now required of modern HVAC systems to provide a comprehensive solution that takes energy efficiency, ventilation, humidity control, and thermal comfort into account [3]. The landscape of HVAC systems has undergone a fundamental shift with the incorporation of smart technology, sensors, and digital controls. Building Automation Systems (BAS) and Building Management Systems (BMS) have become essential because they provide data-driven decision-making, real-time monitoring, and centralized control. A more flexible, responsive, and energy-efficient HVAC system is required, which is driving this trend toward intelligent systems.

It is impossible to overestimate the importance of HVAC control applications in the modern world. The way that buildings and their systems are planned, run, and maintained has to be reevaluated in light of the growing global issues surrounding energy sustainability and climate change. HVAC systems are significant energy consumers in buildings; therefore, lowering their environmental effect and reaching energy efficiency targets depend heavily on them. The potential of sophisticated HVAC control to resolve long-standing issues with conventional control approaches is what makes it significant. More sophisticated approaches to HVAC management, such as model predictive control (MPC) and the incorporation of AI

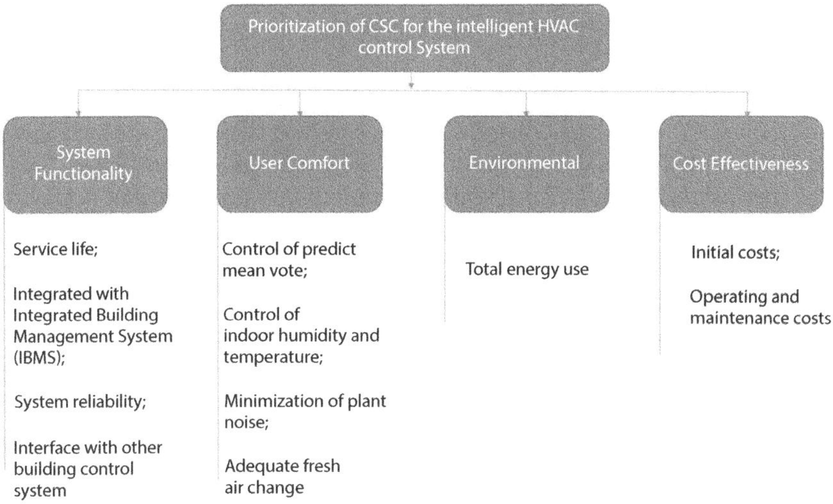

Figure 12.1 Hierarchy of CSC for the intelligent HVAC control system. Source: Figure 12.1 created by author.

and ML, provide a more flexible and reactive solution [4]. This flexibility is essential for improving overall system performance, cutting operating costs, and dynamically optimizing energy consumption. Furthermore, the importance goes beyond simple productivity (Figure 12.1 represents the hierarchy of CSC for the AI enabled HVAC control system). The comfort, productivity, and health of occupants are all directly impacted by the quality of indoor surroundings. Innovative HVAC control solutions support occupant well-being by fostering healthier and cozier interior environments [5]. This is especially important given that people are spending more and more time indoors, whether in homes, businesses, or institutions. The investigation of HVAC control applications in this study is not only a scientific undertaking; it also supports the larger societal objectives of energy conservation, sustainability, and the development of indoor settings that have a good influence on occupants' lives. Comprehending the context and importance lays the groundwork for analyzing control tactics, obstacles, and emerging patterns in the HVAC applications domain. The objective of this study is to provide a valuable contribution to the current discussion and progress in HVAC control by providing practical insights that affect building design, operation, and occupant well-being.

12.2 Challenge in HVAC System

It is still difficult to achieve the best energy efficiency while preserving occupant comfort. Real-time data analysis and sophisticated control algorithms are needed to strike a balance between energy efficiency and thermal comfort. Because renewable energy sources are intermittent, there are issues associated with the growing usage of solar and wind power. HVAC control systems need to ensure constant performance while integrating these sources efficiently [6]. The modification of HVAC systems to engage in demand response initiatives and grid integration poses new difficulties with dynamic load management, predictive modeling, and real-time communication. The integration of personalization elements, occupant feedback mechanisms, and data privacy presents obstacles when designing HVAC management systems that can adjust to the preferences and behaviors of individual occupants.

HVAC control must successfully monitor and manage pollutants, ensure appropriate ventilation rates, and integrate air purification technology in light of the growing awareness of indoor air quality. It is imperative to get smooth communication between various smart building systems, such as security, lighting, and HVAC. While efforts to standardize systems continue, developing universally compatible systems continues to present difficulties. Cybersecurity concerns become increasingly likely as HVAC systems become more digitally integrated and dependent [7]. Cyberattack protection for HVAC control systems is a serious problem that calls for constant attention to detail and strong security protocols. HVAC systems need to adjust to shifting building consumption patterns, occupancy patterns, and dynamic environmental conditions. The difficulty of creating adaptive control techniques that can react well to unforeseen circumstances never goes away.

It is difficult for decision-makers to strike a balance between the upfront expenses of sophisticated HVAC management systems and long-term energy savings and operational effectiveness. To evaluate life-cycle costs, precise metrics and models are required. The design, installation, and maintenance of sophisticated HVAC control technology necessitate the expertise of qualified professionals [8]. For implementation to be successful, skill shortages must be addressed, and appropriate training programs must be offered. The design and execution of HVAC control systems become more sophisticated as a result of keeping up with changing building rules, standards, and regulations. It is a constant struggle to maintain compliance while reaching performance targets. It is crucial to remember that solving these problems frequently calls for interdisciplinary cooperation, technological innovation, and a dedication to continuous research and development within the HVAC sector. Furthermore, because the subject is dynamic, new problems could arise as building practices and technologies advance [9].

12.3 HVAC Control Applications

HVAC control systems have developed over time, moving from simple manually operated systems to highly automated and complex systems. In the past, HVAC systems relied on human temperature and airflow adjustments and simple thermostats. Initially, the goal was to offer a straightforward way to keep interior comfort levels high. Increasingly advanced control tactics were required as buildings and their HVAC needs got increasingly complex. A significant advancement came with the introduction of pneumatic control systems in the middle of the 20th century, which made it possible to control dampers and valves with compressed air [10]. Although pneumatic controls offered a more automated method, they were still not precise or flexible enough for contemporary buildings.

1. Emergence of Building Management Systems (BMS):
The creation of BMS or BAS marked a sea change in the development of HVAC control. BMS combine several building system components—such as HVAC, lighting, security, and more—into a single, centralized control system. This signaled a paradigm change and made it possible to approach building control in an integrated and coordinated manner. BMS popularized the idea of computerized control, paving the way for more dynamic and intelligent reactions to shifting circumstances [11]. It made it possible to optimize energy use and improve overall system efficiency by monitoring and adjusting HVAC parameters based on real-time data.

2. Smart Technologies and Digital Control:
The development of digital control systems and smart technologies has had a significant impact on HVAC control in recent years. HVAC systems are becoming smart, adaptable creatures because of the integration of sensors, data analytics, and networking. This change makes it possible to apply sophisticated control algorithms, do predictive maintenance, and monitor in real time. With the use of MPC, HVAC systems can now predict future circumstances and adjust control parameters appropriately [12]. Furthermore, self-learning capabilities brought about by the integration of

artificial intelligence and machine learning allow HVAC systems to continuously adapt and enhance their performance over time.

3. Energy Efficiency and Sustainability Focus:
The evolution of HVAC control has placed an increased emphasis on optimizing performance for decreased energy consumption and environmental effects since sustainability and energy efficiency have emerged as major concerns. In addition to improving comfort, smart control tactics also work toward these more general objectives, supporting international initiatives to combat climate change and encourage resource conservation. In conclusion, the development of HVAC control has seen a shift from manual to pneumatic systems, and BMS represent the pinnacle of this sophisticated digital technology integration [13]. The newest development in this evolution is the emphasis on smart technologies and a sustainability-driven strategy, which places HVAC control applications at the center of designing built environments that are environmentally conscious, adaptable, and efficient. The foundation for a thorough comprehension of the present situation and anticipated developments in HVAC control applications is laid by this historical viewpoint.

12.4 Advanced Control Strategies Mentioned for HVAC Systems

The built environment relies heavily on HVAC systems to provide energy efficiency, indoor air quality, and thermal comfort. This study explores the intricacies of modeling HVAC systems and the inherent difficulties in putting efficient control strategies into practice. The study includes a thorough investigation of the dynamic relationships between ambient elements, system components, and occupant needs. The study starts with a thorough examination of the current models of HVAC systems, taking into account both dynamic and steady-state representations. The complex behaviors of various components, including ductwork, temperature zones, and air-handling devices, are captured by mathematical equations that are based on the fundamentals of fluid dynamics, thermodynamics, and control theory [14]. The study identifies and groups the main difficulties in modeling and controlling HVAC systems. These issues range from integrating control algorithms and effectively capturing transitory situations to addressing occupant behavior uncertainties and taking into consideration the influence of external environmental elements. Examining the drawbacks and compromises included in various modeling strategies is a crucial component of this research. Control problems are examined in detail and include problems with system stability, responsiveness, and control strategy flexibility under different circumstances [15]. The study evaluates the efficacy of combining machine learning and advanced control methods, including MPC, to overcome conventional control constraints. The modeling framework incorporates environmental considerations, such as how temperature changes and external circumstances affect HVAC system performance. The study also emphasizes how crucial it is to create models that can take into account a variety of system configurations, building kinds, and operational conditions. This study offers a comprehensive viewpoint on the modeling and control issues associated with HVAC systems, providing scholars, practitioners, and business experts with relevant insights [13]. Future developments in HVAC technology can be guided toward more resilient, adaptable, and energy-efficient systems by tackling the issues that have been highlighted. This will eventually help to create sustainable and comfortable indoor environments.

MPC forecasts system behavior in the future by using mathematical models. It takes into account things like occupancy schedules, weather forecasts, and the building's thermal properties. By optimizing control operations across a predetermined time horizon, MPC enables the system to react proactively to conditions that change. It works especially well in situations where the ability to forecast outcomes is critical, such as energy optimization and comfort maintenance in buildings with variable loads.

Multiple indoor units and variable-speed compressors are used by VRF systems to adjust the flow of refrigerant. This maximizes energy use and permits accurate temperature control in various zones. Zoning flexibility provided by VRF systems enables customized control in different parts of a building. As loads change, the variable-speed operation adjusts to improve both energy economy and occupant comfort. Sensors are used in occupancy-based control to determine whether people are present in particular regions. By communicating with the HVAC system, these sensors allow the system to modify temperature set points and ventilation rates concerning occupancy [16]. By matching HVAC operations to real occupancy patterns, this technique lowers energy waste in vacant buildings. Based on pollution levels and occupancy in real time, DCV modifies ventilation rates. Indoor air quality is often monitored with CO_2 sensors. In vacant or sparsely populated spaces, DCV maximizes ventilation to satisfy occupant needs without over-ventilating. This method helps to enhance indoor air quality and save electricity.

Using machine learning, adaptive learning algorithms evaluate inputs in real-time and past data. With time, these algorithms adjust HVAC management techniques based on patterns they identify, thereby optimizing system operations. They can adapt set points according to occupant preferences, foresee changes in building conditions, and continuously increase system efficiency. Weather forecasts are integrated into HVAC operations through weather-responsive control [13, 17]. The system modifies its set points and operating parameters in response to predicted meteorological factors, including variations in temperature, sun radiation, and humidity. This proactive strategy maximizes energy efficiency and HVAC performance (Figure 12.2 represents the energy efficient inputs to thermal models).

FDD systems employ algorithms to identify anomalies and diagnose problems while continuously monitoring the performance of HVAC systems. Early problem detection (EDD) allows for timely maintenance, minimizes downtime, and assures peak system performance. FDD is essential to preserving system dependability and averting significant malfunctions [18]. Different HVAC system components use local decision-making in distributed control systems. Every component can independently make decisions based on local circumstances rather than depending on a centralized controller [19]. In big and intricate HVAC installations, this method improves fault tolerance, scalability, and flexibility. Smart thermostats enable remote HVAC control via mobile devices; they are frequently included in home automation systems. To adjust to user preferences and behavior, certain smart thermostats use machine learning algorithms.

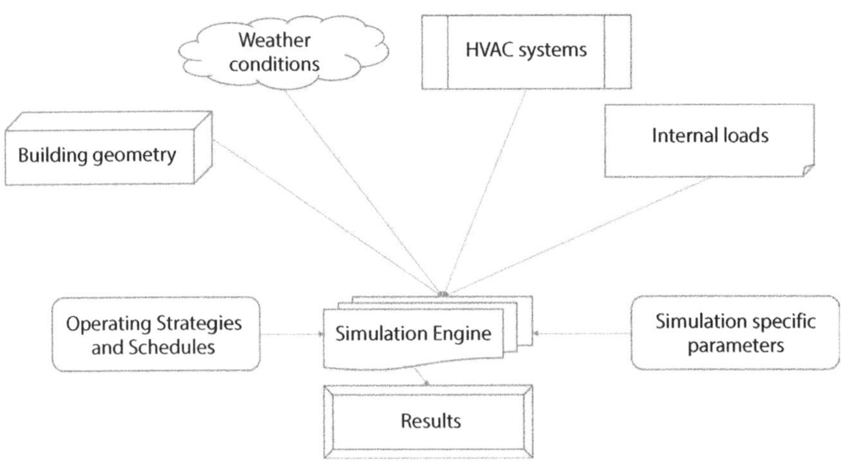

Figure 12.2 Input to thermal models. Source: Figure 12.2 created by author.

They improve user convenience, shed light on trends in energy consumption, and promote energy conservation.

Hybrid HVAC systems maximize overall performance by combining various technologies. For instance, a system may combine geothermal or radiant heating elements with conventional air systems. In order to create a comprehensive and cooperative approach to building management, multimodal integration entails integrating HVAC with other building systems, such as lighting. This improves the efficiency, flexibility, and adaptation of the system to various building requirements [20].

Together, these cutting-edge control techniques enable the development of intelligent, flexible, and energy-efficient HVAC systems that satisfy the changing requirements of contemporary structures and their inhabitants. These strategies' capabilities are further enhanced by the integration of IoT, machine learning, and artificial intelligence technologies, which also present potential for ongoing optimization and improvement.

12.5 Energy Efficiency and Sustainability in HVAC System

HVAC (heating, ventilation, and air-conditioning) systems are crucial in determining how built environments use energy. This study investigates how to integrate intelligent control strategies to improve HVAC systems' energy efficiency, meeting the need for responsive and sustainable building operations. The study looks into how cutting-edge technology like predictive control, machine learning, and artificial intelligence can be used to maximize HVAC efficiency and reduce energy use. The study starts with a thorough analysis of conventional HVAC control strategies and how poorly they can adjust to changing circumstances and a wide range of customer preferences [21]. Expanding on this basis, the paper presents intelligent control strategies, stressing their capacity to independently adjust to evolving environmental conditions, occupancy trends, and user actions. The incorporation of machine learning algorithms enables proactive modifications to HVAC set points and predictive modeling of thermal loads [22]. The creation and application of intelligent control algorithms specifically suited for HVAC systems are important facets of the study. To predict system requirements, these algorithms make use of historical patterns, real-time sensor data, and weather forecasts. The efficiency of these sophisticated controls is evaluated in the study concerning overall system performance, interior comfort, and energy savings. Additionally, the study takes into account the real-world implementation difficulties that come with intelligent HVAC control, like cybersecurity, scalability, and model interpretability [23]. To guarantee that intelligent controls are seamlessly integrated into a variety of HVAC applications, strategies for overcoming these obstacles are investigated. To verify how well intelligent control strategies work in actual situations, simulations are run. The findings show that considerable energy reductions are possible without sacrificing occupant comfort. An understanding of the superiority of intelligent control approaches—especially in dynamic and uncertain environments—can be gained through comparisons with traditional control methods. This research adds to the growing body of knowledge in the area of intelligent HVAC control by providing a road map for the operation of sustainable and energy-efficient buildings. The results highlight how intelligent control strategies can significantly reduce environmental impact and advance the objectives of energy conservation in the built environment [24]. The conclusions of the study are intended to direct practitioners, scholars, and legislators toward the acceptance of intelligent HVAC control as a crucial facilitator of ecologically friendly and energy-efficient building practices.

When designing, running, and maintaining HVAC (heating, ventilation, and air-conditioning) systems, energy efficiency and sustainability are crucial factors to take into account [25]. These ideas are emphasized by recent advancements in HVAC system utilization to cut down on energy use, lessen environmental effects, and support more environmentally friendly building techniques. The following are important components of HVAC systems' sustainability and energy efficiency, along with current trends and practices.

High-efficiency HVAC equipment, such as energy-efficient heat exchangers, variable-speed compressors, and sophisticated control systems, is becoming more and more common. This equipment is meant to maximize energy efficiency by modifying its functioning in response to actual demand. Advanced algorithms are used by smart HVAC controls, which are frequently linked with BAS, to maximize system performance [26]. These controllers make real-time modifications for energy efficiency based on many criteria, including occupancy, outside weather conditions, and past usage trends. Demand-response programs, which allow power providers to signal HVAC systems to modify operations during periods of peak demand, are increasingly integrating HVAC systems. In addition to helping to balance the electrical grid, this frequently gives building owners financial benefits.

In order to prepare incoming new air, ERV systems recover and repurpose energy from exhausted air. In addition to maintaining interior air quality and increasing overall energy efficiency, this lessens the strain on heating and cooling systems [27]. Geothermal heat pumps use the ground's constant temperature to generate heat and coolness. This renewable energy source provides a sustainable substitute with less energy usage, reducing dependency on conventional HVAC systems. The use of solar panels to power HVAC systems is growing in popularity. By offsetting electricity use, solar-powered systems can increase the sustainability and environmental friendliness of HVAC operations [28]. Reducing the heating and cooling load on HVAC systems is a result of improved building insulation and envelope design. This tendency is consistent with energy-saving sustainable building techniques.

More and more building owners and developers are pursuing green building certifications (like LEED and BREEAM), which frequently have strict guidelines for the sustainability and efficiency of HVAC systems. A dedication to environmental responsibility is demonstrated by compliance with these certifications. HVAC systems can adapt their operation to the number of occupants and consumption patterns thanks to the integration of occupancy sensors and zoning systems [29]. Energy can be saved by conditioning vacant or sparingly occupied spaces more effectively. HVAC systems are increasingly using machine learning and data analytics. Large volumes of data are analyzed by these technologies to spot trends, forecast system behavior, and maximize energy efficiency through control strategy optimization.

Life-cycle analysis is a tool that building owners and operators are using more and more to evaluate the long-term environmental effects of HVAC systems. Proactive maintenance techniques also guarantee that systems run as efficiently as possible, which lowers energy waste. Stricter energy efficiency requirements for HVAC systems are being implemented by governments and regulatory agencies across the globe [30, 31]. By offering incentives and rebates, building owners are incentivized to adhere to these criteria, advancing the adoption of sustainable technologies. Heat pumps, especially those powered by electricity, are becoming more and more well-liked as cost-effective substitutes for air-conditioning and heating. These systems use electricity, which is frequently produced from renewable energy sources, to deliver effective heating and cooling solutions [32]. The use of hybrid HVAC systems, which integrate many technologies (such as conventional air-conditioning systems with geothermal or radiant heating components), is growing in popularity. These systems are flexible, enabling buildings to maximize energy use and adjust to changing conditions. Machine learning algorithms examine past HVAC system data to find trends that can point to possible equipment breakdowns. Facility managers can anticipate maintenance needs and schedule repairs ahead of time to save downtime and avoid significant system failures. By making sure HVAC equipment runs at peak efficiency, this proactive strategy improves energy efficiency.

Individual occupant preferences are taken into consideration by adaptive thermal comfort models, which then modify HVAC settings appropriately. These models take into account variables like activity levels, clothing choices, and individual comfort preferences rather than rigidly sticking to set points for temperature [33]. By cutting out on needless heating and cooling, this adaptive technique improves occupant pleasure while simultaneously conserving energy.

Decentralized HVAC systems minimize the amount of ductwork required by distributing heating and cooling sources across a structure. This method reduces heat loss during transportation and enables more accurate control in each zone. Decentralized HVAC systems reduce the inefficiencies of central HVAC systems, hence increasing energy efficiency and offering customized comfort in designated building regions. Building systems such as HVAC, lighting, and security are all integrated into a single platform by iBEMS. All building components' energy usage may be thoroughly monitored, analyzed, and controlled thanks to this integration. iBEMS improves building performance, sustainability, and energy management by maximizing the cooperation of many systems. Reducing energy use using real-time feedback systems for residents promotes more environmentally friendly behavior [32]. Residents can make educated decisions to cut down on energy waste by having access to information about their usage patterns. This involvement immediately includes tenants in reaching energy-saving targets and cultivates a sense of responsibility.

A green façade or green roof involves adding greenery to the outside of a structure. By offering thermal insulation, these components lessen the HVAC systems' overall cooling burden. Green roofs and façades promote environmental sustainability, energy efficiency, and better indoor air quality by establishing a more sustainable building envelope. Systems for storing thermal energy use power that is produced during off-peak hours [34, 35]. Storing energy can be used for heating or cooling during times of high demand, which lessens the need for grid electricity. This methodology facilitates load shifting, optimizes the overall efficiency of the HVAC system, and adds to a more robust and sustainable energy infrastructure. The ventilation rates of HVAC systems with dynamic airflow control mechanisms are adjusted in real time according to occupancy and demand [36]. By ensuring that ventilation is in line with actual requirements, indoor air quality is maintained and energy efficiency is promoted. In vacant or sparsely populated spaces, dynamic airflow control minimizes energy waste by optimizing the dispersion of conditioned air.

The whole cost of purchasing, running, and maintaining an HVAC system over the course of its lifetime is evaluated via life-cycle cost analysis. It takes into account the initial outlay, energy usage, upkeep expenses, and other elements. Building owners can make well-informed decisions that strike a balance between initial expenditures and the advantages of sustainability and energy savings by assessing the long-term costs. Buildings certified as zero net energy (ZNE) strive to produce as much energy on the premises as they need [37]. Demand-side management, renewable energy sources, and extremely efficient technologies are all integrated into HVAC systems, which are essential to reaching ZNE goals. ZNE buildings stand for dedication to environmental responsibility and sustainability [38, 39]. HVAC systems that have been electrified use electricity for cooling and heating, which is frequently produced utilizing renewable energy sources. This pattern supports initiatives to lessen dependency on fossil fuels and decarbonize energy sources. Because electrified HVAC systems use cleaner, greener energy, they help achieve sustainability goals. Blockchain technology is being investigated for safe and transparent HVAC system energy management. It can make data sharing, smart contracts, and secure transactions easier. Blockchain helps to ensure the integrity of data in the energy management ecosystem and supports sustainability goals by improving the efficiency and reliability of energy-related processes.

The goal of research and development is to find ecologically benign refrigerants that have a decreased potential to cause global warming. Evaporative cooling and passive design techniques are examples of sustainable cooling solutions that try to lessen dependency on conventional high-energy cooling techniques. Sustainability, environmental stewardship, and reducing the impact of climate change on HVAC operations are given top priority in these advances.

Recent developments in HVAC system operation and design demonstrate a dedication to environmental stewardship, sustainability, and energy efficiency. The diverse methodologies and inventive tactics utilized in contemporary HVAC operations underscore a comprehensive strategy that is beyond mere energy efficiency. This method incorporates cutting-edge design techniques, renewable energy sources, smart technology, and environmental responsibility in addition to occupant enjoyment. Buildings that are becoming more robust and sustainable are increasingly relying on HVAC systems [37, 40, 41]. Utilizing renewable energy sources in conjunction with cutting-edge technologies like blockchain and machine intelligence shows a proactive approach to solving environmental issues.

Furthermore, the emphasis on decentralized systems, green building components, and occupant participation highlights how comprehensive these improvements are. Future developments in regulations and continued innovation are expected to affect the HVAC sector as it continues to evolve. These developments continue to emphasize lowering energy use and minimizing environmental effects, which strengthens the industry's resolve to provide a resilient and sustainable built environment. HVAC systems are at the forefront of making a positive, sustainable, and adaptive impact on the environment in the future because of the convergence of technology, sustainability, and a holistic viewpoint.

12.6 Human-Centric HVAC Control

Give the tenants of the building the comfort and happiness they deserve. Make use of cutting-edge sensors to keep an eye on temperature, humidity, and air quality within buildings. To maintain the highest levels of comfort, use adaptive control algorithms that dynamically modify HVAC parameters in response to changes in occupant preferences [42, 43]. Give residents authority over their local surroundings. Give them access to user interfaces and smart thermostats that let them customize their comfort levels. Include smartphone apps that let tenants adjust the HVAC system from a distance, offering a smooth and personalized experience.

Utilize energy wisely by cooling just areas that are in use. Use occupancy sensors to identify spaces that are occupied. Include zoning systems that create distinct zones within the building so that HVAC systems can modify their operations in response to changes in occupancy patterns. Zones that are seldom or never occupied can have less cooling, which will maximize energy efficiency. Go beyond predetermined cutoff points and take personal preferences for thermal comfort into account. Adopt models of adaptive thermal comfort that account for individual preferences, activity levels, and clothes, among other personal aspects. This methodology guarantees that HVAC systems adjust dynamically to the fluctuating comfort requirements of inhabitants [44]. Give your initial opinion on the indoor climate. Make use of real-time monitoring systems with sensors to check the humidity, temperature, lighting, and quality of the air. By giving residents immediate feedback via smartphone apps or visual displays, you can raise awareness and empower them to make wise decisions. Recognize and anticipate comfort needs in advance. Use machine learning algorithms to examine past data, such as patterns of comfort and tenant behavior. Then, predictive models may foresee comfort needs and modify HVAC settings beforehand, guaranteeing a proactive and customized approach.

Foster a sense of well-being by introducing them to nature. When planning an HVAC system, consider biophilic design aspects like natural lighting, vistas of greenery, and natural materials. It has been demonstrated that biophilic design features improve mental health and foster an atmosphere that is more focused on people [45, 46]. Verify that indoor air quality satisfies health requirements. Install sensors to track variables related to air quality, such as particle matter, carbon dioxide (CO_2) concentrations, and volatile organic compounds (VOCs). To ensure the best possible interior air quality, use ventilation control strategies to modify airflow and introduce fresh outdoor air as needed.

Involve tenants in maximizing the efficiency of the HVAC system. Provide systems so that tenants can comment on how comfortable they are. Utilize mobile apps, feedback buttons, or surveys to get data on preferences and satisfaction. Examine this input to continuously adjust the HVAC system's settings and enhance overall performance. Energy efficiency and occupant comfort should be balanced. Investigate the usage of personal comfort systems that use less energy, such as localized heating and cooling options [47]. By giving each person customized thermal comfort, these technologies reduce the need to heavily condition entire spaces while maximizing energy efficiency. For greater well-being, match lighting to circadian rhythms. Include dynamic lighting systems that change throughout the day in terms of intensity and color temperature [48]. Align lighting with the cycles of natural daylight to maintain circadian rhythms, which will improve occupant health and provide a more human-centered atmosphere.

Give HVAC systems the ability to instantly adjust to changing circumstances. Use dynamic HVAC parameter adjustments based on occupancy patterns, outside weather, and user preferences by implementing responsive control algorithms. These systems can proactively adjust settings for comfort and energy efficiency in anticipation of changes. The well-being and preferences of building occupants are given priority through the complex integration of technology and tactics known as human-centric HVAC control [31, 49]. These systems help to create interior environments that are not only energy-efficient but also supportive of occupant health, satisfaction, and general quality of life by taking into account individual comfort demands, offering user control, and utilizing real-time data and adaptable algorithms. It is anticipated that the capabilities of human-centric HVAC control will continue to improve with the further development of smart technologies and data analytics.

12.7 Indoor Environmental Quality

To guarantee the health, comfort, and productivity of inhabitants, indoor environmental quality (IEQ) control refers to the management and optimization of numerous environmental parameters within indoor environments. The phrase refers to a variety of elements, and managing these specifications is essential to establishing comfortable and healthful interior spaces. IEQ control in the context of HVAC systems includes tactics and technology that deal with the following crucial elements: The cleanliness and purity of the air inside a building are referred to as air quality. It entails controlling the amounts of contaminants, including gases, particle matter, and VOCs. Sensors are used by intelligent HVAC systems to track many aspects of indoor air quality [50]. To ensure ideal air quality, sophisticated control algorithms modify ventilation rates, air filtration, and purification procedures depending on real-time data. The degree to which occupants are content with the thermal environment—which includes elements like temperature, humidity, and air movement—is known as thermal comfort [51]. Adaptive temperature and humidity management strategies are used by HVAC systems with intelligent control mechanisms. To achieve individualized comfort levels and maximize energy savings, machine learning algorithms can learn the preferences of the occupants and modify the thermal environment. Designing and managing artificial and natural lighting to meet lighting needs, minimize glare, and maintain circadian cycles is known as lighting quality. Control systems for integrated lighting and HVAC work together to provide lighting scenes that enhance the surrounding temperature. Adaptive lighting and daylight harvesting systems help reduce energy use and enhance visual comfort. Controlling sound levels and preventing loud noises indoors are related to acoustic comfort. It is possible to design and operate HVAC systems to reduce noise pollution. Furthermore, clever control algorithms can optimize air distribution and ventilation to minimize noise while preserving sufficient airflow [52]. The whole effect of interior environment reflects on productivity, cognitive function, and occupant health. Air quality, illumination, and thermal comfort are just a few of the elements that intelligent HVAC systems integrate to take occupant well-being into account. Customized control elements, including temperature controls for workstations, enhance worker comfort and efficiency. Critical modification on ventilation rates in a place according to activities, impacts the indoor air quality, occupant comfort and productivity. Adaptive ventilation is made possible by sensors and algorithms, which maximize the introduction of fresh outdoor air while consuming the least amount of energy. This guarantees that the rates of ventilation correspond with the real occupancy and concentrations of pollutants. IEQ metrics are continuously monitored, and building occupants receive real-time feedback [53]. Intelligent HVAC systems give residents real-time feedback on the environment through the use of sensors and communication technology. Through user interfaces, occupants can also adjust some parameters, which increase their sense of control and enjoyment. To put it briefly, IEQ control in HVAC systems refers to the use of clever technologies and tactics to establish and preserve interior spaces that support comfort, productivity, and health. HVAC systems may adjust dynamically to changing conditions and individual preferences by combining sensors, data analytics, and adaptive control algorithms. This enhances the quality of interior environments. IEQ has a major impact on occupant comfort, productivity, and health, making it a crucial factor to take into account when designing and managing built environments. In this regard, creating and preserving ideal indoor conditions is mostly dependent on HVAC systems. To improve IEQ, this research focuses on integrating intelligent control strategies into HVAC systems [1]. It addresses the various facets of air quality, thermal comfort, lighting, and acoustics.

12.8 Case Studies

In order to maximize the efficiency of HVAC systems, intelligent control in HVAC systems integrates cutting-edge technologies, automation, and clever algorithms. The following case studies illustrate the use of intelligent control in practical HVAC applications:

Case Study: The Crystal, London, Construct a smart, sustainable building that features cutting-edge energy and HVAC technologies. Intelligent building control is exemplified by Siemens' sustainable cities effort, The Crystal, located in London. A complex BMS is used at The Crystal to connect the building's many functions, such as the HVAC, lighting, and security. Sensor data are analyzed by intelligent algorithms to maximize energy efficiency while preserving occupant comfort. Ground-source heat pumps and solar panels are examples of renewable energy sources that are integrated into the structure. By minimizing dependency on traditional energy sources, intelligent control systems dynamically modify HVAC operations based on the availability of renewable energy. The Crystal's HVAC systems use predictive maintenance algorithms in conjunction with Internet-of-Things sensors. By analyzing equipment data, these algorithms can predict possible problems, which enable proactive maintenance and minimize downtime. The Crystal's incorporation of renewable energy sources and automated HVAC control allows for high levels of sustainability and energy efficiency. The structure is a showcase for urban sustainability, with smart systems that improve the economy and the environment.

Case Study: Helsinki Central Library Oodi, Construct a cutting-edge, energy-efficient library with sophisticated environmental management and HVAC systems. Helsinki Central Library Oodi is renowned for its dedication to sustainability and modern design. Occupancy sensors are used by the library to count the number of patrons in various sections. By modifying ventilation and temperature settings in response to occupancy levels in real time, intelligent HVAC controllers maximize energy efficiency across various zones. Oodi integrates natural ventilation techniques by using smart control systems to modify airflow and open windows in response to outside weather. This method reduces the need for mechanical ventilation while improving interior air quality. Lighting systems are also under intelligent control; they change their settings according to occupancy and natural light. Energy-efficient and cozy environments are the result of the cooperation between HVAC and lighting systems. Helsinki Central Library Oodi is a prime example of how sustainable architecture and smart HVAC control combine to provide a comfortable and energy-efficient public area. Adaptive control and real-time data integration improve the usability of the library while also contributing to environmental sustainability.

Case Study: Microsoft's Project Natick, Using intelligent control in submerged data center modules, data centers can be cooled more energy-efficiently. In order to attain more energy-efficient cooling, Microsoft's Project Natick tested the viability of submerging data center modules in the water. The subterranean data Centre's conditions were monitored and adjusted by the project using intelligent control technologies. The sensors that monitored temperature, humidity, and other environmental parameters were installed in the submerged data center modules. To maximize energy efficiency, intelligent control algorithms modified the cooling systems in response to real-time data. To forecast environmental conditions and analyze past data, machine learning techniques were used. As a result, the system was able to proactively modify HVAC operations to maintain ideal server conditions by anticipating changes. Through the application of intelligent control in unusual conditions, the study revealed the potential for energy-efficient cooling in data centers. In comparison to conventional air-conditioned data centers, Microsoft was able to obtain a more economical and sustainable cooling solution by utilizing the ocean's steady temperature.

Case Study: The Edge, Amsterdam, Construct a highly energy-efficient and sustainable office building by using intelligent building technology and HVAC control. Intelligent control technologies for HVAC and building management are integrated into The Edge in Amsterdam, which is frequently referred to be the "smartest" office skyscraper in the world. The Edge monitors a number of variables, such as occupancy, temperature, and lighting conditions, through the use of an advanced BMS that incorporates IoT devices and sensors. Real-time data are gathered by the BMS to support intelligent control techniques. Through a smartphone app, tenants have control over their workstation settings. Users of the app may schedule meeting rooms, change the lighting, and even control the temperature. Intelligent control algorithms that adapt to user preferences enable this degree of personalization. The structure has dynamic lighting systems that change according to occupancy and ambient light levels. In a similar vein, the HVAC system maximizes energy use by adjusting to the number of occupants and their preferences. The Edge's sophisticated control systems have allowed for exceptional sustainability and energy efficiency. The building has produced a workspace that is not only effective but also supportive of occupant well-being and productivity thanks to its creative approach to personalized comfort and real-time modifications based on occupancy and environmental circumstances.

These case studies show how occupant comfort, sustainability, and energy efficiency may all be significantly increased with sophisticated HVAC control. Through the utilization of real-time data, predictive analytics, and user-centric design principles, these projects demonstrate the versatility and potential applications of intelligent HVAC control.

12.9 Conclusion

In conclusion, the complexities inherent in modeling and managing HVAC systems arise from their intricate dynamics, featuring interconnected components and influenced by uncertainties from external factors like weather and occupancy. The integration of renewable energy sources and advanced technologies further complicates the formulation of effective control strategies, necessitating sophisticated algorithms for energy optimization. The lack of comprehensive sensor networks in traditional HVAC systems underscores the importance of robust sensing technologies and data analytics. Intelligent control systems, employing machine learning algorithms and reinforcement learning, play a pivotal role in sustainable building management by creating adaptive predictive models and refining control strategies based on system responses and occupant feedback. Integration with smart grids enhances efficiency, particularly during low-demand or high-renewable periods. Intelligent control extends to IEQ, utilizing smart technologies and data analytics for real-time monitoring and adaptive responses. Case studies illustrate transformative impacts, demonstrating significant energy savings and improved comfort in both commercial and residential settings. Ultimately, intelligent solutions drive sustainability goals by optimizing energy efficiency, enhancing occupant comfort, and improving overall building management.

References

1. Bearg, D.W., *Indoor Air Quality and HVAC Systems* (1st ed.), Rout-ledge, 1993. https://doi.org/10.1201/9780203751152/indoor-air-quality-hvac-sys-tems-david-bearg.
2. Al Horr, Y., Arif, M., Kaushik, A., Mazroei, A., Katafygiotou, M., Elsarrag, E., Occupant productivity and office indoor environment quality: A review of the literature. *Build. Environ.*, 105, 369–389, 2016.
3. Arif, M., Katafygiotou, M., Mazroei, A., Kaushik, A., Elsarrag, E., Impact of indoor environmental quality on occupant well-being and comfort: A review of the literature. *Int. J. Sustain. Built Environ.*, 5, 1, 1–11, 2016.
4. Ghosh, B., Lal, H., Srivastava, A., Review of bioaerosols in indoor environment with special reference to sampling, analysis and control mechanisms. *Environ. Int.*, 85, 254–272, 2015.
5. Leung, D.Y., Outdoor-indoor air pollution in urban environment: challenges and opportunity. *Front. Environ. Sci.*, 2, 69, 2015.
6. Luengas, A., Barona, A., Hort, C., Gallastegui, G., Platel, V., Elias, A., A review of indoor air treatment technologies. *Rev. Environ. Sci. Bio/Technol.*, 14, 3, 499–522, Sep. 2015, doi: 10.1007/s11157-015-9363-9.
7. Tham, K.W., Indoor air quality and its effects on humans—A review of challenges and developments in the last 30 years. *Energy Build.*, 130, 637–650, 2016.
8. Tran, V.V., Park, D., Lee, Y.-C., Indoor air pollution, related human diseases, and recent trends in the control and improvement of indoor air quality. *Int. J. Environ. Res. Public Health*, 17, 8, 2927, 2020.
9. Chenari, B., Carrilho, J.D., da Silva, M.G., Towards sustainable, energy-efficient and healthy ventilation strategies in buildings: A review. *Renew. Sustain. Energy Rev.*, 59, 1426–1447, 2016.

10. Frontczak, M. and Wargocki, P., Literature survey on how different factors influence human comfort in indoor environments. *Build. Environ.*, 46, 4, 922–937, 2011.
11. Khan, A.H. and Karuppayil, S.M., Fungal pollution of indoor environments and its management. *Saudi J. Biol. Sci.*, 19, 4, 405–426, 2012.
12. Sundell, J., *et al.*, Ventilation rates and health: multidisciplinary review of the scientific literature: Ventilation rates and health. *Indoor Air*, 21, 3, 191–204, Jun. 2011, doi: 10.1111/j.1600-0668.2010.00703.x.
13. Vardoulakis, S., *et al.*, Impact of climate change on the domestic indoor environment and associated health risks in the UK. *Environ. Int.*, 85, 299–313, 2015.
14. Chua, K.J., Chou, S.K., Yang, W.M., Yan, J., Achieving better energy-efficient air conditioning–a review of technologies and strategies. *Appl. Energy*, 104, 87–104, 2013.
15. Li, W., Koo, C., Hong, T., Oh, J., Cha, S.H., Wang, S., A novel operation approach for the energy efficiency improvement of the HVAC system in office spaces through real-time big data analytics. *Renew. Sustain. Energy Rev.*, 127, 109885, 2020.
16. Rhee, K.-N. and Kim, K.W., A 50 year review of basic and applied research in radiant heating and cooling systems for the built environment. *Build. Environ.*, 91, 166–190, 2015.
17. Sathvara, P.B., Anuradha, J., Sanjeevi, R., Tripathi, S., Rathod, A.B., Spatial Analysis of Carbon Sequestration Mapping Using Remote Sensing and Satellite Image Processing, in: *Multimodal Biometric and Machine Learning Technologies*, pp. 71–83, Wiley Scrivener Publishing LLC, 2023, doi: 10.1002/9781119785491.ch4.
18. Afroz, Z., Shafiullah, G., Urmee, T., Higgins, G., Modeling techniques used in building HVAC control systems: A review. *Renew. Sustain. Energy Rev.*, 64–84, Mar. 2018, doi: 10.1016/j.rser.2017.10.044.
19. Afram, A. and Janabi-Sharifi, F., Theory and applications of HVAC control systems – A review of model predictive control (MPC). *Build. Environ.*, 72, 343–355, Feb. 2014, doi: 10.1016/j.buildenv.2013.11.016.
20. Blum, D., *et al.*, Field demonstration and implementation analysis of model predictive control in an office HVAC system. *Appl. Energy*, 318, 119104, Jul. 2022, doi: 10.1016/j.apenergy.2022.119104.
21. Asim, N., *et al.*, Sustainability of heating, ventilation and air-conditioning (HVAC) systems in buildings—An overview. *Int. J. Environ. Res. Public Health*, 19, 2, 1016, 2022.
22. Rasmussen, B.P., Price, C., Koeln, J., Keating, B., Alleyne, A., HVAC System Modeling and Control: Vapor Compression System Modeling and Control, in: *Intelligent Building Control Systems, Advances in Industrial Control*, J.T. Wen and S. Mishra (Eds.), pp. 73–103, Springer International Publishing, Cham, 2018, doi: 10.1007/978-3-319-68462-8_4.

23. Serale, G., Fiorentini, M., Capozzoli, A., Bernardini, D., Bemporad, A., Model Predictive Control (MPC) for Enhancing Building and HVAC System Energy Efficiency: Problem Formulation, Applications and Opportunities. *Energies*, 11, 3, 631, Mar. 2018. doi: 10.3390/en11030631.
24. Zhang, Y., *et al.*, Can commonly-used fan-driven air cleaning technologies improve indoor air quality? A literature review. *Atmos. Environ.*, 45, 26, 4329–4343, 2011. doi: 10.1016/j.atmosenv.2011.05.041.
25. Felgueiras, C., Santos, R., Fonseca, L.M., Caetano, N.S., Buildings sustainability: the HVAC contribution. *J. Clean Energy Technologies-JOCET*, 4, 5, 375–379, 2016.
26. Khalid, F., Dincer, I., Rosen, M.A., Development and analysis of sustainable energy systems for building HVAC applications. *Appl. Therm. Eng.*, 87, 389–401, 2015.
27. Khazaii, J., *Energy-Efficient HVAC Design: An Essential Guide for Sustainable Building*, Springer International Publishing, Cham, 2014, doi: 10.1007/978-3-319-11047-9.
28. Kim, J.T. and Yu, C.W.F., Sustainable development and requirements for energy efficiency in buildings – The Korean perspectives. *Indoor Built Environ.*, 27, 6, 734–751, Jul. 2018. doi: 10.1177/1420326X18764618.
29. Ligade, J. and Razban, A., Investigation of Energy Efficient Retrofit HVAC Systems for a University: Case Study. *Sustainability*, 11, 20, 5593, 2019.
30. Vakiloroaya, V., Samali, B., Fakhar, A., Pishghadam, K., A review of different strategies for HVAC energy saving. *Energy Convers. Manage.*, 77, 738–754, 2014.
31. Sanjeevi, R., Haruna, M., Tripathi, S., Singh, B., Jayaraman, A., Impacts of Global Carbon Foot Print on the Marine Environment. *Int. J. Eng. Res. Technol.*, 5, 51–54, May 2017.
32. Yang, Z., Ghahramani, A., Becerik-Gerber, B., Building occupancy diversity and HVAC (heating, ventilation, and air conditioning) system energy efficiency. *Energy*, 109, 641–649, 2016.
33. Vujanović, M., Wang, Q., Mohsen, M., Duić, N., Yan, J., Recent progress in sustainable energy-efficient technologies and environmental impacts on energy systems. *Appl. Energy*, 283, 116280, 2021. Accessed: Jan. 30, 2024. [Online]. Available: https://www.sciencedirect.com/science/article/pii/S030626192031669X.
34. Bakar, N.N.A., *et al.*, Energy efficiency index as an indicator for measuring building energy performance: A review. *Renew. Sustain. Energy Rev.*, 44, 1–11, 2015.
35. Rathod, A.B., Anuradha, J., Sathvara, P.B., Tripathi, S., Sanjeevi, R., Vegetational Change Detection Using Machine Learning in GIS Technique: A Case Study from Jamnagar (Gujarat). *J. Data Acquisition Process.*, 38, 1, 1046–1061, 2023, doi: 10.5281/zenodo.7700655.

36. Pérez-Lombard, L., Ortiz, J., Coronel, J.F., Maestre, I.R., A review of HVAC systems requirements in building energy regulations. *Energy Build.*, 43, 2–3, 255–268, 2011.
37. Wang, Y., Feng, H., Xi, X., Monitoring and autonomous control of Beijing Subway HVAC system for energy sustainability. *Energy Sustain. Dev.*, 39, 1–12, 2017.
38. Xiao, F. and Wang, S., Progress and methodologies of lifecycle commissioning of HVAC systems to enhance building sustainability. *Renew. Sustain. Energy Rev.*, 13, 5, 1144–1149, 2009.
39. Sanjeevi, R., Rathod, A.B., Sathvara, P.B., Tripathi, A., Anuradha, J., Tripathi, S., VEGETATIONAL CARTOGRAPHY ANALYSIS UTILIZING MULTI-TEMPORAL NDVI DATA SERIES: A CASE STUDY FROM RAJKOT DISTRICT (GUJARAT), INDIA. *Tianjin Daxue Xuebao (Ziran Kexue yu Gongcheng Jishu Ban)/ J. Tianjin Univ. Sci. Technol.*, 55, 04:2022, 490–497, 2022. doi: 10.17605/OSF.IO/UGJYM.
40. Abbasi, T., Sanjeevi, R., Anuradha, J., abbasi, S.A., Impact of Al\textsuperscript3+ on sludge granulation in UASB reactor, Apr. 01, 2013, [Online]. Available: https://nopr.niscpr.res.in/handle/123456789/19284.
41. Abdelrahman Aly, A., Al-Omran, A.M., Sallam, A.S., Al-Wabel, M., II, Al-Shayaa, M.S., Vegetation cover change detection and assessment in arid environment using multi-temporal remote sensing images and ecosystem management approach. *Solid Earth*, 7, 2, 713–725, Apr. 2016. doi: 10.5194/se-7-713-2016.
42. Zhang, Y., Chen, J., Liu, H., Chen, Y., Xiao, B., Li, H., Recent advancements of human-centered design in building engineering: A comprehensive review. *J. Build. Eng.*, 84, 108529, 2024.
43. Sathvara, P.B., Anuradha, J., Tripathi, S., Sanjeevi, R., Impact of climate change and its importance on human performance, in: *Insights on Impact of Climate Change and Adaptation of Biodiversity*, 1st, pp. 1–9, Kripa Drishti Publications, 2023.
44. Tiwari, A., Varandani, V., Mandali, S., Arsenault, J., Design of a Human-Centric Auto-Climate Control System for Electric Vehicles. *SAE Int. J. Adv. Curr. Pract. Mobility*, 748–761, 2022.
45. Pioppi, B., Piselli, C., Crisanti, C., Pisello, A.L., Human-centric green building design: the energy saving potential of occupants' behaviour enhancement in the office environment. *J. Build. Perform. Simul.*, 13, 6, 621–644, Nov. 2020, doi: 10.1080/19401493.2020.1810321.
46. Abbasi, T., Sanjeevi, R., Makhija, M., Abbasi, S.A., Role of Vitamins B-3 and C in the Fashioning of Granules in UASB Reactor Sludge. *Appl. Biochem. Biotechnol.*, 167, 2 348–357, May 2012, doi: 10.1007/s12010-012-9691-y.
47. Moreno-Cano, M.V., Santa, J., Zamora-Izquierdo, M.A., Skarmeta, A.F., Future Human-Centric Smart Environments, in: *Modeling and Processing for Next-Generation Big-Data Technologies, vol. 4, Modeling and Optimization in Science and Technologies*, vol. 4, F. Xhafa, L. Barolli, A. Barolli, P. Papajorgji

(Eds.), pp. 341–365, Springer International Publishing, Cham, 2015, doi: 10.1007/978-3-319-09177-8_14.

48. Meoni, A., Vittori, F., Piselli, C., D'Alessandro, A., Pisello, A.L., Ubertini, F., Integration of structural performance and human-centric comfort monitoring in historical building information modeling. *Autom. Constr.*, 138, 104220, 2022.
49. Martin, K.B. and Mills, W., Human centric building control system [J]. *Int. J. Eng. Technol.*, 8, 4, 25–28, 2017.
50. Ramakrishnan, S., Jayaraman, A., Rajesh Kumar, A., B. S., Prashantkumar, C., Dushyant Singh, S., tripathi, *indoor air quality monitoring system*, p. 6310836, Sep. 25, 2023, Accessed: Jan. 30, 2024. [Online]. Available: https://www.registered-design.service.gov.uk/find/6310836.
51. Deng, M., Wang, X., Li, D., Menassa, C.C., Digital ID framework for human-centric monitoring and control of smart buildings. *Build. Simul.*, 15, 10, 1709–1728, Oct. 2022, doi: 10.1007/s12273-022-0902-3.
52. Sarbu, I. and Sebarchievici, C., Aspects of indoor environmental quality assessment in buildings. *Energy Build.*, 60, 410–419, 2013.
53. Mujan, I., An\djelković, A.S., Munćan, V., Kljajić, M., Ružić, D., Influence of indoor environmental quality on human health and productivity-A review. *J. Clean. Prod.*, 217, 646–657, 2019.

13

Enhancing UAV Navigation in Partially Observable 2D Environments: An Optimized Obstacle Avoidance Approach

Jun Jet Tai[1] and Swee King Phang[2]*

[1]*Institute of Future Transport and Cities, Coventry University, Priory Street, CV1 5FB, Coventry, United Kingdom*
[2]*Digital Innovation & Smart Society Impact Lab, Taylor's University, 1 Jalan Taylor's, Subang Jaya, Malaysia*

Abstract
The field of Obstacle Avoidance and Navigation (OAN) algorithms is constantly being researched, with innovations having far-reaching uses. These methods can be generally split into two variants – offline methods and online methods. Offline methods tend to require a prior known map to function but provide goal-reaching guarantees, while online methods forego this need but may result in getting stuck on local minima. Many OAN schemes that currently exist are usually a compromise of either computational efficiency or algorithm robustness. However, when integrating Obstacle Avoidance Navigation (OAN) algorithms into commercially used robots, like multirotor unmanned aerial vehicles (UAVs), the algorithms need to be efficient and robust while also being light enough computationally to operate directly on the UAV. This section introduces one such algorithm, the Closest Obstacle Avoidance and A* (COAA*) algorithm, which addresses the limitations associated with both offline and online algorithms while ensuring it reaches the most optimal solution. This algorithm has been tested on the Heavy Lift Experimental (HLX) UAV at Taylor's University, with the simulation results closely mirroring those obtained in real-world scenarios, underlining its potential for broad application.

***Keywords*:** Permanent COAA, obstacle avoidance and navigation, path planning, multirotor robots

**Corresponding author*: sweeking.phang@taylors.edu.my

Arindam Mondal and Souvik Ganguli (eds.) Controller Design for Industrial Applications, (257–286)

13.1 Introduction

Unmanned aerial vehicles (UAVs), often referred to as "drones," are autonomous flying machines. UAVs like quadcopters, hexacopters, and octacopters, which have multiple rotors, are prime examples. They serve as outstanding platforms for handling payloads and can function in both organized and disorganized settings [1]. This is attributed to the configuration of their propulsion systems which enable the execution of highly precise maneuvers [2]. Furthermore, attributed to their low-cost and scalable designs, usage of UAVs has seen proliferation in many sectors, including photography, cinematography [3], agriculture [4], security [5], delivery [6], search and rescue [7], and various warfare scenarios [8]. In addition, with the rapid advancement of technologies such as 5G networking [9], computer vision [10], and Internet-of-Things (IoT) information backbones [11], UAVs are starting to become an integral part of modern society.

The most common method of operating multi-rotor systems is through Global Positioning System (GPS) enabled waypoint flight over open areas. This enables users to operate these systems easily for a vast majority of tasks. However, for GPS-denied, indoor environments, these UAVs still require trained human operators to assist with the UAV's navigation via a radio control module. With the rapidly growing number of UAVs being deployed, the eventuality of self-navigation within closed environments is unavoidable. For that to happen, scalable and easy-to-implement algorithms for Obstacle Avoidance and Navigation (OAN) must be available. This is especially evident for systems that are meant to work within UAV delivery networks in a city-like environment, or for search and rescue within post-disaster sites or forested regions.

Current systems that possess this capability employ two approaches: offline methods with a preexisting global map or online methods that demand significant computational resources. While both approaches are functional, they may not be practical for a universal algorithm. For instance, offline navigational algorithms work well in city environments but falter when faced with dynamic obstacles like moving vehicles. Additionally, deploying numerous UAVs and routing them offline would require an enormous amount of computational power. On the other hand, for search and rescue UAVs operating in forested regions, generating a global map and then computing a smoothing spline for each rescue operation may not be feasible. Therefore, in such cases, online methods are preferred. However, these methods necessitate high-power and high-cost processors. Hence, there is a need for algorithms that can efficiently run on low-power and low-cost processors, which would significantly facilitate the expansion of UAVs across various markets.

This book chapter seeks to explore the challenges of designing scalable OAN algorithms for operating on UAVs, considering various existing methods such as artificial potential fields [12], A* path search algorithms [13], and dynamics modeling of the UAV [14]. Then, an approach for performing OAN using a combination of various methods is proposed. This method is then extensively tested in simulation, as well as on the Heavy Lift Experimental (HLX) UAV from Taylor's University, Malaysia.

13.2 Background

OAN algorithms are one of the most studied research topics in the field of robotics. While we categorize them into an umbrella term of OAN here, they are generally formed of two distinct components. The first is a pathfinding algorithm, which tends to be the predominant form of algorithm in the offline setting. On the other hand, many online algorithms tend to use obstacle avoidance algorithms, owing to their low computational requirements.

13.2.1 Preliminaries

OAN operates based on a known local map M_L, which is an underexplored variant of an unknown global map M_G.. In this chapter, both maps are represented as occupancy grids $M \in \{0,1\}^{m \times n}$. In the beginning, M_L is initialized as all zeros. As the environment is explored more by the robot, M_L is updated in areas where the robot has been to match M_G, such that as $t \to \infty$, $M_L \to M_G$. This paradigm of map representation is the classic case in Simultaneous Localization and Mapping (SLAM) techniques. Issues concerning inaccurate odometrical data as well as difficulties in obtaining closed loop maps are generally not the concern of OAN, and instead, a task that SLAM is intended to solve.

Canonically, OAN's purpose is to get the robot from its current position P_P to a target position P_T. During traversal of the environment, previously unobserved obstacles may appear, and the OAN algorithm must be able to accommodate these obstacles into its plan. In addition, OAN algorithms should consider vehicle dynamics, as certain maneuvers cannot be accomplished via UAVs. An assumption we make is that the UAV can observe obstacles within a radius R_{obs} of itself, and the information from this observation is immediately added to the local map. This is not necessarily a limiting factor and is easily accomplished with off-the-shelf Lidar systems or 360-degree cameras.

13.2.2 Pathfinding Algorithms

Pathfinding algorithms remain one of the oldest algorithms meant for robotics. The most well-known and well-studied of these algorithms is the A* pathfinding algorithm [15], which itself is an extension of Dijkstra's pathfinding algorithm [16]. A canonical example of a pathfinding algorithm working is shown in Figure 13.1. These algorithms work by viewing the environment as a series of nodes, and then assigning each node a cost associated with the amount of steps required to reach the target. The cost for each node is computed via dynamic programming, and the optimal path is then found by recursively traversing to adjacent nodes with the least cost. In the case of an occupancy grid, every unoccupied cell in the grid is represented as a node. The immediate problem here is that a known global map must be known beforehand. An alternative, more subtle problem is that these algorithms do not consider the influence of the physical dynamics of the robots they are controlling, resulting in velocity and acceleration profiles that are not continuous or may even be impossible for any robot to follow at fixed speeds. For this reason, various works have been done to smooth out the obtained path, from performing smoothing splines to optimal trajectory tracking. The main issue with these techniques is that they tend to be computationally very intensive, requiring the computation to first be done in an offline manner, defeating the notion of a lightweight pathfinding algorithm.

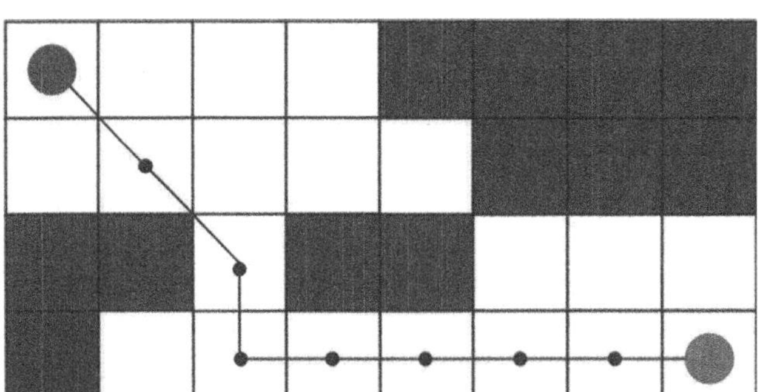

Figure 13.1 Canonical example of an A* pathfinding algorithm [15].

13.2.3 Obstacle Avoidance Algorithms

Obstacle avoidance refers to algorithms that allow robots to avoid obstacles while still maintaining a general heading toward a potentially moving target, with no guarantees on whether it will reach that target. The most common form of obstacle avoidance is simply using an artificial potential field [17]. In this algorithm, obstacles exert an implied repulsive force on the robot, while the target applies an attractive force on the robot. The resulting vector—a sum of the repulsive and attractive force—is subsequently used as a target acceleration vector for the robot. The magnitude of this force varies with distance, with further away obstacles exerting a lower amount of force to have less influence on the robot's trajectory. This is a widely used algorithm that is widely applicable to [12, 18].

On top of having no guarantees for goal reaching, one downside of the artificial potential field method is the failure to produce paths that avoid unnecessary deceleration for UAVs [19]. This is immediately obvious when thinking of an environment with a target, a single obstacle, and the robot, where the positions of all three are colinear. In this case, the attraction and repulsive vector both point in opposing directions, canceling each other out, resulting in the UAV slowing down instead of simply curving around the obstacle. One method to bypass the drawbacks of potential field methods is to utilize artificial vector fields [20]. Vector fields maintain most of the ease of implementation of potential fields, with the ability to form smoother trajectories.

More sophisticated methods of obstacle avoidance include Rapidly exploring Random Trees (RRT) [21–23], which find paths around obstacles by computing a space-filling tree of possible paths around the obstacle to find an optimal one. Sampling-based algorithms can also be used; popular ones include Sampling-based Incremental Planning (SIP) or Probabilistic Roadmaps (PRMs) [24, 25]. For algorithms that consider the robot's dynamics more explicitly, dynamic window approaches are ideal. These algorithms repeatedly generate paths that fit within a dynamically feasible trajectory based on the robot's dynamics capabilities and have parallels with model predictive control (MPC) in that the best trajectory is treated as an optimization problem at every step [26].

13.2.4 Learned Obstacle Avoidance and Navigation

So far, the work discussed here encompasses heuristic and exact methods. There is also a branch of research that utilizes learned, non-heuristic methods to perform OAN, predominantly using neural networks [27, 28].

Some examples utilize deep reinforcement learning for path planning [29], while others utilize supervised learning frameworks for obstacle avoidance [30]. Learned techniques are alluring because they can potentially offer end-to-end solutions for OAN directly from sensor measurement. The immediate downside, however, is the lack of interpretability in the resulting algorithm, as well as the difficulty in scaling the trained models to new sensor suites or different platforms. For these reasons, heuristic or exact methods are the predominant form of OAN algorithms so far, especially in applied settings.

13.3 Overview of the Proposed Method

Here, we propose a heuristic, exact method for performing OAN that is easily understood and configured. Our proposed method is called Closest Obstacle Avoidance and A*, where functionally, it works by aiming to avoid the closest obstacle and is guided by an A* algorithm to formulate paths toward the target position. The work here concerns itself with the 2D setting, where the height of the UAV is held constant. However, an extension to the 3D setting of our algorithm is trivial. The flow chart for COAA* is shown in Figure 13.2, and each component is described in subsequent sections.

13.3.1 Notation

COAA* operates on several variables. The current list of waypoints is denoted as $\boldsymbol{p}_W = \{\boldsymbol{p}_{T_1}, \boldsymbol{p}_{T_2}, \ldots, \boldsymbol{p}_{T_n}\}$, where $\boldsymbol{p} \in \mathbb{R}^2$ defines a position in space. The current location of the UAV is denoted as $\boldsymbol{p}_P \in \mathbb{R}^2$. The local and global maps are denoted as M_L and M_G, respectively, and the UAV's radius of observation is r_{obs}.. We denote vector pointing from entity A to entity B as $\boldsymbol{v}_{A2B}$ and its unit vector as $\boldsymbol{u}_{A2B}$, where A and B can be either the UAV's current position denoted as P,, the target denoted as T, or an obstacle denoted as O. For instance, a vector pointing from the UAV to the target would be written as $\boldsymbol{v}_{P2T}$.

13.3.2 Obstacle Avoidance and Target Heading

In the interest of computational simplicity, the obstacle avoidance component utilizes a blended potential and vector field approach. Consider the case of the UAV with a known target location and a local map with a single occupied cell (a single obstacle). We denote this obstacle position as $\boldsymbol{p}_{O_i}$, denoting the i-th nearest obstacle to the UAV. These positions are

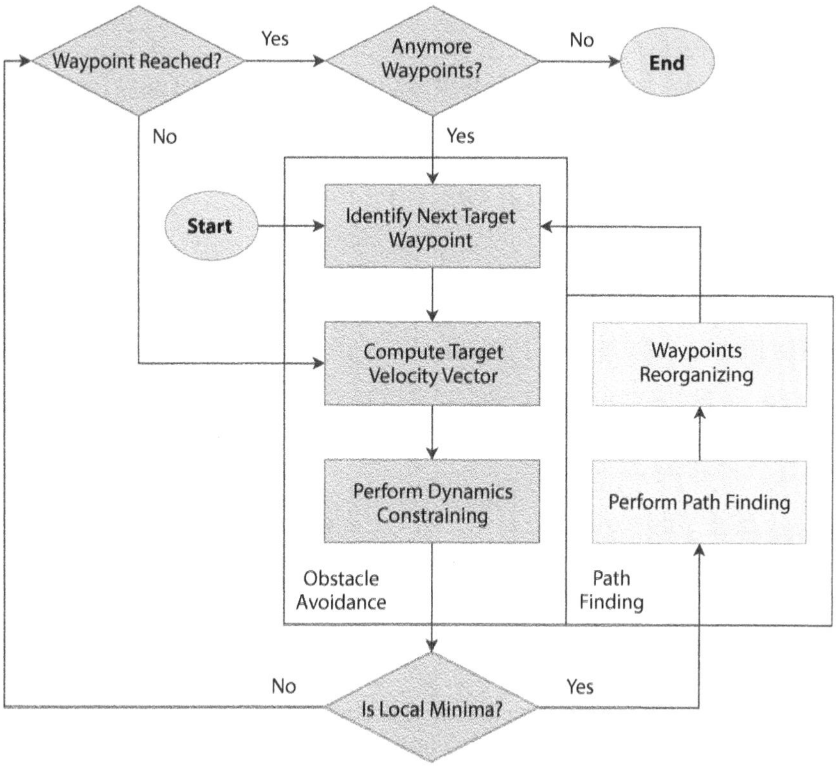

Figure 13.2 Overall architecture of the Closest Obstacle Avoidance and A* (COAA*) algorithm.

sufficient for computing the relevant unit vectors, shown in Equations 13.1 to 13.3, for constructing the vector field.

$$\boldsymbol{u}_{O_i 2T} = \frac{\boldsymbol{p}_T - \boldsymbol{p}_{O_i}}{|\boldsymbol{p}_T - \boldsymbol{p}_{O_i}|} \tag{13.1}$$

$$\boldsymbol{u}_{P2O_i} = \frac{\boldsymbol{p}_{O_i} - \boldsymbol{p}_P}{|\boldsymbol{p}_{O_i} - \boldsymbol{p}_P|} \tag{13.2}$$

$$\boldsymbol{u}_{T2P} = \frac{\boldsymbol{p}_T - \boldsymbol{p}_P}{|\boldsymbol{p}_T - \boldsymbol{p}_P|} \tag{13.3}$$

We then define a vector field that induces a heading vector for the UAV as the unit vector $\boldsymbol{u}_{\mathrm{UAV}_i}$, which is dependent on the angle θ_i between the unit vectors $\boldsymbol{u}_{O_i 2P}$ and $\boldsymbol{u}_{T2O_i}$, as well as the heading to target $\boldsymbol{u}_{T2P}$, as shown in Equations 13.4 and 13.5.

$$\theta_i = -atan2\left(\boldsymbol{u}_{P2O_{ix}}, \boldsymbol{u}_{P2O_{iy}}\right) + atan2\left(\boldsymbol{u}_{O2T_{ix}}, \boldsymbol{u}_{O2T_{iy}}\right) \tag{13.4}$$

$$\boldsymbol{u}_{UAVi} = \begin{cases} \begin{bmatrix} \boldsymbol{u}_{P2O_{iy}} \\ -\boldsymbol{u}_{P2O_{ix}} \end{bmatrix}, & \text{if } 0 < \theta_i < \dfrac{\pi}{2} \\ \begin{bmatrix} -\boldsymbol{u}_{P2O_{iy}} \\ \boldsymbol{u}_{P2O_{ix}} \end{bmatrix}, & \text{if } -\dfrac{\pi}{2} < \theta_i < 0 \\ \begin{bmatrix} \boldsymbol{u}_{T2P_x} \\ \boldsymbol{u}_{T2P_y} \end{bmatrix}, & \text{otherwise} \end{cases} \tag{13.5}$$

This defines the vector field around the position of obstacle i, $\boldsymbol{p}_{o_i}$ like that shown in Figure 13.3.

The unit vector of the vector field induced by obstacle i on the UAV $\boldsymbol{u}_{\mathrm{UAV}_i}$ from Equation 13.5 is only computed for obstacles that are within a certain range of influence D_{infl}. To avoid limit cycles where the UAV continually switches vector fields between two obstacles, the heading vector in Equation 13.5 is computed for the two nearest obstacles to the UAV,

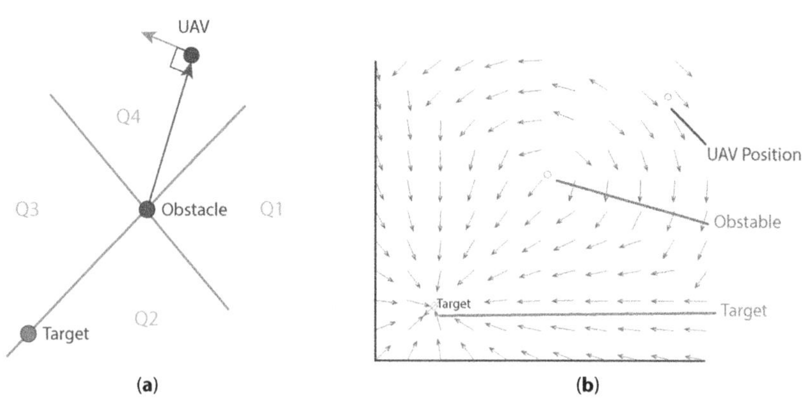

Figure 13.3 Illustration of the vector field around a single obstacle. (a) The quadrants defined by Equation 13.5 and (b) the vector field around an obstacle given the UAV's position.

$\boldsymbol{p}_{o_1}$ and $\boldsymbol{p}_{o_2}$ and a weighted average of both vectors are taken depending on the Euclidean distances between the UAV and the obstacles as shown in Equation 13.6.

$$\boldsymbol{u}_{\text{UAV}_C} = \frac{\boldsymbol{u}_{\text{UAV1}} \cdot D_2 + \boldsymbol{u}_{\text{UAV2}} \cdot D_1}{\left| \boldsymbol{u}_{\text{UAV1}} \cdot D_2 + \boldsymbol{u}_{\text{UAV2}} \cdot D_1 \right|} \tag{13.6}$$

where D_1 and D_2 denotes the Euclidean distance of $\boldsymbol{v}_{P2O_1}$ and $\boldsymbol{v}_{P2O_2}$, respectively. When only one obstacle exists within D_{infl}, we set $\boldsymbol{u}_{\text{UAV}_C} = \boldsymbol{u}_{\text{UAV1}}$,; and when there are none, we set $\boldsymbol{u}_{\text{UAV}_C} = \boldsymbol{u}_{P2T}$.

To avoid having the UAV flying through gaps smaller than an allowable threshold distance D_{prox}, we augment $\boldsymbol{u}_{\text{UAV}_C}$ into $\boldsymbol{u}_{\text{UAV}_T}$ according to Equations 13.7 and 13.8 by smoothly flipping the vector depending on the distance to the obstacle.

$$\boldsymbol{u}_{\text{UAV}_T} = \frac{\boldsymbol{u}_{\text{UAVC}} - \lambda \cdot \boldsymbol{u}_{\text{UAVC}}}{\left| \boldsymbol{u}_{\text{UAVC}} - \lambda \cdot \boldsymbol{u}_{\text{UAVC}} \right|} \tag{13.7}$$

$$\lambda = \begin{cases} \dfrac{D_{\text{prox}}}{D_1}, & \text{if } D < D_{\text{prox}} \\ 0, & \text{otherwise} \end{cases} \tag{13.8}$$

Equations 13.7 and 13.8 form the condition shown in Equation 13.9, effectively forming a push-back vector whenever the UAV is within D_{prox}.

$$\frac{\boldsymbol{u}_{\text{UAV}T}}{\boldsymbol{u}_{\text{UAV}C}} = \begin{cases} 1, & if\ D_{\text{prox}} < D \\ 0, & if\ D_{\text{prox}} = D \\ -1, & if\ D_{\text{prox}} > D \end{cases} \tag{13.9}$$

To promote safer flight, we allow the UAV to have higher flight speeds when far away from obstacles and utilize slower speeds when performing proximity flights. This is achieved by allowing the magnitude to be defined using a potential field. We adopt the sigmoid function defined in Equation 13.10.

$$V_a = V_{\max} \cdot \frac{1}{1+e^{k(-D_1+D_{\text{prox}})}} \tag{13.10}$$

where $V_{\max}$ defines the maximum allowable flight speed of the UAV, and k represents a scaling factor with higher values causing a steeper transition from $V_{\max}$ to 0 velocity and D_{prox} defining the distance from the obstacle from when the transition starts. Visually, the Sigmoid function produces a potential field that is like Figure 13.4. The velocity target of the UAV based on the potential field and vector field is thus shown in Equation 13.11.

$$\boldsymbol{V}_{\text{UAV}_C} = \boldsymbol{u}_{\text{UAV}} \cdot V_a \tag{13.11}$$

Finally, to account for discontinuities in the velocity profile when switching between obstacles, we utilize a first-order low-pass filter defined as in Equation 13.12.

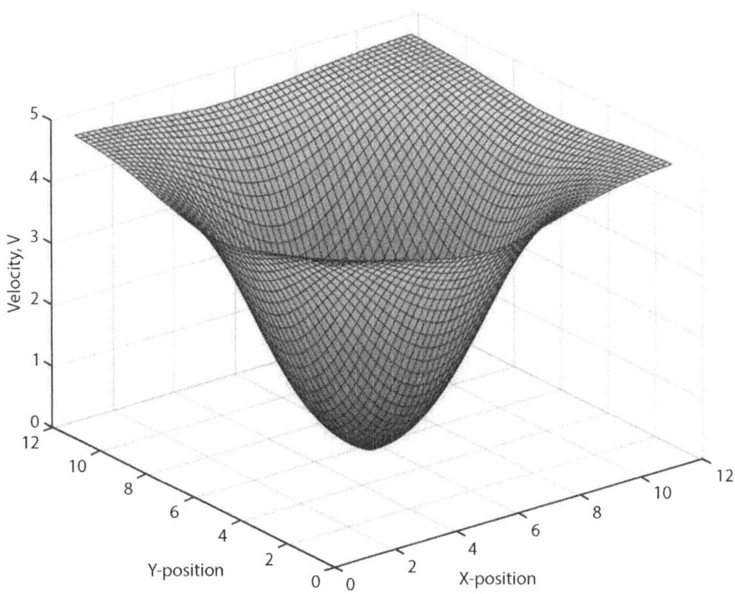

Figure 13.4 Illustrated potential field centered on an obstacle located at (5, 5), with $V_{\max} = 5$, $k = 1$, and $D_{\text{prox}} = 3$.

$$\boldsymbol{V}_{\mathrm{UAV}t} = (\boldsymbol{V}_{UAVC} - \boldsymbol{V}_{UAVt-1})\left(1 - e^{-\frac{\Delta t}{k_t}}\right) + \boldsymbol{V}_{UAVt-1} \tag{13.12}$$

where $\boldsymbol{V}_{\mathrm{UAV}}$ defines the current velocity of the UAV, k_t is the system time constant, and Δt represents the loop time of the algorithm.

When performing this obstacle avoidance and target heading, the known target corresponds to the first waypoint on the list of waypoints $\boldsymbol{p}_w$. Whenever the UAV's distance is less than a predetermined target-reached distance D_{WP}, the waypoint is considered reached and the next waypoint in the list is reselected as the target waypoint.

13.3.3 A* Search Integration

Up until this point, the algorithm has no mechanism for getting rid of local minima—whenever the UAV approaches an obstacle gap, the current plan of action is to just slow to a halt. This behavior is induced by Equation 13.9. This subsection aims to tackle the identified challenge, which necessitates resolving two subsidiary issues. Initially, it involves enabling the UAV to recognize when it is nearing or has encountered a local minimum. Following that, the focus shifts to assisting the UAV in escaping from this local minimum.

The first challenge can be solved using several methods ranging from look-ahead searches to looking at the UAV's velocity profile within a moving average window. Applying look-ahead search implies simulating the UAV flight internally for a certain range at every time step to identify, ahead of time, whenever the UAV is about to reach a minimum. Here, we simply adopt the simpler approach where we look at the velocity profile of the UAV over a time frame of t_{f}. More concisely, we define the time-averaged velocity of the UAV as in Equation 13.13.

$$\bar{V}_{\mathrm{f}} = \frac{1}{t_{\mathrm{f}}}\int_{t-t_{\mathrm{f}}}^{t} | V_{\mathrm{UAV}T} |_2 \, dt \tag{13.13}$$

Whenever $\bar{V}_{\mathrm{f}}$ reaches below a certain threshold $\bar{V}_{\mathrm{f}} < V_{\mathrm{GG}}$, we flag the UAV as getting stuck at a local minimum. We can perform the same calculation for the look-ahead search, where the search is done over future simulated time steps.

To get the UAV out of local minima, we apply a modified A* algorithm. The A* algorithm belongs to a family of techniques known as graph search, wherein a graph is represented as a series of nodes connected by edges, and the goal is to find the best path from a source node to a target node—in this case, the last position in the waypoints list p_w. Here, we view the local map M_L as a series of nodes, where every empty space on the grid is a valid node, and each node is connected to adjacent edges in eight directions—the four cardinal directions and their diagonals. We adopt a formulation that includes a "danger" cost, adopted from Kasim *et al.*, the ground cost at each node is associated with how close the node is to an obstacle. This formulation allows the search algorithm to prioritize wide open spaces, where the UAV can fly at faster speeds as defined in Equation 13.10. Since the A* algorithm only runs on the local map, it is generally expected that the UAV will eventually encounter obstacles again in previously unseen regions. However, our formulation of running A* whenever the UAV gets stuck circumvents this issue and allows the algorithm to be very lightweight by running A* only whenever the UAV gets stuck, as opposed to all the time. By this intuition, each consecutive run of the A* algorithm is run with more information about the world. Therefore, as $t \to \infty$, the UAV is guaranteed to reach the global minimum.

As is, the A* algorithm is incompatible with the obstacle avoidance algorithm because the A* algorithm produces a series of nodes that go from the starting position to the ending location. Canonically, this means that the path is formulated as a list of waypoints $p_W = \{p_{T_1}, p_{T_2}, \ldots, p_{T_n}\}$, where each consecutive waypoint is adjacent to the next. While it is possible to follow this regime of flight, it results in the algorithm being overly sensitive to smaller, previously unseen obstacles on the flight path. We address this challenge by first performing a split-and-merge (SaM) algorithm on the A* algorithm-generated flight path. This algorithm recursively splits the path up into a series of straight-line segments until this series of straight-line segments appropriately approximates the original path. More formally, the SaM algorithm has pseudocode shown in Algorithm 1 and a visual procedure shown in Figure 13.5.

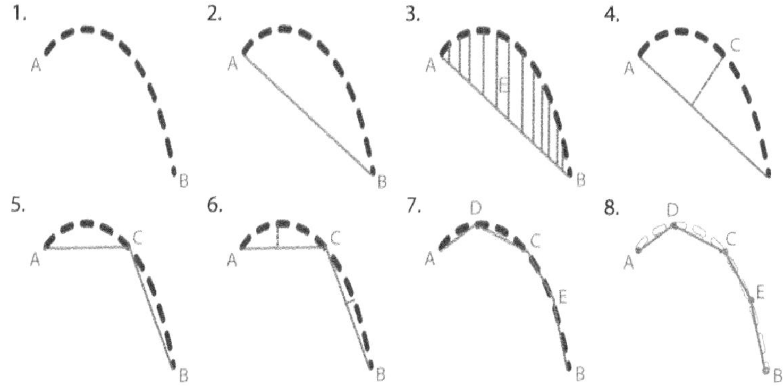

Figure 13.5 Illustration of the split-and-merge algorithm, where a series of waypoints forming a path is split up into an approximation using only 5 points.

Algorithm 1: Split-and-merge algorithm for path splitting.

Input: A list of waypoints $p_i = \{p_1, p_2, \ldots, p_n\}$
Initialize: An index list $I = \{1, n\}$.

while not end:

  for pair of consecutive points i_1, i_2 ***in I:***

    form linear equation $Y = mX + C$ from p_{i_1} and p_{i_2}
    calculate sum of squares between curves:

$$E = \sum_{k=i_1}^{i_2} \left(p_{k_y} - mp_{k_x} + C \right)^2$$

    if $E > \eta_E$:
    insert $i_{\text{mid}} = \text{round}\left(\frac{i_2 - i_1}{2}\right)$ into I between i_1 and i_2:

$$I = \{\ldots, i_1, i_{\text{mid}}, i_2, \ldots\}$$

    continue

  if $i_2 \equiv n$:

    end

13.4 Implementation and Deployment

13.4.1 UAV Platform

To evaluate COAA*'s efficacy in a naïve real-world scenario, the algorithm described in section 3 was implemented on an actual UAV. The UAV itself utilized an Intel UP Squared single-board computer (SBC), connected via UART to a Pixhawk 4 flight controller, both components are shown in Figure 13.6. Perception was enabled via a 2.1-mm ESP USB camera providing mono-visual camera data to the ORB SLAM 2 library (*Mur-Artal et al.*). The Ubuntu 18.04.02 operating system was utilized on the SBC, and COAA* was written in the C++ programming language compiled with MAVROS bindings for communicating with the flight controller.

The UAV platform itself is a custom-built TAROT 650 class quadcopter, shown in Figure 13.7. This platform was chosen for its high endurance and high payload nature, more than sufficient for the given purpose. The 2.1-mm USB ESP camera was gimbal mounted.

13.5 Results

13.5.1 Simulation Performance vs. Real-World Performance

The first set of experiments was done to address the similarities between the simulated and real-world domains. A simulation environment was constructed in MATLAB with a kinematic model for the UAV. The maps were designed with three primary goals—easy to replicate in the real world, does not require pathfinding, and can produce multiple flight scenarios with various hyperparameters. Two different maps were constructed with a waypoint located 15 m ahead of the UAV's takeoff point. These maps are

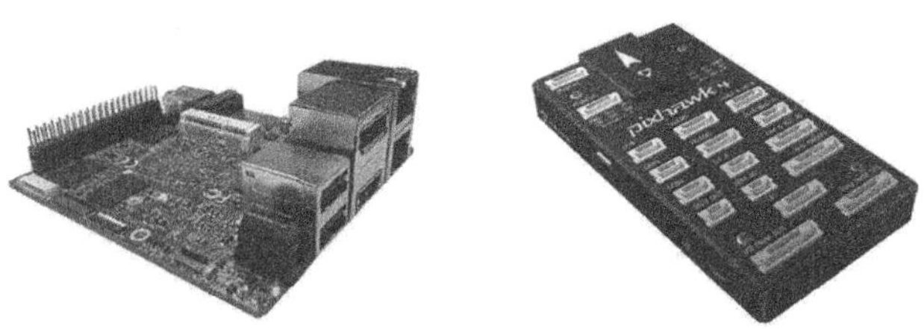

(a) (b)

Figure 13.6 Components used to implement COAA*. (a) Intel® UP Squared companion computer; (b) pixhawk 4 flight controller.

Figure 13.7 The TAROT 650 platform used in the physical experiments.

shown in Figure 13.8. This is the constrained limit within our allowable flight area of 28 m. The maps were flown using different sets of hyperparameters, shown in Table 13.1. Figure 13.9 further shows the images of actual UAV flights, along with GPS-gathered data of the flight path performed for Map 1. Instead of actual obstacles, the use of traffic cones was used to represent obstacles. The 2D trajectories of simulated UAV flights against real UAV flights are shown in Figure 13.10, with the corresponding velocity profiles displayed in Figure 13.11.

Ultimately, both simulated and real-world trajectories show reasonably good correlations except for some drift errors. The source of this drift error is evident from the velocity plots—the actual UAV is struggling to follow the commanded velocity by COAA*'s obstacle avoidance target heading module. As far as performance is concerned, this result can be expected. COAA* is derived without any notion of UAV dynamics modeling.

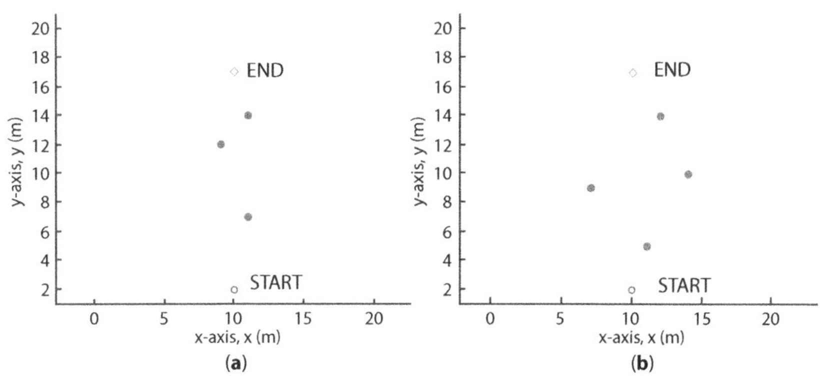

Figure 13.8 Two different maps were used to test real-world vs. simulated performance. (a) Map 1; (b) map 2.

Table 13.1 Two sets of hyperparameters used in real-world and simulated tests.

Hyperparameter	Map 1	Map 2
Proximity allowance, D_{prox} (m)	3	3
Maximum velocity, V_{max} (m/s)	3	1.5
Acceleration time constant, k_t (s)	0.5	0.8

Figure 13.9 Images gathered from actual flight tests. (a) The UAV performing obstacle avoidance while trying to progress; (b) GPS image of the flight path taken.

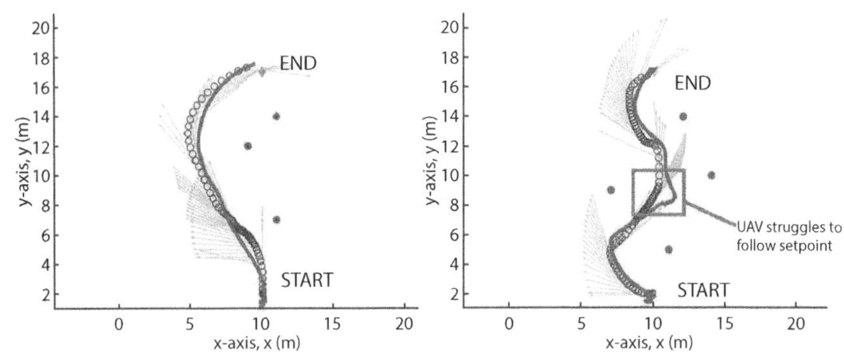

Figure 13.10 2D trajectory of the UAV in the simulated world versus real-world scenarios for both maps using different hyperparameter sets. (a) Map 1, hyperparameter set 1; (b) map 2, hyperparameter set 2.

This results in an accumulation of error when the commanded velocity is unreachable by the onboard flight controller. However, as can be observed from the graphs, the error itself is not severe and can be fine-tuned further. If perfect tracking of the velocity profile is desired, one can always swap out the velocity smoothing module of COAA* in favor of a more sophisticated algorithm, such as MPC or dynamics-constrained velocity profiling.

13.5.2 Sensitivity Analysis of Varying Hyperparameters

The simple comparison between real-world performance and the simulation performance does not tell the full story of the algorithm's performance in more complex environments. To this end, a series of hyperparameter sweeps were performed with the objective of analyzing the algorithm performance from the perspective of various metrics. More concisely, Table 13.2 lists the possible hyperparameters that COAA* contains. With the exception of moving window interval, t_f, we select a range of values for each hyperparameter to define a hyperparameter search grid, shown in Table 13.3. We evaluate this grid of values on four maps representing a range of environments, illustrated in Figure 13.12 and Table 13.4. Several metrics were selected to quantify performance. These metrics are based on actual desired flight characteristics. In no order of importance, the metrics are shown in Table 13.5. A did-not-finish (DNF) scenario occurs when the UAV gets less than 50% of the proximity allowance to any obstacle at any given time, or when A* fails.

The outcome of the sensitivity analysis involves a complex framework, specifically a six-factor, three-level setup, leading to 1,728 distinct interactions examined under four different scenarios or maps. A variety of analytical methods can be applied to this extensive dataset. In this context, a detailed analysis is conducted using a 6-by-3-way analysis of variance (ANOVA), applied at every level for each scenario, to determine the impact of each factor on the outcome metrics. The ANOVA identifies how significantly each factor affects a response metric, quantified by a P-value, with lower P-values indicating a stronger likelihood that the factor significantly impacts the metric. Subsequent to the ANOVA, the study employs a Pearson's correlation analysis to further investigate the relationships within the data. This results in a Pearson's correlation coefficient, which is a measure of the strength and type of association between two variables. A positive correlation denotes a positive correlation between a hyperparameter and the corresponding metric, where a higher value denotes a greater interaction. Conversely, negative values indicate negative interactions.

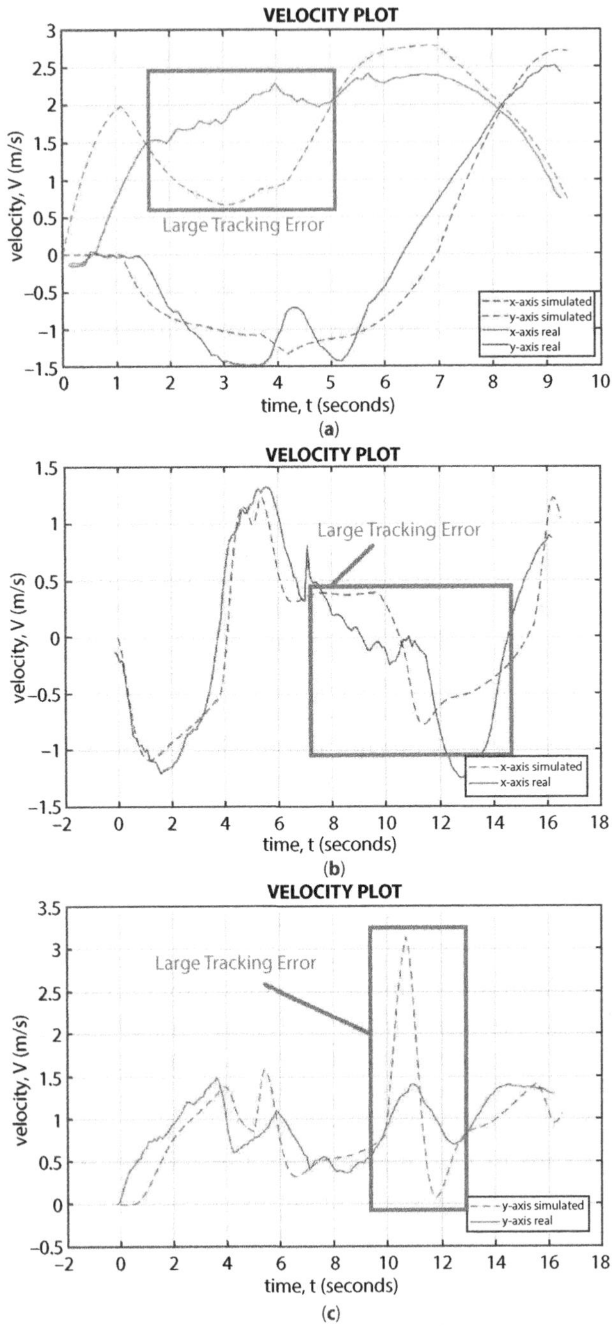

Figure 13.11 Velocity profiles for real-world flights versus simulated flights. (a) x- and y-axis simulated versus real velocity profile for map 1 with hyperparameter set 1; (b) x- and (c) y-axis simulated versus real velocity profile for map 2 with hyperparameter set 2.

Table 13.2 All top-level hyperparameters that determine COAA*'s performance.

Hyperparameter	Symbol	Units	Description
Observation radius	R_{obs}	m	The UAV's observation radius.
Proximity Allowance	D_{prox}	m	Minimum allowable distance between UAV and obstacle.
Maximum velocity	V_{max}	m/s	Maximum absolute commanded flight velocity.
Acceleration time constant	k_t	s	Time constant for acceleration control.
Waypoint reached threshold	D_{WP}	m	Distance between UAV to waypoint to declare waypoint reached.
Moving window interval	t_f	s	Size of moving window for average velocity calculation.
Stuck velocity threshold	V_{GG}	m/s	Minimum average velocity threshold before flagging DIS trigger

The result of these two analytical methods is the visualization of two things—hyperparameter influence significance and type of influence. The results of these studies are shown in Tables 13.6 to 13.9. These tables show the interaction of individual hyperparameters (columns) against the responding metrics (rows). The color of the cell indicates the significance of influence (broken down in Table 13.10), while the value in the cell is Pearson's correlation coefficient. The tables provide a reliable and understandable method for visualizing the overall effect of hyperparameter change on algorithm performance.

From the results, the most influential parameters to algorithm success, in order of descending importance, are the proximity allowance D_{prox}, maximum velocity V_{max}, acceleration time constant k_t, and observation radius R_{obs}. Waypoint reached threshold D_{WP} and stuck velocity threshold V_{GG}. Some interesting points to note include all variations of experiments reaching a DNF whenever $D_{prox} = 5$. Intuitively, this makes sense—when the UAV is prohibited from navigating through gaps smaller than 5 m, it is unlikely that the UAV will progress at all.

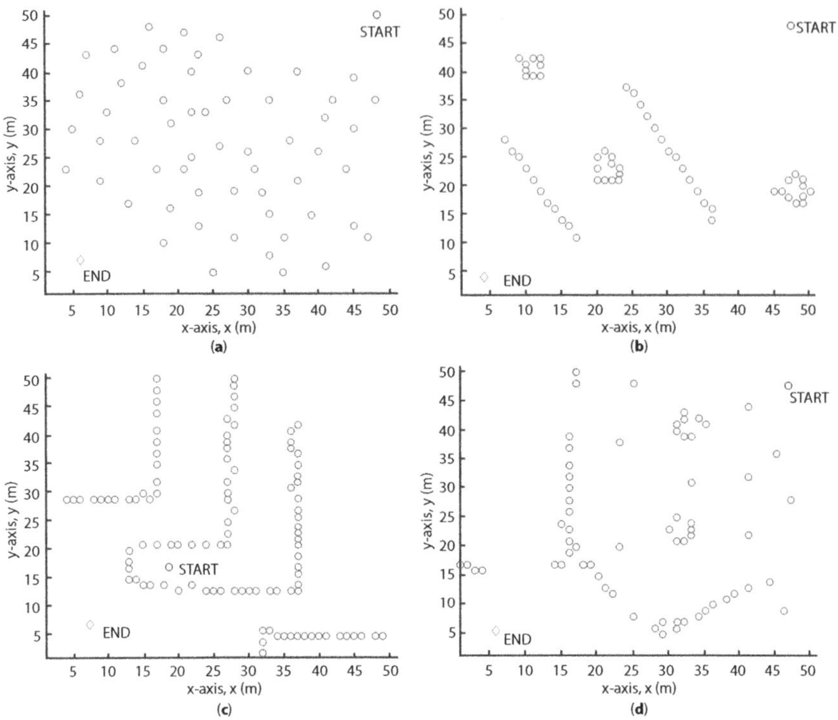

Figure 13.12 Various maps used in the sensitivity analysis. (a) "Sparse". (b) "Two walls". (c) "Corridor". (d) "Mixed".

Table 13.3 The grid of hyperparameters for a sensitivity analysis.

Hyperparameter	Symbol	Range
Observation radius	R_{OBS}	[5, 10, 15, 20]
Proximity allowance	D_{prox}	[1, 3, 5]
Maximum velocity	V_{max}	[3, 5, 7, 10]
Acceleration time constant	k_t	[0.2, 1.0, 2.0, 5.0]
Waypoint reached threshold	D_{WP}	[1, 3, 5]
Stuck velocity threshold	V_{GG}	[0.2, 0.5, 1]

Table 13.4 Map description and purpose for the sensitivity analysis.

Map name	Description	Simulated space
Sparse	Obstacles randomly scattered throughout the map space.	Forested regions
Two Wall	Almost-parallel walls and blocks staggered from the target location to the UAV position.	Urban landscapes
Corridor	Approximately equally spaced connected walls.	Indoor corridors
Mixed	A mixture of all the above.	A complex environment traversing multiple environments

Table 13.5 Metrics studied in the sensitivity analysis.

Performance metric	Symbol	Units	Max.	Min.	Flight characteristic indicator
Mission completion time	T_c	s	$+\infty$	0	Flight speed; Lower is better
Maximum proximity allowance breach	D_{PAB}	-	1	0	Safety; Lower is better
No. of A* runs	n_{A*}	-	$+\infty$	0	Robustness; Lower is better

13.5.3 Repeated Performance Example

This section demonstrates the performance of the algorithm in a more complex, real-world map based on a properly configured set of hyperparameters, shown in Table 13.11. The runs were repeated multiple times and modified between runs to simulate the operation of the UAV in an industrial setting. The starting map used in this experiment is shown in Figure 13.13. In this scenario, we aim to get the UAV from a starting point

Table 13.6 Pearson's correlation for hyperparameter influence against algorithm performance for the sparse map.

	R_{obs}	D_{prox}	V_{max}	k_t	D_{WP}	V_{GG}
T_c	0.0359	0.6271	-0.5290	0.1403	0.0591	0.0483
D_{PAB}	0.0550	0.8056	-0.0700	0.1663	0.0179	0.0020
n_{A*}	0.0367	0.5749	-0.1558	0.0138	0.0839	0.1334

Table 13.7 Pearson's correlation for hyperparameter influence against algorithm performance for the two-wall map.

	R_{OBS}	D_{prox}	V_{max}	k_t	D_{WP}	V_{GG}
T_c	-0.0428	0.7132	-0.5104	0.0248	0.0046	-0.0824
D_{PAB}	-0.0347	-0.3313	0.1357	0.3028	-0.0301	-0.0077
n_{A*}	-0.1747	0.7829	-0.1481	-0.1449	-0.0373	0.0222

Table 13.8 Pearson's correlation for hyperparameter influence against algorithm performance for the corridor map.

	R_{OBS}	D_{prox}	V_{max}	k_t	D_{WP}	V_{GG}
T_c	0.0753	0.8795	-0.3546	-0.1388	-0.1113	-0.0122
D_{PAB}	-0.0852	-0.3452	0.0410	-0.3275	-0.0090	0.0448
n_{A*}	-0.0524	0.8345	-0.1565	-0.2183	-0.0882	0.0607

to an ending point repeatedly while studying the change in behavior in between runs.

A demonstration of all runs is shown in Figures 13.14 and 13.15. In the beginning, the UAV is unaware of its surroundings and spends a large

Table 13.9 Pearson's correlation for hyperparameter influence against algorithm performance for the mixed map.

	R_{OBS}	D_{prox}	V_{max}	k_t	D_{WP}	V_{GG}
T_c	-0.0236	0.7994	-0.4176	-0.0456	-0.0736	-0.0897
D_{PAB}	-0.0177	0.5972	0.0618	0.0831	-0.0096	-0.0109
n_{A*}	-0.0644	0.7380	-0.1456	-0.0871	-0.0374	0.0544

Table 13.10 Legend used for the tables of P-value against the associated colors.

P-value							
1.0 to 0.5	0.5 to 0.25	0.25 to 0.1	0.1 to 0.05	0.05 to 0.01	0.01 to 0.005	0.005 to 0.001	Less than 0.001

Table 13.11 Hyperparameters used for the example of repeated performance.

Hyperparameter	**Symbol**	**Value**
Observation radius	R_{obs}	10
Proximity allowance	D_{prox}	2
Maximum velocity	V_{max}	5
Acceleration time constant	k_t	0.2
Waypoint reached threshold	D_{WP}	3
Stuck velocity threshold	V_{GG}	1

portion of time learning about its environment and finding a way out of the starting room. Eventually, the UAV reaches its goal in a room on the opposite end of the map. It accomplishes this sub-optimally, through a more restrictive section of the map.

In the second run, the UAV is aware of the map from the first run and leverages this information to immediately fly toward the ending goal. In addition, leveraging information gathered from the first run, it reaches the target location while avoiding the tighter corridor in favor of the one with a wide gap, prioritizing safer and faster flights.

We follow up in the third run by walling off the previous two gaps, forcing more exploration of the global map. As can be seen, the UAV attempts to take its previously known path but realizes that the previously open corridor has been blocked off. This results in it adapting to the scenario on the fly and performing A* searches halfway down the traversal. Running the same routine again results in the same behavior seen in the first two runs—the UAV leverages past information to make a more informed flight path, this time flying straight to the target smoothly using the newfound path.

Finally, when we move the ending point to the starting point and start the UAV from a completely different section of the map with the same information, the UAV can utilize all previously learned information to perform navigation from the new starting point to the new ending point, without performing any additional A* searches.

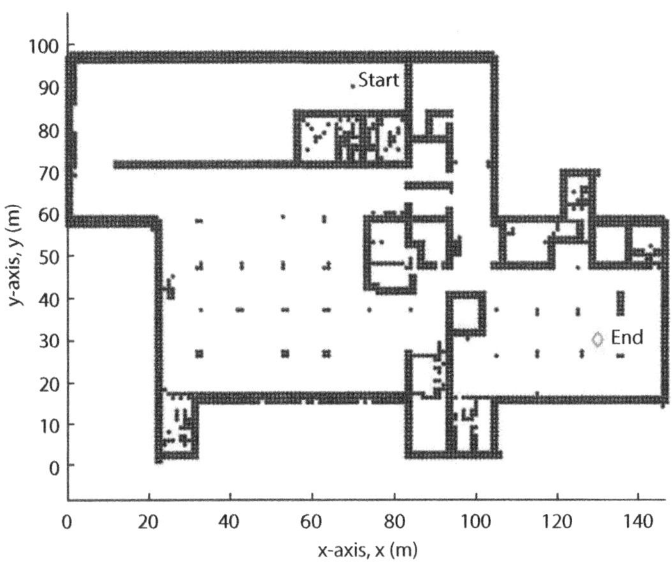

Figure 13.13 The floor plan of an actual factory used in the repeated performance example, showing the starting and ending points.

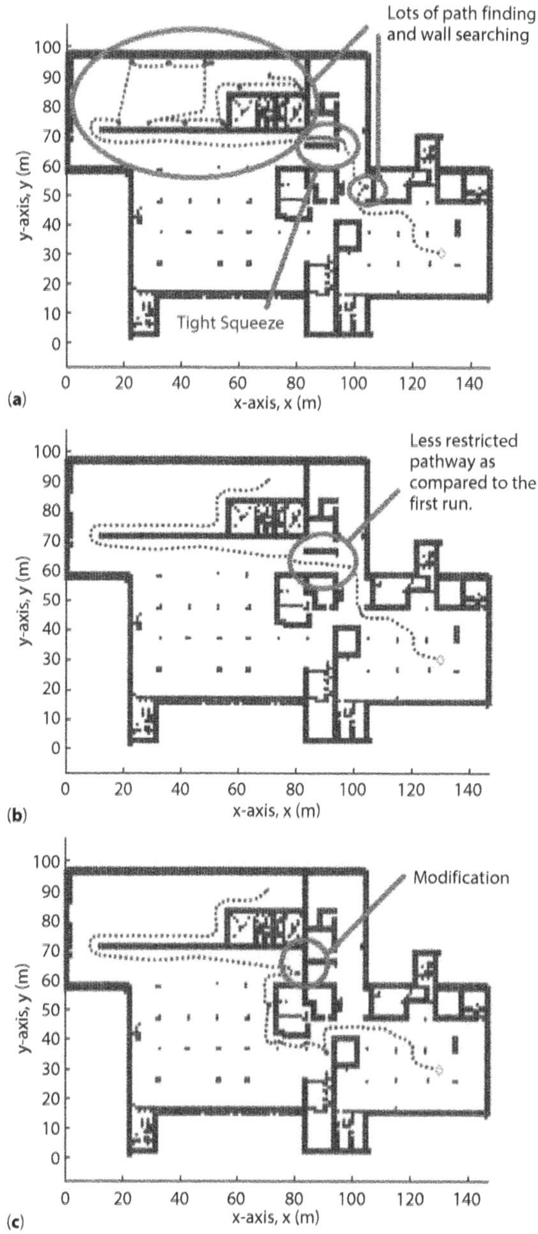

Figure 13.14 First three runs of the repeated runs experiment. (a) The first run in the map, showing the many points where the A* search happened and a traversal through a tight corridor. (b) The second run, showing a much smoother run utilizing past known information and utilization of a wider corridor for faster and safer flight. (c) The third run, where the map has been modified outside of the UAV's knowledge, resulting in the COAA* needing to adapt to the new map.

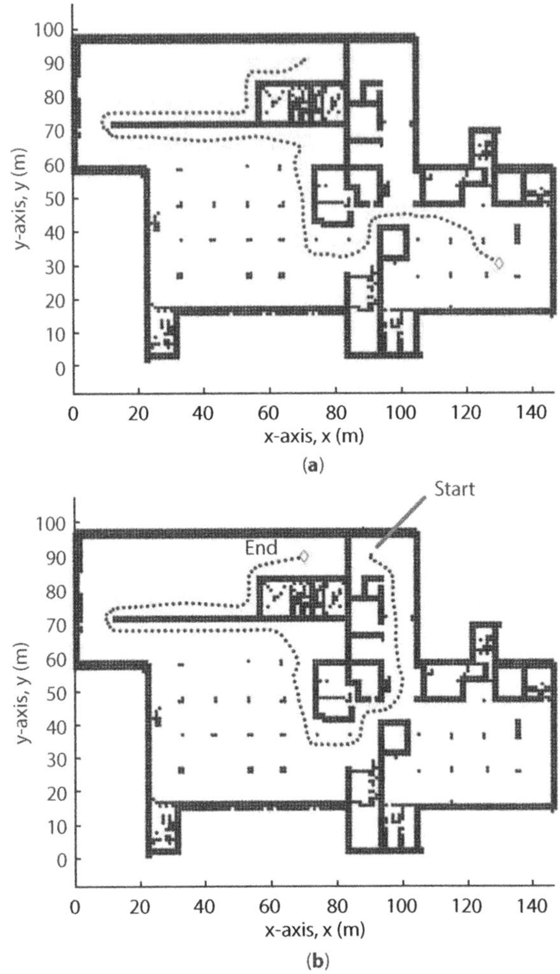

Figure 13.15 Final two runs of the repeated runs experiment. (a) A repeated run in the modified map, showing the UAV adapting to the change and immediately traversing to the end. (b) A run where the starting position is now the new ending position, and the starting position has been shifted into a different room.

13.6 Conclusion

In this work, we introduced various existing OAN algorithms, along with the strengths and weaknesses of each algorithm. To that end, we proposed a novel lightweight and online OAN algorithm that is efficient enough to run onboard a UAV companion computer, robust enough to adapt to environmental changes, and scalable enough for components to be swapped out for more sophisticated ones if needed. The algorithm, termed COAA*, relies on an occupancy grid and outputs direct UAV velocity targets. COAA* was tested in the real world in toy examples, with performances lining up with simulated examples reasonably well. The algorithm was further run using a range of hyperparameters in a grid sensitivity analysis, and the influence of changing various hyperparameters on the performance of the algorithm has been highlighted. Further research would entail utilizing more complex components for each part of the architecture. Most notably, a look-ahead search for stuck detection or Model Predictive Control (MPC) for velocity constraints could significantly enhance the system's robustness and adaptability, enabling more efficient and reliable operation under dynamic and challenging conditions. More obviously, an extension to 3D domains would be a very promising direction that is left unexplored.

References

1. Achtelik, M., Bachrach, A., He, R., Prentice, S., Roy, N., Stereo vision and laser odometry for autonomous helicopters in GPS-denied indoor environments, in: *Unmanned Systems Technology XI*, vol. 7332, Orlando, FL, USA, pp. 336–345, SPIE, 2009, April.
2. Lee, H. and Kim, H.J., Trajectory tracking control of multirotors from modelling to experiments: A survey. *Int. J. Control Autom. Syst.*, 15, 281–292, 2017.
3. Mademlis, I., Mygdalis, V., Nikolaidis, N., Pitas, I., Challenges in autonomous UAV cinematography: An overview, in: *2018 IEEE International Conference on Multimedia and Expo (ICME)*, IEEE, pp. 1–6, 2018, July.
4. Phang, S.K., Chiang, T.H.A., Happonen, A., Chang, M.M.L., From Satellite to UAV-based Remote Sensing: A Review on Precision Agriculture. *IEEE Access*, 11, 127057–127076, 2023.
5. Abro, G.E.M., Zulkifli, S.A.B., Masood, R.J., Asirvadam, V.S., Laouti, A., Comprehensive review of UAV detection, security, and communication advancements to prevent threats. *Drones*, 6, 10, 284, 2022.
6. Dissanayaka, D., Wanasinghe, T.R., De Silva, O., Jayasiri, A., Mann, G.K., Review of Navigation Methods for UAV-Based Parcel Delivery, in: *IEEE Transactions on Automation Science and Engineering*, 2023.
7. Silvagni, M., Tonoli, A., Zenerino, E., Chiaberge, M., Multipurpose UAV for search and rescue operations in mountain avalanche events. *Geomatics Nat. Hazards Risk*, 8, 1, 18–33, 2017.
8. Lushenko, P., Review of the special issue 'robotics autonomous systems and warfare,' Small Wars and Insurgencies 31, 4 June 2020. *Small Wars Insurgencies*, 32, 3, 550–570, 2021.

9. Ullah, H., Nair, N.G., Moore, A., Nugent, C., Muschamp, P., Cuevas, M., 5G communication: An overview of vehicle-to-everything, drones, and healthcare use-cases. *IEEE Access*, 7, 37251–37268, 2019.
10. Tai, J.J., Innocente, M.S., Mehmood, O., FasteNet: a fast railway fastener detector, in: *Proceedings of Sixth International Congress on Information and Communication Technology: ICICT 2021*, London, vol. 1, Springer Singapore, Singapore, pp. 767–777, 2021, September.
11. Ray, P.P. and Nguyen, K., A review on blockchain for medical delivery drones in 5G-IoT era: Progress and challenges, in: *2020 IEEE/CIC International Conference on Communications in China (ICCC Workshops)*, IEEE, pp. 29–34, 2020, August.
12. Chen, Y.B., Luo, G.C., Mei, Y.S., Yu, J.Q., Su, X.L., UAV path planning using artificial potential field method updated by optimal control theory. *Int. J. Syst. Sci.*, 47, 6, 1407–1420, 2016.
13. Liu, X. and Gong, D., A comparative study of A-star algorithms for search and rescue in perfect maze, in: *2011 International Conference on Electric Information and Control Engineering*, IEEE, pp. 24–27, 2011, April.
14. Wang, B., Dong, X., Chen, B.M., Lee, T.H., Phang, S.K., Formation flight of unmanned rotorcraft based on robust and perfect tracking approach, in: *2012 American Control Conference (ACC)*, IEEE, pp. 3284–3290, 2012.
15. Duchoň, F., Babinec, A., Kajan, M., Beňo, P., Florek, M., Fico, T., Jurišica, L., Path planning with modified a star algorithm for a mobile robot. *Procedia Eng.*, 96, 59–69, 2014.
16. Wang, H., Yu, Y., Yuan, Q., Application of Dijkstra algorithm in robot path-planning, in: *2011 Second International Conference on Mechanic Automation and Control Engineering*, IEEE, pp. 1067–1069, 2011, July.
17. Khatib, O., The potential field approach and operational space formulation in robot control, in: *Adaptive and Learning Systems: Theory and Applications*, pp. 367–377, Springer US, Boston, MA, 1986.
18. Sun, J., Tang, J., Lao, S., Collision avoidance for cooperative UAVs with optimized artificial potential field algorithm. *IEEE Access*, 5, 18382–18390, 2017.
19. Mac, T.T., Copot, C., Hernandez, A., De Keyser, R., Improved potential field method for unknown obstacle avoidance using UAV in indoor environment, in: *2016 IEEE 14th International Symposium on Applied Machine Intelligence and Informatics (SAMI)*, IEEE, pp. 345–350, 2016, January.
20. Chen, H., Chang, K., Agate, C.S., UAV path planning with tangent-plus-Lyapunov vector field guidance and obstacle avoidance. *IEEE Trans. Aerosp. Electron. Syst.*, 49, 2, 840–856, 2013.
21. Wei, K. and Ren, B., A method on dynamic path planning for robotic manipulator autonomous obstacle avoidance based on an improved RRT algorithm. *Sensors*, 18, 2, 571, 2018.
22. Chen, L., Mantegh, I., He, T., Xie, W., Fuzzy kinodynamic RRT: a dynamic path planning and obstacle avoidance method, in: *2020 International*

Conference on Unmanned Aircraft Systems (ICUAS), IEEE, pp. 188–195, 2020, September.

23. Yang, F., Fang, X., Gao, F., Zhou, X., Li, H., Jin, H., Song, Y., Obstacle avoidance path planning for UAV based on improved RRT algorithm. *Discrete Dyn. Nat. Soc.*, 2022, 1–9, 2022.
24. Karaman, S. and Frazzoli, E., Incremental sampling-based algorithms for optimal motion planning. *Rob. Sci. Syst. VI*, 104, 2, 267–274, 2010.
25. Castro, L., II, Chaudhari, P., Tůmová, J., Karaman, S., Frazzoli, E., Rus, D., Incremental sampling-based algorithm for minimum-violation motion planning, in: *52nd IEEE Conference on Decision and Control*, IEEE, pp. 3217–3224, 2013, December.
26. Kunchev, V., Jain, L., Ivancevic, V., Finn, A., Path planning and obstacle avoidance for autonomous mobile robots: A review, in: *Knowledge-Based Intelligent Information and Engineering Systems: 10th International Conference, KES 2006*, Bournemouth, UK, October 9-11, 2006, Springer Berlin Heidelberg, 2006.
27. Back, S., Cho, G., Oh, J., Tran, X.T., Oh, H., Autonomous UAV trail navigation with obstacle avoidance using deep neural networks. *J. Intell. Rob. Syst.*, 100, 1195–1211, 2020.
28. Singla, A., Padakandla, S., Bhatnagar, S., Memory-based deep reinforcement learning for obstacle avoidance in UAV with limited environment knowledge. *IEEE Trans. Intell. Transp. Syst.*, 22, 1, 107–118, 2019.
29. Tai, J.J., Wong, J., Innocente, M., Horri, N., Brusey, J., Phang, S.K., *PyFlyt–UAV Simulation Environments for Reinforcement Learning Research*, 2023, arXiv preprint arXiv:2304.01305.
30. Sedaghat-Pisheh, H., Rivera, A.R., Biaz, S., Chapman, R., Collision avoidancc algorithms for unmanned aerial vehicles using computer vision. *J. Comput. Sci. Colleges*, 33, 2, 191–197, 2017.

14

Fast Inner and Outer Dynamics Control of Multi-Rotor UAVs with Novel SIPIC and RPT Controllers Design

Swee King Phang[1]* and Jun Jet Tai[2]

[1]*Digital Innovation & Smart Society Impact Lab, Taylor's University, 1 Jalan Taylor's, Subang Jaya, Malaysia*
[2]*Institute of Future Transport and Cities, Coventry University, Priory Street, CV1 5FB, Coventry, United Kingdom*

Abstract
The industry application of racing-grade multi-rotors remains largely unexplored, with a primary focus on optimizing system actuation while minimizing overshoot and oscillations. The control of these machines is commonly a series of cascaded PID control loops with predominantly hand-tuned control parameters. This chapter introduces a new control scheme codenamed steady-state integral proportional integral controller (SIPIC) derived from steady-state integral techniques used in motor speed controllers, aiming to establish an innovative technique of control specifically for racing-grade multi-rotors. The proposed controller's performance is evaluated using a multi-rotor nonlinear dynamic model simulator. Additionally, the outer dynamics of the multi-rotor is governed by the robust and perfect tracking (RPT) controller to ensure accurate tracking of position, velocity, and acceleration set points with minimal errors even in the presence of disturbances. The controllers demonstrated remarkable robustness and performance. This chapter also showcases their intuitive and straightforward tuning process.

***Keywords*:** Unmanned aerial vehicles, multi-rotor UAV, SIPIC, RPT, controller design

**Corresponding author*: sweeking.phang@taylors.edu.my

Arindam Mondal and Souvik Ganguli (eds.) Controller Design for Industrial Applications, (287–308)

14.1 Introduction

Multi-rotor systems, comprising rotors in a parallel plane with fixed pitch, have garnered significant attention due to their mechanical simplicity compared to traditional helicopters. Despite their inherent instability, the orientation of these rotors imparts exceptional agility, enabling them to execute quick and precise maneuvers [1]. As a result, multi-rotor systems find versatile applications as robotics platforms in both unstructured and structured environments [2], excelling in payload manipulation [3].

In recent times, high-performance, racing-grade quadrotors have emerged as a prominent niche within the field. Beyond their captivating freestyle flying category, these vehicles exhibit remarkable translational velocities that have practical implications in rescue missions, civil defense, surveillance, and emergency scenarios. Consequently, the racing quadrotor industry is experiencing a surge, with enthusiasts assembling customized quad rotors using readily available components from diverse manufacturers [4].

Numerous control schemes for inner and outer-loop control of these vehicles have been developed. Classical, modern, and artificial intelligence-based methods, such as $H\infty$ PID [5], LQR and LQG [6], learned neural network–based control [7], and nonlinear control [8], have demonstrated effectiveness in governing the attitude of multi-rotors.

Despite the availability of sophisticated control methods, the racing quadrotor industry still predominantly favors the traditional cascaded proportional-proportional integral derivative (PPID) controller. Its widespread adoption can be attributed to its ease of implementation and architectural simplicity, requiring minimal knowledge of the actual system's mathematical model [9]. However, achieving desired performance levels often necessitates manual tuning by end users.

This book chapter aims to explore the challenges of tuning controllers for racing quadrotors, considering various methods such as the Ziegler-Nichols step response tuning rules [10] and IMC-PID [11] tuning. Additionally, we will present an alternative approach, the steady-state integral analytic method [4], which has proven effective in speed control of direct current motors, and discuss its potential application to multi-rotor systems, together with a robust and perfect tracking (RPT) controller for computing set points for outer-loop control [12].

By examining the Newton-Euler mathematical model, we propose a novel steady-state integral method to design a general controller specifically tailored to multi-rotor systems. This approach allows for the joint development of proportional and integral gains, considering a single system constant. By eliminating the derivative component of a conventional PID controller, the resulting controller requires less computation per cycle yet offers enhanced robustness and performance, which can be simply tuned with just a single parameter.

This chapter seeks to shed light on the unique characteristics and control challenges of multi-rotor systems, with a particular focus on the racing quadrotor industry. The proposed controller design methods present a promising avenue for improving performance while streamlining the tuning process.

14.2 Dynamic Model of Quadrotor UAV

The block diagram representation of the mathematical control system dynamics for a multi-rotor system, specifically a quadrotor, is illustrated in Figure 14.1. It forms the foundation for the control algorithm to be discussed. The dynamics model of the system comprises two parts: the linear acceleration dynamics is governed by an outer-loop controller, while the angular acceleration dynamics is governed by an inner-loop controller. In most implemented circumstances, the outer-loop controller steps at a much lower loop rate than the inner-loop controller, typically by an order of magnitude. This architecture results in a system whereby the three rotational axes are handled by the inner-loop attitude controller, while all three linear axes are governed by the outer-loop controller.

Figure 14.2 provides an overview of the physical dynamics of a quadrotor model. The inputs to the drone, δ_n, represent normalized motor control signals transmitted from the onboard flight control hardware. The output of the system comprises the linear velocity $V_b = \left[u, v, w\right]^T$, linear position $P_g = \left[x, y, z\right]^T$, Euler angles $\Theta = \left[\varphi, \theta, \psi\right]^T$, and angular velocity $\omega = \left[p, q, r\right]^T$. These outputs will be utilized for feedback control in later stages, with the control of linear position/velocity typically termed position control, while angular position/velocity control is coined attitude control. It should be noted that the mathematical derivation assumes a quadrotor UAV with motors oriented in an X configuration, with rotor numbers 1, 2, 3, and 4, indicating front-right, back-right, back-left, and front-left motors, respectively.

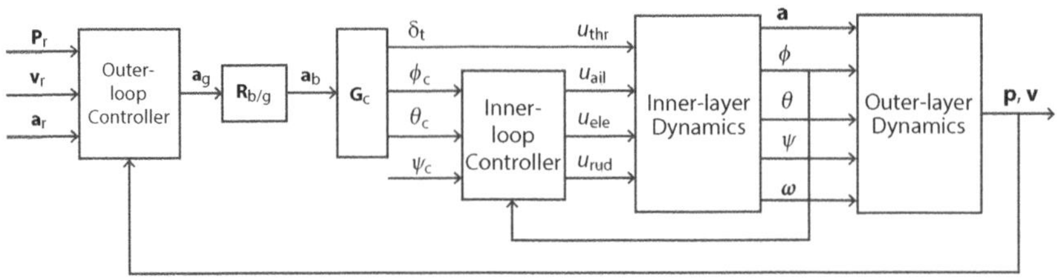

Figure 14.1 Dual-loop control structure of the quadrotor UAV [4].

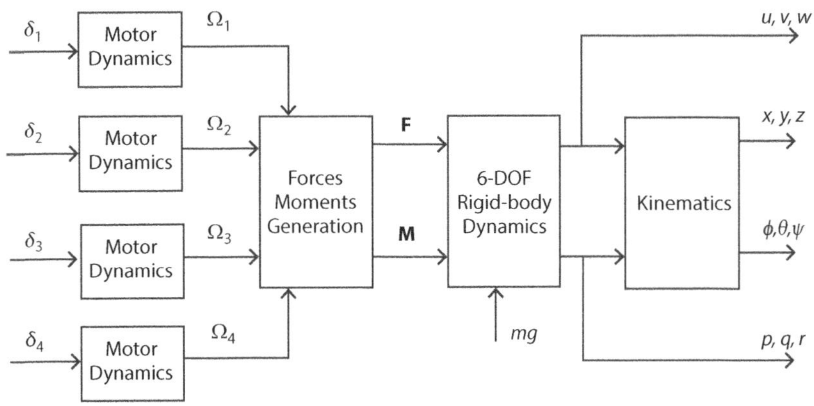

Figure 14.2 Physical system dynamics of the quadrotor UAV model, starting with motor commands on the left and resulting in systems states on the right [4].

14.2.1 Kinematic and Rigid-Body Dynamics

One important aspect of robot dynamics modeling is the frames in which the robot is moving. In our derivation, the movement of the quadrotor UAV will be described in two frames: the ground frame and the body frame. The ground frame is a frame of reference that is affixed on the ground in an XYZ-NED configuration, where the x-axis points North, y-axis East, and z-axis downward. The body frame is fixed on the UAV's body, with the x-axis pointing toward the nose of the UAV, the y-axis toward the right of the UAV, and the z-axis toward the belly of the UAV.

Equations 14.1 and 14.2 depict the well-known frame-of-reference conversion equations for converting motions in one frame of reference to another. These equations are well documented [13] and, therefore, will not be discussed in this book chapter.

$$\dot{P}_g = R_{g/b} V_b \tag{14.1}$$

$$\dot{\theta} = S^{-1} \omega \tag{14.2}$$

where the $R_{g/b}$ is the ground frame to body frame 3D rotation matrix, and the S^{-1} is the lumped transformation matrix, given by

$$R_{g/b} = \begin{bmatrix} c_\theta c_\psi & s_\theta s_\phi c_\psi - c_\phi s_\psi & s_\theta c_\phi c_\psi + s_\phi s_\psi \\ c_\theta s_\psi & s_\theta s_\phi c_\psi + c_\phi s_\psi & s_\theta c_\phi c_\psi - s_\phi s_\psi \\ -s_\theta & c_\theta s_\phi & c_\theta c_\phi \end{bmatrix} \tag{14.3}$$

$$S^{-1} = \begin{bmatrix} 1 & t_\theta s_\phi & t_\theta c_\phi \\ 0 & c_\phi & -s_\phi \\ 0 & s_\phi / c_\theta & c_\phi / c_\theta \end{bmatrix} \tag{14.4}$$

with s_*, c_*, t_* corresponds to $\sin(*), \cos(*), \tan(*)$, respectively.

Translational and rotational dynamics of any rigid body can be formulated using Newton-Euler's formalism as

$$m\dot{V}_b + \omega \times (mV_b) = F \tag{14.5}$$

$$J\dot{\omega} + \omega \times (J\omega) = M \tag{14.6}$$

where F and M are the resultant forces and moments in three axes acting on the quadrotor's center of mass. Here, we assume that the quadrotor is symmetrical. Thus, the inertia matrix of the UAV, J, is assumed to be a diagonal matrix

$$J = \begin{bmatrix} J_x & 0 & 0 \\ 0 & J_y & 0 \\ 0 & 0 & J_z \end{bmatrix} \tag{14.7}$$

14.2.2 Forces and Moments Generation

The quadrotor is affected by external forces and moments in addition to those generated by the UAV itself. These forces and moments are primarily derived from gravity, rotor thrust, and fuselage drag [10, 14]. In practical applications, the drone's air speed can be approximated to be proportional to its ground speed measured via an onboard GPS. These forces and moments provide the input necessary for the quadrotor to move longitudinally or laterally. The gravitational force acting in the ground-frame z-axis, directly downward, can be expressed as

$$F_{gravity} = R_{b/g} \begin{bmatrix} 0 \\ 0 \\ mg \end{bmatrix} = \begin{bmatrix} -mgs_\theta \\ mgc_\theta s_\phi \\ mgc_\theta c_\phi \end{bmatrix} \tag{14.8}$$

The main source of the forces and moments experienced by the quadrotor is derived from aircraft propulsion. As conventional multi-rotors have smaller propellers where the hinge of the rotor is rigid, it is typically assumed that there are no rotor flapping dynamics. The thrust T_i and torques Q_i produced by the rotor $i = n$ can be modeled as

$$T_n = \frac{1}{4\pi^2} C_T \rho (2r)^4 \Omega_n^2 \tag{14.9}$$

$$Q_n = \frac{1}{4\pi^2} C_Q \rho (2r)^5 \Omega_n^2 \tag{14.10}$$

where C_T and C_Q are the thrust and torque coefficients of the propeller, ρ is the density of the air, and r is propeller radius. The aerodynamic coefficients can be assumed constant since the propellers are fixed pitch, leading to the following simplified expressions:

$$T_n = k_T \Omega_n^2 \tag{14.11}$$

$$Q_n = k_Q \Omega_n^2 \tag{14.12}$$

The identification of the constant k_T and k_Q can be done experimentally. A detailed experiment to estimate the constants can be found in [10] and therefore will not be further discussed here.

Finally, the penultimate force and moment vectors acting on the center of mass of the quadrotor because of the four motors can be formulated as

$$F_{rotor} = \begin{bmatrix} 0 \\ 0 \\ -(T_1 + T_2 + T_3 + T_4) \end{bmatrix} \tag{14.13}$$

$$M_{rotor} = \begin{bmatrix} \frac{\sqrt{2}}{2} l(-T_1 + T_2 + T_3 - T_4) \\ \frac{\sqrt{2}}{2} l(T_1 + T_2 - T_3 - T_4) \\ Q_1 - Q_2 + Q_3 - Q_4 \end{bmatrix} \tag{14.14}$$

To achieve racing-grade performance, a quadrotor traveling at up to 20 m/s must be considered. For this consideration, drag force due to air speed is assumed to follow the following formula:

$$F_{drag} = \begin{bmatrix} -b_w & 0 & 0 \\ 0 & -b_w & 0 \\ 0 & 0 & 0 \end{bmatrix} \tag{14.15}$$

where b_w is the drag coefficient, identified through a wind tunnel test of the quadrotor.

14.2.3 Motor Dynamics

Brushless electric motors operating out of startup range can be modeled as a second-order system. However, the electrical dynamics of these systems tend to be much faster than their mechanical dynamics. Therefore, motor dynamics can sufficiently be modeled as a first-order system [15]:

$$\dot{\Omega}_n = \frac{1}{\tau_m}\left[k_m\left(\delta_n - \delta_n^*\right) - \Omega_n\right] \tag{14.16}$$

where k_m and τ_m are the steady-state gain and time constants of the motor, which can be obtained experimentally. δ_n^* defines the motor's normalized input trim, while δ_n is the normalized motor speed command, which remains constant at $\delta_n \in \{0,1\}$.

14.3 Control Strategy for High-Speed Maneuver

The dynamic structure of the quadrotor is separated into orientation and position control loops, where the orientation control runs at a much higher rate than the position control and therefore the terms inner loop and outer loop. Combining Equations 14.6, 14.7, and 14.14 allows for the attainment of a nonlinear relationship between the inputs to the system and its angular accelerations:

$$\begin{bmatrix} J_x\dot{p} \\ J_y\dot{q} \\ J_z\dot{r} \end{bmatrix} = u + \begin{bmatrix} \left(J_y - J_z\right)rq \\ \left(J_z - J_x\right)pr \\ \left(J_x - J_y\right)pq \end{bmatrix} \tag{14.17}$$

where u is the normalized moment commanded to the system. r, p, q represents the angular rate of the quadcopter and can be measured directly onboard IMU, an augmented input to the system, $\bar{u}$, can be formulated as

$$\bar{u} = u + \begin{bmatrix} (J_y - J_z)rq \\ (J_z - J_x)pr \\ (J_x - J_y)pq \end{bmatrix} \tag{14.18}$$

This is essentially a feedback linearization of the system. It is a reasonable substitution if the quadrotor provides a low-noise estimation of its angular rates. This is typically achieved via an extended Kalman filter (EKF).

14.3.1 SIPIC Orientation Controller Design

To adopt steady-state control theory for quadcopter controller design, we assume that the quadcopter is influenced by a constant external disturbance, T, Equations 14.17 and 14.18 are reformulated as

$$\begin{bmatrix} J_x\dot{p} \\ J_y\dot{q} \\ J_z\dot{r} \end{bmatrix} = J\dot{\omega} = \bar{u} + T \tag{14.19}$$

Equation 14.19 shows a linearized inner-loop system for all three axes, where each axis is completely decoupled because of the feedback linearization of Equation 14.18. Therefore, it is safe to implement individual controllers for each axis by considering the quadrotor rate controller:

$$\bar{u} = K_p e_\omega + K_i q_i \tag{14.20}$$

Here, e_ω and q_i represent the angular velocity error from the set point and its integral state, respectively. In steady-state conditions, it is optimal for the quadrotor to accurately reach the designated set point with no residual steady-state error, where the angular velocity and the error in angular position tend toward zero, i.e., $\dot{\omega} \to 0, e_\omega \to 0$. Additionally, it is preferable for the integral action q_i to entirely offset any deficiency in proportional action, influenced by K_i parameter. Therefore, the output of the controller in Equation 14.20 at steady-state is

$$\bar{u}_{ss} = K_i q_{ss} = -T \tag{14.21}$$

where q_{ss} is the steady-state integral controller output. This can be substituted into Equation 14.19 to produce

$$K_i q_{ss} = \bar{u} - J\dot{\omega} \tag{14.22}$$

The integral state component mentioned in reference [16] offers a practical criterion, ensuring that the final error value of a system reaches zero. This equation is appropriately documented, and its Laplace domain representation is expressed as follows:

$$\frac{q_{ss}}{s} - Q(s) = As^n f(s) + B \tag{14.23}$$

where A and B are arbitrary constants, and $f(s)$ is the system function described as

$$K_i(q_{ss} - q_i) = sq_i \tag{14.24}$$

Applying Laplace transform on Equations 14.20 and 14.22 assuming zero initial values, we have

$$q_i = \left(\frac{1}{s}\right)\left(K_p e_\omega - J\omega s\right) \tag{14.25}$$

The suggested integral component block diagram of the rate proportional-integral (PI) controller is depicted in Figure 14.3. The controller can be further simplified by solving for the controller's integral component in the time domain, where we can obtain

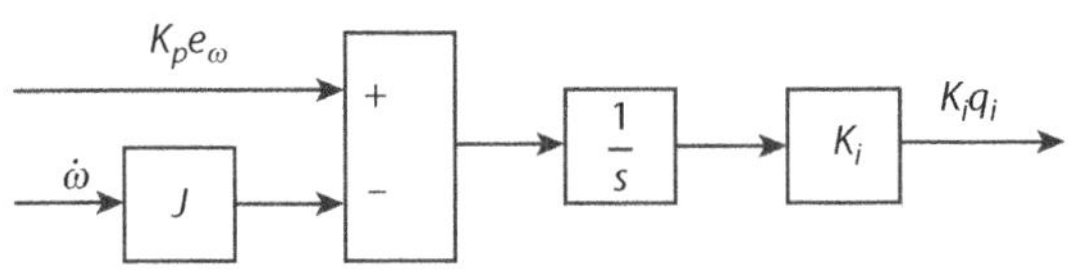

Figure 14.3 PI integral component proposed [4].

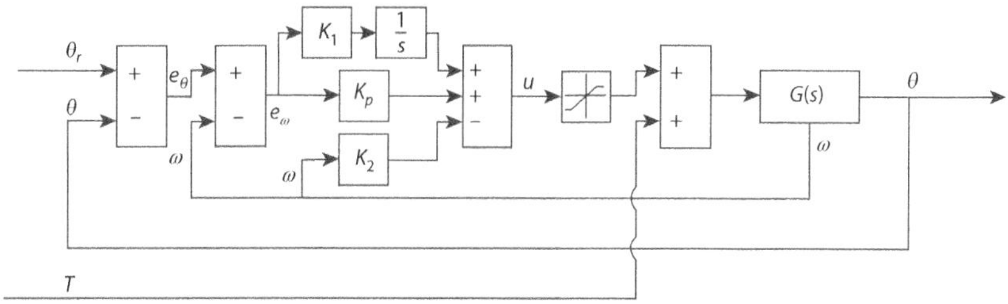

Figure 14.4 Finalized attitude control block diagram [4].

$$u = K_p e_\omega + K_i K_p \int e_\omega dt - K_i J \omega \tag{14.26}$$

The dynamics controller resulting from Equation 14.26 is illustrated in Figure 14.5. The associated transfer function of the closed-loop controller is:

$$\theta = \frac{\frac{1}{J} Ts + \frac{1}{J} K_p (s + K_i)\theta_r}{s^3 + s^2\left(\frac{J_m}{J} K_i + \frac{1}{J} K_p\right) + s\left(\frac{1}{J} K_p K_i + \frac{1}{J} K_p\right) + \frac{1}{J} K_p K_i} \tag{14.27}$$

In an ideal scenario, where system identification is performed perfectly, $J_m = J$. Here, θ_r represents the angular reference for the quadrotor system. Referring to Figure 14.4, the constants are defined as follows: $K_1 = K_p K_i$ and $K_2 = K_i J_m$.

14.3.2 Inner-Loop Stability Analysis

Analysis of any linear time-invariant system can be done by looking at the poles of the system transfer function. This section explores two distinct scenarios of the system: one under the ideal condition where $J_m = J$, and another in a less-than-ideal scenario where $J_m \approx J$.

First, when $J_m = J$, the closed-loop characteristic equation can be obtained as

$$c(s) = (s + K_i)\left(s^2 + \frac{K_p}{J} s + \frac{K_p}{J}\right) \tag{14.28}$$

The system exhibits stability under the conditions $K_i > 0$ and $K_p / J > 0$. Typically, for a small-scale quadrotor, the moment of inertia J falls within 10^{-3} to 10^{-6} kgm². Since the third pole, located at $s = -K_i$, is considerably more negative than the other two poles, this system will behave more like a second-order system. Notably, oscillation in the system starts when $K_p < 4J$.

On the other hand, when $J_m \approx J$, perfect characteristic equation factorization is no longer achievable. When assessing the system's stability through a pole plot perspective, two factors are considered: the imaginary component and the sign of the real component in the s-plane. To investigate the latter factor, a Routh table is constructed:

Routh Table Formulated from Equation 14.27.

s^3	1	$\frac{1}{j}K_pK_i + \frac{1}{J}K_p$
s^2	$\frac{Jm}{J}K_i + \frac{1}{J}K_p$	$\frac{1}{J}K_pK_i$
s^1	A	0
s^0	$\frac{1}{J}K_pK_i$	0

where

$$A = \left(\frac{J_m}{J}K_i + \frac{1}{J}K_p\right)\left(\frac{1}{J}K_pK_i + \frac{1}{J}K_p\right) - \frac{1}{J}K_pK_i \qquad (14.29)$$

All constants in Column 1 are positive with the exception of A. $A > 0$ for all system poles to be on the negative s-plane, conveniently expressed as

$$J_m > J\frac{1}{K_i + 1} - \frac{K_p}{K_i} \qquad (14.30)$$

or written alternatively as

$$\frac{J_m}{J} > \frac{1}{K_i + 1} \tag{14.31}$$

While this is not an absolute requirement, it does ensure the stability of the overall system. From a practical perspective, the system is asymptotically stable if $J_m > J$. In cases where $J_m < J$, increasing the integral gains is enough to regain system stability.

14.3.3 RPT Position Controller Design

The concept of RPT control, introduced in [17], aligns seamlessly with our racing quadrotor controller design. In theory, a system under the control of the RPT method possesses the capability to track any provided reference with arbitrarily fast settling time, even when influenced by external disturbance and varying initial conditions. The design principle of the RPT controller starts with a linear time-invariant system:

$$\begin{aligned} \dot{x} &= Ax + Bu + Ew \\ y &= C_1 x + D_1 w \\ h &= C_2 x + D_2 u + D_{22} w \end{aligned} \tag{14.32}$$

Given variables x, u, w, y, h representing the state of the system, control input to the system, external disturbance, measured output from the system, and controlled output of the system, respectively, the aim of our controller is to establish a control law with the structure:

$$\dot{v} = A_c(\varepsilon)v + B_c(\varepsilon)y + G_0(\varepsilon)r + \ldots + G_{k-1}(\varepsilon)r^{k-1} \tag{14.33}$$

$$\dot{u} = C_c(\varepsilon)v + D_c(\varepsilon)y + H_0(\varepsilon)r + \ldots + H_{k-1}(\varepsilon)r^{k-1} \tag{14.34}$$

with v being the controller state and u being the input state. With an appropriately chosen $\varepsilon > 0$, it is proven in [12] that

1. The resulting closed-loop system is stable when given a finite input.
2. Letting $e(t,\varepsilon)$ be tracking error, under any initial condition x_0, there exists:

$$\|e\|_p = \left(\int_0^{\infty} \left| e(t)^p \right| dt \right)^{1/p} \rightarrow 0, \text{ as } \varepsilon \rightarrow 0 \tag{14.35}$$

For cases of non-zero reference, additional control inputs are derived from their derivatives. Consequently, the RPT formulation encompasses references to the form

$$r(t) = p_1 t^k + p_2 t^{k-1} + \ldots + p_{k+1} \tag{14.36}$$

According to [18], quadrotor UAVs have differentially flat outer dynamics, meaning that all their inputs and state variables can be expressed as algebraic functions of flat outputs and their derivatives. An appropriate flat output is given by

$$\sigma = [x, y, z, \psi]^T \tag{14.37}$$

Evidently, the position outputs, namely, $x, y,$ and $z,$ are completely independent, implying that the quadrotor may be treated as a single mass point with constrained velocity and acceleration along each axis of the UAV. The assumption is advantageous primarily when deriving outer-loop control laws and the corresponding references. Consequently, a standalone controller built on a dual-layer integrator model is sufficient for tracking the corresponding reference. The nominal system for the $x, y,$ and $z-$axis can then be expressed simply as a double integral system, as shown below:

$$\dot{x} = \begin{bmatrix} 0 & 1 \\ 0 & 0 \end{bmatrix} x + \begin{bmatrix} 0 \\ 1 \end{bmatrix} u, \quad y = x \tag{14.38}$$

where x represents velocity and position state variables and u is the desired acceleration action. To ensure zero steady-state error, integral action is widely used. The system in Equation 14.38 will then be augmented as

$$\dot{x} = \begin{bmatrix} 0 & -1 & 0 & 0 & 1 & 0 \\ 0 & 0 & 1 & 0 & 0 & 0 \\ 0 & 0 & 0 & 1 & 0 & 0 \\ 0 & 0 & 0 & 0 & 0 & 0 \\ 0 & 0 & 0 & 0 & 0 & 1 \\ 0 & 0 & 0 & 0 & 0 & 0 \end{bmatrix} x + \begin{bmatrix} 0 \\ 0 \\ 0 \\ 0 \\ 0 \\ 1 \end{bmatrix} u$$

$$y = x \qquad (14.39)$$

$$h = \begin{bmatrix} 1\,0\,0\,0\,0\,0 \end{bmatrix} x$$

where $x = \left[\int p_e \; p_r \; v_r \; a_r \; p \, v \right]^T$ with p_r, v_r, a_r representing the position, velocity, and acceleration references and p, v are the measured position and velocity with $p_e = p_r - p$ being the error of the position. A linear feedback control law can then be acquired following RPT controller design in [12] with the following form:

$$u = \begin{bmatrix} \frac{k_i \omega_n^2}{\varepsilon^3} & \frac{\omega_n^2 + 2\zeta\omega_n k_i}{\varepsilon^2} & \frac{2\zeta\omega_n + k_i}{\varepsilon} & 1 & -\frac{\omega_n^2 + 2\zeta\omega_n k_i}{\varepsilon^2} & -\frac{2\zeta\omega_n + k_i}{\varepsilon} \end{bmatrix} x \qquad (14.40)$$

where ε is a design variable to modify the closed-loop settling time and ω_n, ζ, k_i are parameters for moving the desired closed-loop pole locations.

In theory, with sufficiently small ε, the RPT controller has the potential to deliver arbitrarily fast responses. Nevertheless, in practical applications, considering the constraints imposed by the physical dynamics of the quadrotor and the bandwidth of its inner loop, it is prudent for overall system safety to ensure that outer-loop bandwidth is only a fraction of the inner-loop dynamics.

14.4 Simulation and Flight Results

The nonlinear mathematical model of a quadrotor UAV identified in section 2 is constructed in a MATLAB environment. The proposed controller is implemented based on a linearized model and implemented in the simulator. It is important to note that the UAV's physical parameters are derived from the T-Lion quadrotor UAV from the National University of Singapore, as indicated in [8].

Tests are conducted for both $J_m = J$ and $J_m \approx J$, including a gust input disturbance, T. Figures 14.5 and 14.6 display the step responses of quadrotor angular positions in both scenarios. For the case with $J_m \approx J$ we assigned $J_m = J/2$ to assess controller robustness.

For both cases, the controller successfully tracks the reference with angular acceleration disturbance, with only minor chattering observed. When J is not accurately identified, increased chatter is observed, but the controller still successfully suppresses it, maintaining almost no angular error. The comparison of angular errors in these cases is illustrated in Figure 14.7. Despite the suboptimal J, the largest error remains less than 0.05 radians in the presence of angular acceleration disturbances.

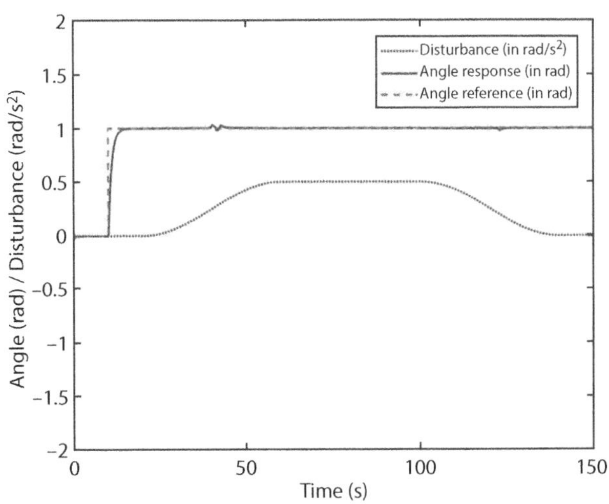

Figure 14.5 Inner-loop system step response for $J_m = J$.

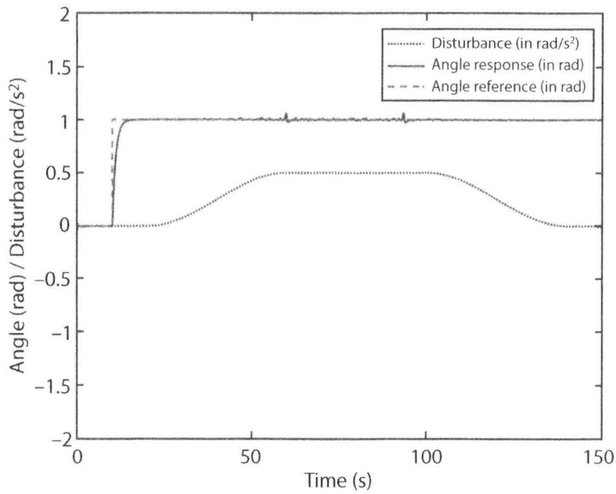

Figure 14.6 Inner-loop system step response for $J_m = J/2$.

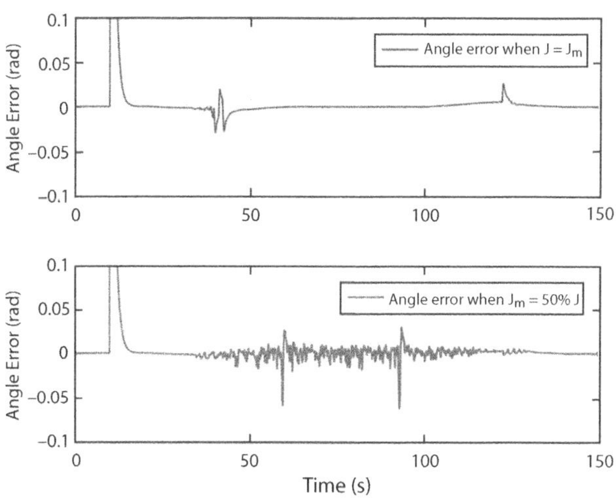

Figure 14.7 Angular position error for both cases.

The RPT position controller on the quadrotor UAV was verified with an actual flight testing of T-Lion, with the reference on position and velocity given in both x and the y-axis. The experiment was carried out with an onboard GPS for position measurements. The velocity of the quadrotor can be estimated through an EKF using the onboard inertial measurement unit (IMU). The experiment was conducted to command the quadrotor to navigate within a 60 m–by–40 m region, at a height of 20 m. x -position and velocity control performance is shown in Figure 14.8, while the y -position and velocity control performance are shown in Figure 14.9. Both responses show a tremendous performance of the quadrotor for velocity and position tracking, as guaranteed by the RPT controllers.

14.5 Conclusions

A novel SIPIC controller, developed through steady-state integral methods, has been introduced for the dynamic system of a multi-rotor UAV system. The tuning approach for this controller is derived from mathematical methods, characterized by its intuitive nature. Additionally, an outer-loop RPT controller, specifically designed for position control, was tested through actual flight trials to complement the inner-loop simulation results.

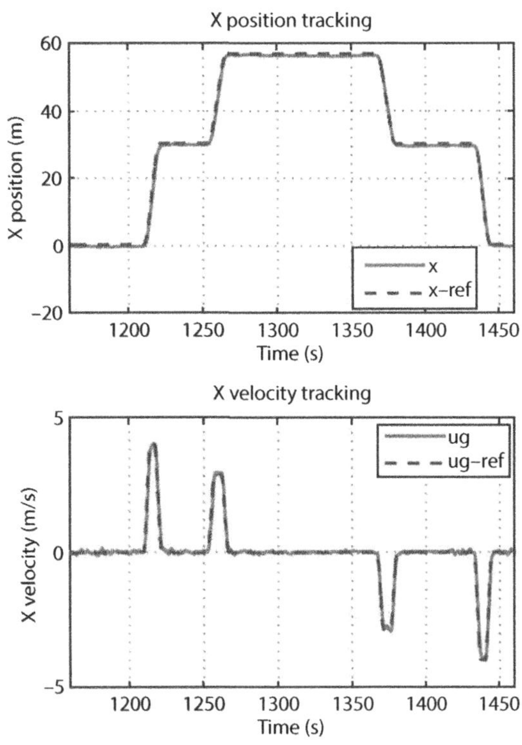

Figure 14.8 Performance of the RPT controller on x -axis.

The proposed controllers were rigorously evaluated using a quadrotor simulator that accurately represents the nonlinear dynamics of quadrotors. Results demonstrate the efficacy and robustness of the proposed controllers. The SIPIC controller exhibited strong performance despite the presence of external disturbances, effectively tracking set points during simulated scenarios. Furthermore, the integration of the RPT controller for outer-loop position control was validated through actual flight trials, contributing to the overall system's stability and precise set-point tracking.

The combination of inner-loop simulation and outer-loop actual flight trials ensures a comprehensive evaluation, emphasizing the efficacy and real-world utility of the suggested control approach for multi-rotor systems. The intuitive tuning methodology ensures a reliable and adaptable control solution for real-world implementation.

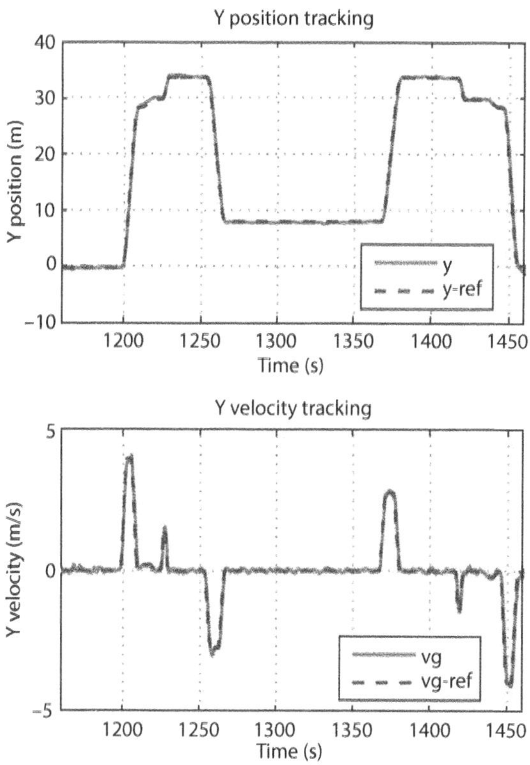

Figure 14.9 Performance of the RPT controller on y -axis.

References

1. Mellinger, D., Michael, N., Kumar, V., Trajectory generation and control for precise aggressive maneuvers with quadrotors. *Int. J. Rob. Res.*, 31, 5, 664–674, 2012.
2. Wang, K., Phang, S.K., Ke, Y., Chen, X., Gong, K., Chen, B.M., Vision-aided tracking of a moving ground vehicle with a hybrid UAV, in: *2017 13th IEEE International Conference on Control & Automation (ICCA)*, IEEE, pp. 28–33, 2017.
3. Phang, S.K., Chiang, T.H.A., Happonen, A., Chang, M.M.L., From Satellite to UAV-based Remote Sensing: A Review on Precision Agriculture. *IEEE Access*, 11, 127057–127076, 2023.
4. Tai, J.J., Phang, S.K., Hoo, C.L., Application of steady-state integral proportional integral controller for inner dynamics control loop of multi-rotor UAVs, in: *2018 Fourth International Conference on Advances in Computing, Communication & Automation (ICACCA)*, IEEE, pp. 1–6, 2018.
5. Ortiz, J.P., Minchala, L.I., Reinoso, M.J., Nonlinear robust H-Infinity PID controller for the multivariable system quadrotor. *IEEE Lat. Am. Trans.*, 14, 3, 1176–1183, 2016.
6. Purnawan, H. and Purwanto, E.B., Design of linear quadratic regulator (LQR) control system for flight stability of LSU-05. *J. Phys. Conf. Ser.*, 890, 1, 012056, 2017.
7. Burman, P., *Quadcopter stabilization with neural network (Doctoral dissertation)*, University of Texas Libraries, The University of Texas at Austin, 2016.
8. Peng, K., Lin, F., Phang, S.K., Chen, B.M., Nonlinear flight control design for maneuvering flight of quadrotors in high speed and large acceleration, in: *2018 International Conference on Unmanned Aircraft Systems (ICUAS)*, IEEE, pp. 212–221, 2018.
9. Ren, J., Liu, D.X., Li, K., Liu, J., Feng, Y., Lin, X., Cascade PID controller for quadrotor, in: *2016 IEEE International Conference on Information and Automation (ICIA)*, IEEE, pp. 120–124, 2016.
10. Phang, S.K., Li, K., Yu, K.H., Chen, B.M., Lee, T.H., Systematic design and implementation of a micro unmanned quadrotor system. *Unmanned Syst.*, 2, 02, 121–141, 2014.
11. Azar, A.T. and Serrano, F.E., Robust IMC–PID tuning for cascade control systems with gain and phase margin specifications. *Neural Comput. Appl.*, 25, 983–995, 2014.
12. Wang, B., Dong, X., Chen, B.M., Lee, T.H., Phang, S.K., Formation flight of unmanned rotorcraft based on robust and perfect tracking approach, in: *2012 American Control Conference (ACC)*, IEEE, pp. 3284–3290, 2012.
13. Phang, S.K. and Chen, X., Autonomous tracking and landing on moving ground vehicle with multi-rotor UAV. *J. Eng. Sci. Technol. (JESTEC)*, 16, 2795–2815, 2021.

14. Phang, S.K., Li, K., Wang, F., Chen, B.M., Lee, T.H., Explicit model identification and control of a micro aerial vehicle, in: *Unmanned Aircraft Systems (ICUAS), 2014 International Conference on*, IEEE, pp. 1048–1054, 2014.
15. Zaccarian, L., *DC motors: dynamic model and control techniques*, Lecture notes, Roma, Italy, 2012.
16. Hoo, C.L., Haris, S.M., Chung, E.C.Y., Mohamed, N.A.N., Steady-state integral proportional integral controller for pi motor speed controllers. *J. Power Electron.*, 15, 1, 177–189, 2015.
17. Chen, B.M., Lee, T.H., Venkataramanan, V., Hard disk drive servo systems, in: *Advances in industrial control series*, Springer, New York, 2002.
18. Mellinger, D. and Kumar, V., Minimum snap trajectory generation and control for quadrotors, in: *Proceedings of the IEEE International Conference on Robotics and Automation (ICRA)*, Shanghai, China, pp. 2520–5, 2011.

15

Type 1 Cascaded Fuzzy Logic–Based Autonomous Vehicles Control Applications

Eshan Samanta[1]*, Sagarika Pal[2] and Anupam De[3]

[1]*Polytechnic Institute, Global Institute of Science & Technology, ICARE Complex, Hatiberia, West Bengal, Haldia, India*
[2]*Government Training Institute, National Institute of Technical Teachers' Training & Research, Falguni, Kolkata, West Bengal, India*
[3]*Autonomous Institute, Haldia Institute of Technology, ICARE Complex, Hatiberia, West Bengal, Haldia, India*

Abstract
Artificial intelligence (AI) is now growing with the utmost versatility. There are many unsolved problems resolved by introducing AI since earlier days in this present century. Many researchers have solved problems related to autonomous vehicle control by considering path planning and collision avoidance. These parameters are highly involved in the design of unmanned autonomous vehicles. It has been found that there are some concerned results that actually evolved with a good impact in the case of the simulation model. Meanwhile, during hardware application, it is quite difficult to implement the same due to errors from the sensor data and entire data processing. In this chapter, a novel cascaded fuzzy logic–based approach is discussed for path planning and collision avoidance control mechanisms for an autonomous vehicle. Simulation is performed in a MATLAB environment and validated by designing three separate fuzzy inference systems (FISs). One of the FISs collected data for collision avoidance by considering obstacles and their distance as inputs. The other FISs collected data from LiDAR sensors, and the output from collision avoidance FIS, to decide whether to stop, move, or turn as per the outcome achieved after the final defuzzification.

***Keywords*:** Unmanned, fuzzy logic, artificial intelligence, LiDAR sensor, FIS

**Corresponding author*: write2eshansamanta@gmail.com

Arindam Mondal and Souvik Ganguli (eds.) Controller Design for Industrial Applications, (309–330)

15.1 Introduction

In this age of artificial intelligence, one of the most talked-about topics is the integration of intelligent systems into autonomous cars (AI). In recent times, many prestigious research works have been carried out after considering the control architecture of an autonomous vehicle [3, 10, 29]. Few more interesting works have been explored considering deep learning in autonomous vehicle control [4, 13]. Considering navigation as one of the most important parameters, many papers have been found with the application of fuzzy logic in solving this problem in the case of autonomous vehicles [5, 7, 17, 24]. Here, the authors have controlled the steering mechanism and navigated the vehicle with an optimum output. But, in this case, no solution is provided if any uncertain obstacle arrives, and the vehicles have to decelerate with an immediate effect with zero collateral damage.

Some of the articles have been presented by considering obstacle avoidance as one of the important agendas and solved by implementing various algorithms [8, 15, 21]. There are several papers that highlight Light Detection and Ranging (LiDAR) as an effective method for detecting obstacles in autonomous propulsion systems [1, 11]. Here, the authors have presented and discussed how the best formation of the algorithm can work to detect obstacles using the LiDAR sensor as input to the autonomous vehicle for applying braking and propelling. One of the articles validated the hybrid application of both navigation and obstacle avoidance by introducing a deep learning mechanism [11]. The primary issues in the earlier studies were either navigation or obstacle avoidance. Every item under discussion has created an algorithm to manage the grave concerns raised by the manufacturers of autonomous vehicles to this day. There is currently no known method that can regulate many inputs with various characteristics and zero mistakes at all.

This chapter presents the development of a new class of fuzzy inference systems (FISs) through the use of cascaded intelligent control. Sensor data are used as input, and a triplet is assigned to it. The cascaded intelligent controller's goal is to enhance any unmanned vehicle's path planning and collision avoidance capabilities. By using both the trapezoidal and triangular membership functions (MFs), the output MF was created. A detailed performance analysis with respect to the entire control system developed for the intelligent controller is reported, conducted with the MATLAB environment.

This chapter has two main contributions: 1) A cascaded fuzzy logic-based intelligent controller is proposed for the first time in any autonomous vehicle. 2) We propose an idea of incorporating sensors that reduce errors before the data processing. The rest of the chapter presents the same by framing the sections with the control architectures for autonomous vehicles, perception and sensor fusion for autonomous navigation, intelligent control for path planning and collision avoidance, and case studies of intelligent control in autonomous vehicles. Finally, the main conclusion and contributions have been discussed in section 15.7.

15.2 Mathematical Kinematic Model

In this part, we present a kinematic model for automobiles. Four wheels make up an automobile: two for steering and two for producing linear motion (the two front or rear wheels, depending on whether the vehicle has four-wheel drive). This is in contrast to the robots with differential drives discussed in the previous chapter. We could decide to model the car as having two inputs: acceleration and steering angle. Even though the forward speed and turning rate are the inputs in our somewhat simplified model—which may be a more accurate physical model—we will stick with it. Unless the car is accelerating quickly or traveling at a high speed, this version is sufficient for most applications.

15.2.1 Car Kinematics

Overall, it is rather clear that the vehicle is driven by its back wheels and has front-wheel steering. As seen in Figure 15.1, we give the car a body-attached frame whose origin is at the midpoint of the rear axle and whose x-axis points forward. If the wheels roll without slipping, then the car's immediate velocity is always in the direction of the x-axis linked to the body. The forward speed is denoted by v_{car}.

The two front wheels may initially appear to be positioned similarly to the car's frame. However, this is untrue. Since the origin of the car frame traces out a curve in the plane, all wheels must roll without slipping tangent to a group of circles having a common origin, known as the center of the turning circle. Because the rear wheels share a single axle and remain stable in relation to the automobile chassis, their axles coincide with the radius of concentric circles, and as a result, each rear wheel has an instantaneous velocity that is perpendicular to this radius. The inner wheel of the front wheels travels in a circle that is smaller in radius than the outer wheel.

This type of front-wheel steering, seen in Figure 15.1, is called Ackermann steering. Ackermann steering calls for a somewhat intricate

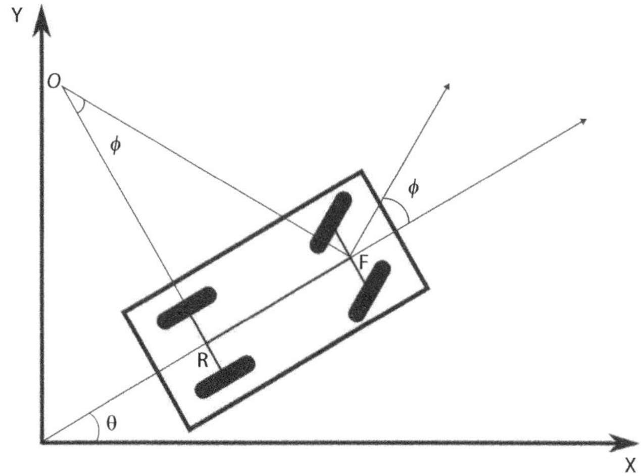

Figure 15.1 Kinematics of a non-holonomic car with a standard Ackermann steering mechanism.

physical mechanism, but thankfully, we can model the system with the help of a single virtual wheel that rolls perpendicular to the line that connects the turning circle's center to this midpoint. The virtual wheel is placed in the middle of the two front wheels. As seen in the image below, we designate ϕ the angle between the virtual wheel's forward motion and the car's x-axis.

The configuration of the automobile can be written as follows: x and y signify the position of the origin of the car frame, ϕ denotes the steering angle, and x and y indicate the car frame's orientation (the angle between the world's x-axis and the car frame's x-axis). The vehicle's inputs are

- $u_1 = v_{car}$, the linear speed of the car and
- $u_2 = \phi$, the turning rate of the virtual front wheel.

15.2.2 Differential Kinematics

An equation describes the car's differential kinematics, $\dot{q} = f(q, u_1, u_2)$.

The two elements $\dot{x}$ and $\dot{y}$ are simple functions of v_{car} and θ, as shown in Equations 15.2 and 15.2.

$$\dot{x} = v_{car} \cos\theta = u_1 \cos\theta \tag{15.1}$$

$$\dot{y} = v_{car} \sin\theta = u_1 \sin\theta \tag{15.2}$$

The steering angle is also straight forward as given in Equation 15.3

$$\dot{\phi} = u_2 \tag{15.3}$$

Note that using this model, we cannot directly control the car's orientation θ. We cannot even directly choose the steering angle ϕ. We have direct control only over $\dot{\phi}$.

The equation for ϕ is slightly more complex. From Equations 15.4 and 15.5 as given below, we observe the following:

$$v_{wheel} \cos\phi = v_{car} \tag{15.4}$$

$$v_{wheel} \sin\phi = v_{tangent} \tag{15.5}$$

Combining these to eliminate v_{wheel}, we obtain Equation 15.6 as shown below.

$$\frac{v_{car}}{\cos\phi} \sin\phi = v_{car} \tan\phi = v_{tangent} \tag{15.6}$$

Now, using $L\dot{\theta} = v_{tangent}$ (where L is the front-to-rear axle baseline distance), we obtain Equation 15.7

$$\dot{\theta} = v_{car} \frac{1}{L} \tan\phi = u_1 \frac{1}{L} \tan\phi \tag{15.7}$$

By combining these terms, the differential kinematics can be expressed in matrix form as shown in Equation 15.8

$$\begin{bmatrix} \dot{x} \\ \dot{y} \\ \dot{\theta} \\ \dot{\phi} \end{bmatrix} = \begin{bmatrix} \cos\theta \\ \sin\theta \\ \frac{1}{L}\tan\phi \\ 0 \end{bmatrix} u_1 + \begin{bmatrix} 0 \\ 0 \\ 0 \\ 1 \end{bmatrix} u_2 \tag{15.8}$$

15.3 Control Architectures for Autonomous Vehicles

When constructing a model like the one outlined, the control architecture of autonomous cars is regarded as one of the most crucial factors. As mentioned before [3], the complexity of self-driving automobile decision-making systems increases with their level of autonomy. The self-driving automobile should be capable of full-time operation of all aspects of driving under various road and environmental situations by the time it reaches the greatest level of autonomy, Level 5. Various designs and connections can be made for the control architecture and its parts, but at a high level, the system structure of a car with at least some self-driving capabilities resembles that shown in Figure 15.2a [20]. According to Montemerlo *et al.* and Urmson *et al.*, each block in Figure 15.2a typically consists of a number of smaller components that are connected by a variety of communication and sensing interfaces [16, 27].

In Figure 15.2a, when creating a fuzzy logic controller–based intelligent autonomous vehicle control system, if the FIS calling theory is put into practice, the system is referred to as cascaded intelligent control. Since the system is constructed with several FIS, the first FIS is considered to be Stage *I*, and the remaining FIS are designated as Stages *II*, *III*, and so on, up to n number of stages. For this reason, another name for this specific system is a multistage-based cascaded fuzzy controller. Considering Figure 15.2a as the parent block diagram in this chapter, we have discussed an intelligent control architecture using a cascaded fuzzy logic controller for motion planner and vehicle control. The design architecture is shown in Figure 15.2a. From Figure 15.2a, it is very clear that the motion planner and vehicle controller are considered the target blocks to develop using a cascaded fuzzy inference system.

Overall, two FISs are required to develop a cascaded design for controlling the entire mechanism one for motion planner and the other for obstacle avoidance. Another hidden FIS is required to ensure that the specific decision is to be considered based on different circumstances. This concept is discussed in detail in the next sections of this chapter.

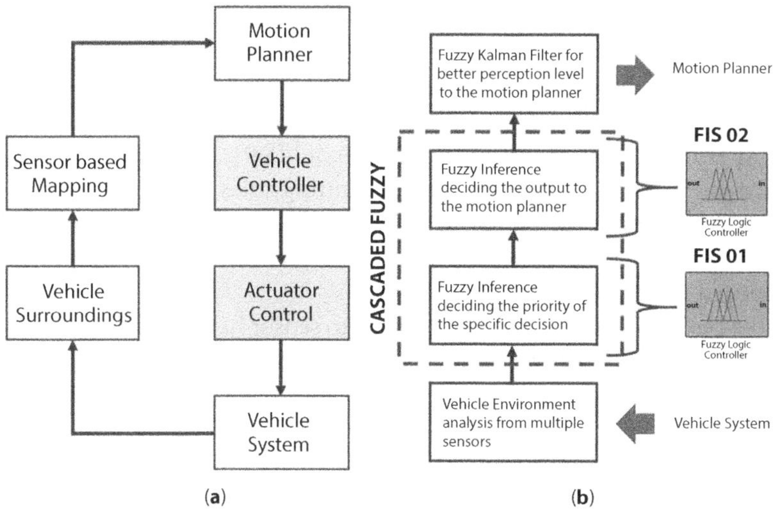

Figure 15.2 (a) Schematic block diagram of the system architecture of an autonomous vehicle from a control perspective. (b) Fusion-based guidance system architecture.

15.4 Perception and Sensor Fusion for Autonomous Navigation

The process of feeding information related to vehicle surroundings to the motion planner and vehicle controller is termed mapping [19]. In Figure 15.2a, the sensor and mapping block are connected to the motion planner following the vehicle controller as well. Previously, navigation of autonomous vehicles was considered to be corrected using fuzzy logic controllers [5, 17, 23, 26]. The application fuzzy controller was explored in many ways for sensor fusion incorporation earlier in the research work using the Kalman filter [25]. Other approaches were also discussed earlier by considering neural networks and reinforcement learning to improve the perception level of sensor fusion in any autonomous unmanned vehicle [9, 28]. For better perception in this chapter, we have proposed an idea of a cascaded fuzzy-based implemented sensor fusion technique for better results. In this concept, two FISs will work together. One will decide what is the priority level and the other will execute the same input along with the other sensor inputs to the motion planner and vehicle controller for error-free navigation using the Kalman filter. The proposed idea of implementing sensor fusion for better perception is shown in Figure 15.2b and discussed in brief. In Figure 15.2b, it is clearly shown that multiple sensors are gathering information and actually deciding which decision to take first. This confirmation is done in the first fuzzy inference system (FIS), FIS 01. After the fundamental defuzzification of FIS 01, simultaneously, FIS 02 processes the actual navigation as per training rules and delivers optimum output to the motion planner for better perception. The entire set of FIS 01 and FIS 02 is considered a cascaded fuzzy logic controller for sensor fusion proposed in this chapter.

The purpose of solving agendas related to sensor fusion technique has been very complex since earlier times [12]. The problem arises when an object is approaching the vehicle at a catastrophic speed, and the navigation system needs to respond if a human comes close to it. So, here in this case if in place of an autonomous system any human being used to drive the vehicle then the vehicle might hit any tree or wall just to save another living things that might be any human or animal due to the presence of wisdom. At that point, is it possible for a modern autonomous system to control the vehicle in the same situation in the same way a human would decide spontaneously [2, 18]?

15.5 Intelligent Control for Path Planning and Collision Avoidance

To solve this problem of AI-based systems, in this chapter, a novel approach for designing fuzzy logic incorporated autonomous vehicle control is discussed in detail in this section. Here, path planning is considered as FIS 02 and fed to the motion planner for more precision. In Figure 15.2b, the decision from FIS 01 ensures input to the path planner and then based on the environment and output from FIS 01 after defuzzification FIS 02 sends a signal to the motion planner. The hybrid application of both the path planner and motion planner is highly accepted on a mass scale globally earlier [6, 14, 22]. Another black box, namely, FIS03, is considered in this work to predict the kinematics to be followed after validating the angular displacement of the steering while ensuring path planning. The function of the FIS01 is to set the steering status, i.e., to move whether left or right, whereas FIS02 will ensure the angular displacement of the steering and the final FIS03 will decide which kinematics has to be followed by the vehicle for any specific environment as per sensor data.

FIS01 and FIS02 together decide the motion of the vehicles, which is why Figure 15.2b only shows FIS01 and FIS02, but no trace of FIS03 is found. Since the vehicle kinematics does not directly fall under the motion planner architecture, FIS03 is omitted from Figure 15.2b.

15.5.1 Fuzzy System Rules

In this proposed concept, three sets of rule configurations have been designed for three different FIS. Table 15.1 explains the rules developed to train the first FIS. Here, LiDAR is designed by considering four different variables: Very Close (VC), Close (C), Far (F), and Very Far (VF). Variables taken for identifying the steering direction are as follows: Left (L), Right (R), and Straight (S). The rules are designed for the first FIS as shown in Table 15.1 is like for example; If LiDAR is VC and Speed is Slow then Steering is Left.

In Table 15.1, it is clearly shown that a total of 12 rules have been set for designing the first FIS that decides the optimum steering direction to be followed. Similarly, in Table 15.2, it is clearly shown that a total of 12 rules have been set for designing the second FIS that decides the optimum control mechanism of steering angular displacement (Angular D) to be followed. Here, Normal (N), Critical (Cr), and Very Critical (VCr) are introduced as output variables for the angular displacement of steering. Table 15.2 explains the decision by the intelligent system to select the steering respective angle during the execution of the obstacle avoidance mechanism to be followed as decided by the first FIS. In this table, the rules are also designed as given in this sample example like '*If LiDAR is Very Close and Steering is turned toward Left then Angular Displacement must be maintained at critical level*' for secured journey and safety as well.

Table 15.1 Rule configuration sets for deciding steering direction.

		Speed		
Steering		**Slow**	**Moderate**	**High**
LiDAR	VC	Left	Right	Right
	C	Left	Right	Right
	F	Straight	Straight	Right
	VF	Straight	Straight	Straight

Table 15.2 Rule configuration sets for making decisions related to the angular displacement of the steering.

Angular D		Steering		
		Left	**Right**	**Straight**
LiDAR	VC	Cr	N	VCr
	C	Cr	Cr	VCr
	F	N	Cr	N
	VF	N	N	N

Table 15.3 Rule configuration sets for making decisions related to actual kinematics to be followed during driving.

Kinematics		Steering		
		Left	**Right**	**Straight**
Angular D	N	DK	DK	CK
	Cr	DK	DK	CK
	VCr	DK	DK	CK
	None	DK	DK	CK

In Table 15.3, it is clearly shown that a total of 12 rules have been set for designing the second FIS that decides the optimum control mechanism of driving kinematics to be followed. Here, differential kinematics (DK) and car kinematics (CK) are introduced as output variables for deciding driving kinematics based on FIS 01 and FIS 02, respectively. Table 15.3 explains the decision by the intelligent system to select the driving kinematics during the execution of the obstacle avoidance mechanism to be followed as decided by the first FIS and the actual path to be followed by the second FIS as well. In this table, rules are also designed as given in this sample example like '*If angular displacement is normal and steering is turned toward the left then differential kinematics must be followed to ensure proper path planning.*'

15.5.2 Membership Function

In this work, three types of MFs have been followed. Triangular MF and trapezoidal MF are introduced in this work to design the entire FIS to develop an intelligent controller for motion planning and obstacle avoidance. In Figure 15.3, a clear pictorial view of LiDAR is given. Here, in Figure 15.3, a trapezoidal MF is considered to ensure a sustainable capture, with a 50-m distance established as the maximum range for this particular cascaded controller.

Figures 15.8 and 15.9 ensures the rules set for this specific intelligent controller through a clear surface view of the same. In Figure 15.8a, the yellow section indicates the steering control while driving toward the right side. The deep blue section ensures the left-side movement. The light blue color ensures the straight direction of the concerned vehicle during normal scenarios and if the road is straight somehow. Figure 15.8b confirms that the vehicle usually maintains a critical or very critical state due to certain variations in the roadways. In Figure 15.11, it is clearly shown that kinematics are not very predictive, and thus, it has been designed using multiple rules and thus showing a different form of surface view in Figure 15.9 compared to Figure 15.8(a and b). In Figure 15.4, it is clearly shown that the speed is designed by considering triangular MF. This is due to the factor of instantaneous decision required to be processed as speed is highly variable. In Figure 15.4, speeds below 30 km/hr are declared as slow, while speeds from 20 km/hr to 60 km/hr are considered moderate.

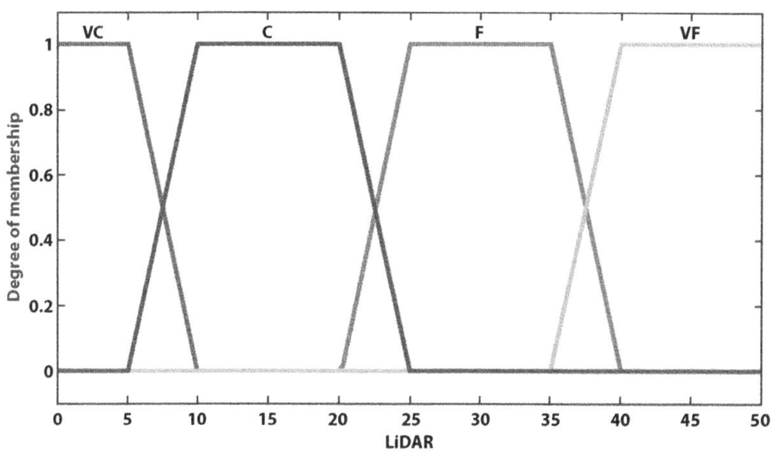

Figure 15.3 First input membership function of FIS01.

In India, there is a speed limit of 50 km/h, which is established to monitor vehicles that exceed this speed. Therefore, the variable range for speed is set from 50 km/h to 100 km/h. In Figure 15.5, the input of vehicle steering is shown. Here, in Figure 15.5, the parameters left and right are set by introducing trapezoidal MF. This is because of the sustainable decision required to be considered when determining the directions of the vehicle.

Straight variables are preferred as triangular MF due to a very specific reason. Since the roadways are not continuously straight at all due to certain curves, and thus, this specific decision is quite instantaneous. The range is -100 to 100 in terms of the angle of rotation and the movement of steering as per direction. In Figure 15.6, angular displacement is shown.

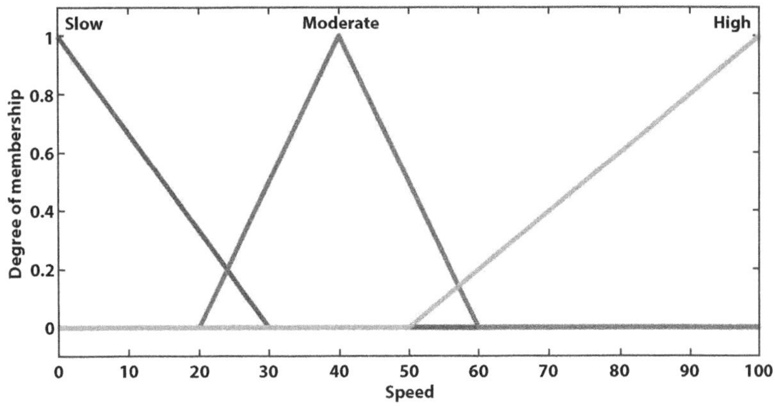

Figure 15.4 Second input membership function of FIS01.

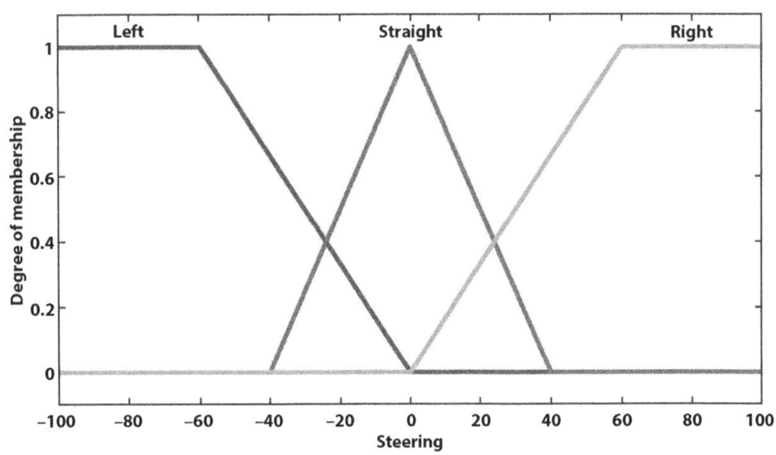

Figure 15.5 Output membership function of FIS01.

This angular displacement is very important during a certain scenario if any certain obstacle arrives while avoiding another regular obstacle. Here, in Figure 15.6, a certain impulse with an angle within the range of 30° is considered a normal variable and declared under triangular MF due to its spontaneous decision. The same concept has been followed in the case of critical variables up to the 60-degree range. On the other hand, a very critical level ensures a huge impulse during an emergency and high priority to avoid any catastrophic accidental fault. This is a lifesaving decision and has to be very sustainable, thus trapezoidal MF is followed. Figure 15.7 indicates the kinematics pattern to be followed by the vehicle to ensure a specific motion to the controller.

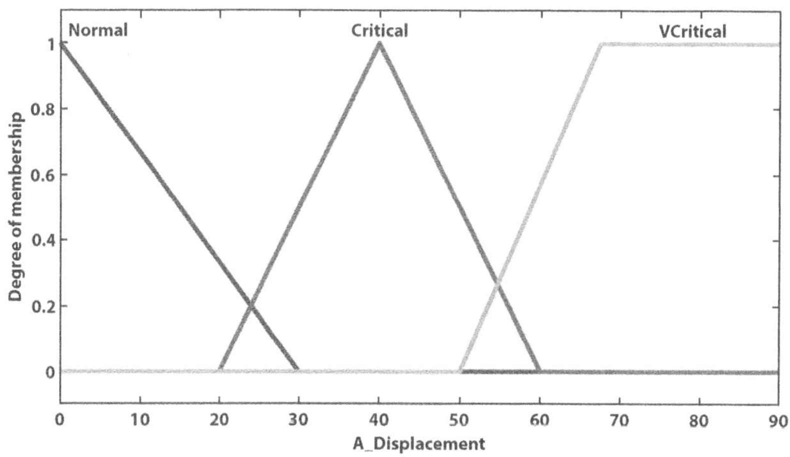

Figure 15.6 Output membership function of FIS02.

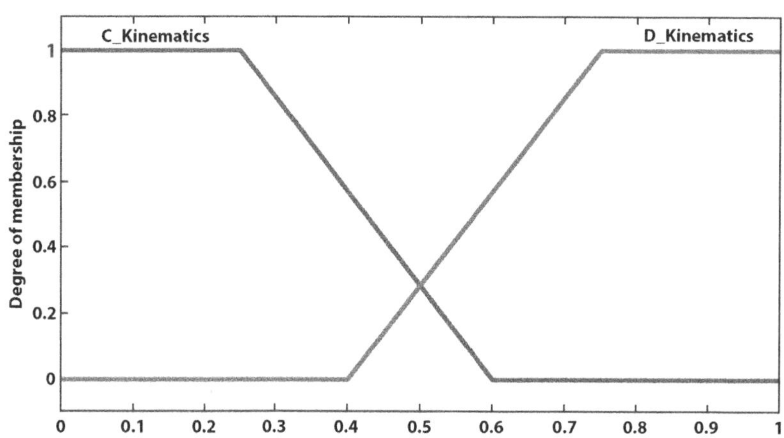

Figure 15.7 Output membership function of FIS03.

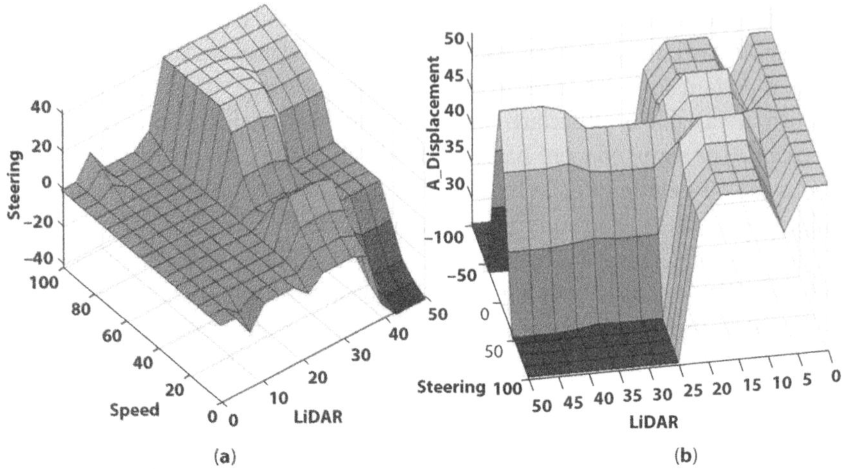

Figure 15.8 (a) Generated surface view of FIS01. (b) Generated surface view of FIS02.

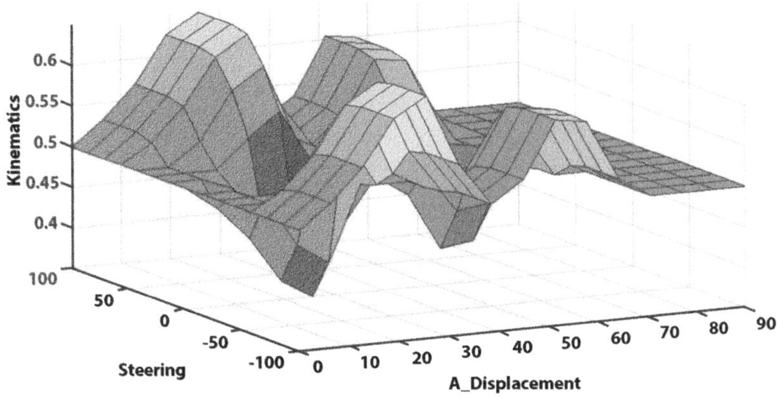

Figure 15.9 Generated surface view of FIS03.

15.5.3 Results and Discussions

In this section, a clear validation has been plotted with a valid justification, as presented in Figures 15.10 and 15.11. In Figure 15.10a, it is very clear that the roadways are not straight, which is why steering ensures a highly dynamic response even after defuzzification. This output is validated by Figure 15.10b, where the signal from the accelerometer introduces uneven turns of the roadways with such a hilarious variation as manually created. Figure 15.11a shows the output characteristics of angular displacement of steering based on time-specific circumstances. As per Figure 15.10b, after 4 seconds when LiDAR is sensing any obstacle, then a pulse is clearly captured. Simultaneously, in Figure 15.11a, there is a change in the output just after 4 seconds and the same is validated as per the learning rules and MFs, as shown in Figure 15.6. After 4 seconds, the speed indicated a moderate level, as shown in Figure 15.10b; at this point, the LiDAR is sensing some obstacles, as depicted in Figure 15.11b. Therefore, as per Figure 15.11a, the angular displacement falls within the range of critical and very critical parameters, as shown in Figure 15.6. The kinematics is validated in Figure 15.11b, where, after 4 seconds, the output seems like differential kinematics is more preferred as compared to car kinematics due to certain turns required due to obstacles predicted by the LiDAR in Figure 15.10b. In case when the steering is normal, the output obtained from Figure 15.11b indicates that the kinematics is very much aligned with car kinematics, as the vehicle is only following certain bends of the roadways but not the obstacles.

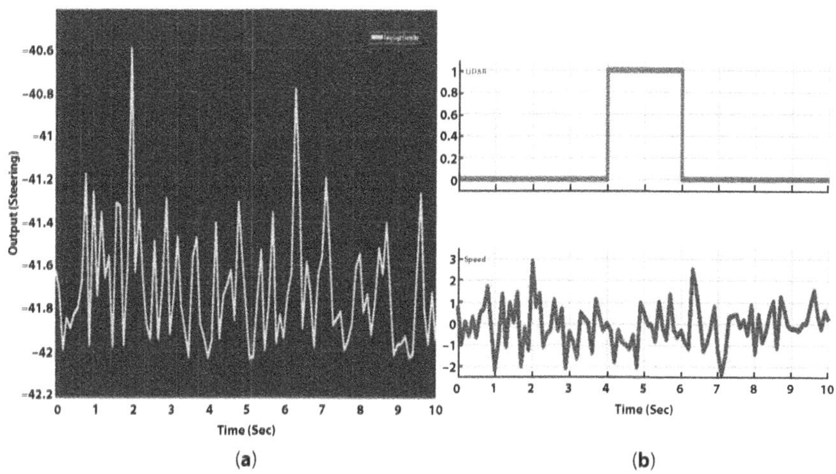

Figure 15.10 (a) Scope analysis of steering. (b) Signal generated from accelerometer and LiDAR in this chapter.

The results obtained in this work have validated both the rules and the MFs, as shown in Figures 15.3 to 15.7. From Figures 15.3 to 15.7, it is very clear that in this work a very specific rules have been considered for validating the developed algorithm to ensure a soothing architecture. In this architecture, three consecutive FISs are followed for predicting steering movement including its angular displacement based on how much critical angle has to be followed. The other thing is to confirm which kinematics to be followed as per the third FIS indeed as explained and discussed.

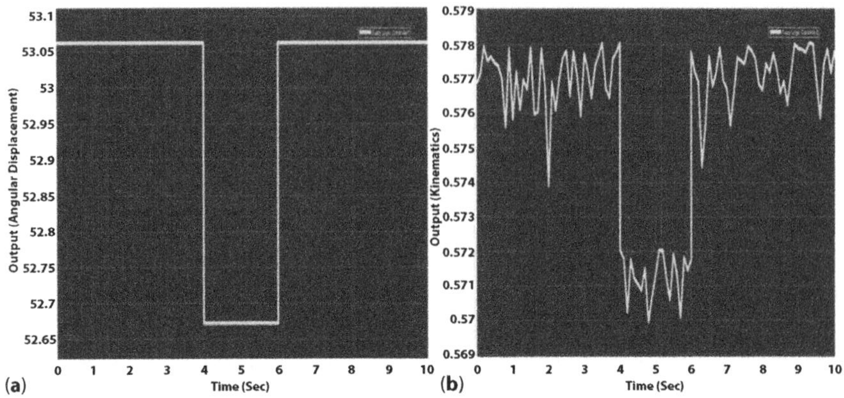

Figure 15.11 (a) Scope analysis of angular displacement. (b) Scope analysis of kinematics.

15.6 Case Studies of Intelligent Control in Autonomous Vehicles

A car that can drive itself is referred to as a self-driving car, autonomous car, driverless car, or robotic car (robocar). Sensors like optical and thermographic cameras, radar, LiDAR, ultrasound/sonar, GPS, odometry, and inertial measurement units are used by self-driving automobiles to understand their environment. Control systems analyze sensory data to build a three-dimensional representation of the environment around the vehicle. The vehicle then determines a suitable navigation route and tactics for navigating around obstacles and track controls (such as stop signs, traffic lights, speed restrictions, and yield signs) based on the model.

Autonomous vehicles are expected to have an impact on the automotive business, as well as the health, welfare, urban planning, traction, insurance, labor market, and other fields once the technology is more developed. Their regulation is becoming a more crucial issue.

Typically, a system created by SAE International (SAE J3016) divides vehicle autonomy into six categories. Level 0 is the lowest level of automation, Level 1 is hands-on/shared control, Level 2 is hands-off, Level 3 is eyes-off, Level 4 is mind-off, and Level 5 is steering wheel optional. These levels can be loosely understood as optional steering wheels.

15.6.1 Google's Driverless Car Technology - First Case Study

Technology for autonomous cars is being developed by Google as part of the Google Driverless Car project. Currently leading the project is Sebastian Thrun, a Google employee and co-inventor of Google Street View. In addition, he is the Stanford Artificial Intelligence Laboratory's director. The robotic vehicle Stanley, created by Thrun's Stanford team, took home the $2 million award from the US Department of Defence along with the 2005 DARPA Grand Challenge title. A team of 15 Google engineers, including DARPA Grand and Urban Challenge veterans Chris Urmson, Mike Montemerlo, and Anthony Levandowski, worked on the system's development. Information from Google Street View and artificial intelligence algorithms are mixed with input from video cameras within the vehicle, a LIDAR sensor on top of the vehicle, radar sensors on the front of the vehicle, and a position sensor attached to the vehicle.

With minimal human interaction, Google gathered 3,500 km of Street View photographs in 2009. As of 2010, Google had tested a number of automobiles using the system, covering 1,609 km (1,000 mi) of autonomous driving and 225,308 km (140,000 km) with sporadic human involvement. In addition to using energy and space on roads more effectively, Google anticipates that the improved accuracy of its automated driving system will contribute to a decrease in the number of accidents and fatalities caused by traffic.

15.6.2 AI-Based Robotaxis - Second Case Study

These are driverless taxis operated by ridesharing companies. These cars developed on electric vehicle technology opening a new branch of study called transportation as a service (TaaS). Baidu announced the radio taxi cars will be available for $77,000. This brings the hope of scalability. Another notable company Waymo expects the hardware cost to go at $0.30 per mile. Currently, robotaxi service providers are testing the car in geo-fencing areas. These areas are labeled as the objective design domain (ODD) in the robotaxi industry. Waymo and Cruise got a license to run their radio taxi in California. Similarly, Baidu and Pony.ai got licenses to run radio taxis in China. The maiden trial was tested in April 2016 by MIT in collaboration with NuTonomy. They worked on Renault Zoes to get initial responses. The encouraging results made Grab a Southeast Asia car-sharing company, tie up with NuTonomy. 2017 turned out to be a major year for the robotaxi industry. In March 2017, Uber tested robotaxis in Pittsburg and Waymo began testing its taxis in Phoenix, Arizona.

The same year, Cruise announced a radio taxi service for its employees. In February 2021, Waymo invited the public to apply to test the radio taxi service in limited areas with its engineers assisting the car. In February 2022, Cruise opened up the robotaxi service in California for the public.

Robotaxi has been implemented with the assistance of Tesla's electric vehicle and autonomous vehicle learning curve. In contrast to its American and Chinese competitors, Tesla lets users learn how to operate the car right away. Different road conditions and safety regulations led to the conclusion. Elon Musk revealed in April 2022 that the business was developing a cutting-edge car for the robotaxi sector. The vehicle will be fully autonomous (FSD) and will not have a pedal or steering wheel. Furthermore, the company works on the principle that cost per mile should be optimum.

15.7 Conclusion

In this chapter, an intelligent approach is discussed to develop a better path planner and obstacle avoidance algorithm to provide a high security-based autonomous vehicle. A novel approach is proposed to incorporate within the autonomous vehicle for providing priority-based decisions using cascaded fuzzy logic technique and reported. The output of the fuzzy logic ensures that if in case two lifesaving incidents occur, then the autonomous vehicle will be able to find the best possible way out by giving priority to saving human loss from both ends. Globally, AI has reached such a certain level that now humans are implementing it somewhere, which might become a serious threat to sustainable existence. Now, the question arises: If AI can solve everything, then how could AI be a responsible factor in cases where wisdom is required?

During autonomous driving, if any anonymous obstacles appear and during steering control another sudden human encounter takes place, then is it possible for the unmanned car to save both and apply its destructive algorithm by considering the case with utmost priority? All these unsolvable questions are only possible for AI to solve when it leads to divine intelligence (DI), which is actually incorporated within humans only. That will be considered the final evolution of this earth.

References

1. Baras, N., Nantzios, G., Ziouzios, D., Dasygenis, M., Autonomous obstacle avoidance vehicle using lidar and an embedded system, in: *2019 8th International Conference on Modern Circuits and Systems Technologies (MOCAST)*, IEEE, pp. 1–4, 2019.
2. Bathla, G., Bhadane, K., Singh, R.K., Kumar, R., Aluvalu, R., Krishnamurthi, R., Kumar, A., Thakur, R.N. and Basheer, S., Autonomous vehicles and intelligent automation: Applications, challenges, and opportunities. *Mobile Inf. Syst.*, 2022, 1, 7632892, 2022.
3. Berntorp, K., Hoang, T., Quirynen, R., Di Cairano, S., Control architecture design for autonomous vehicles, in: *2018 IEEE Conference on Control Technology and Applications (CCTA)*, pp. 404–411, 2018, doi: 10.1109/CCTA.2018.8511371.
4. Deng, Y., Zhang, T., Lou, G., Zheng, X., Jin, J., Han, Q.-L., Deep learning based autonomous driving systems: A survey of attacks and defenses. *IEEE Trans. Ind. Inf.*, 17, 12, 7897–7912, 2021.
5. Driankov, D. and Saffiotti, A., *Fuzzy logic techniques for autonomous vehicle navigation*, vol. 61, Physica-Verlag, Springer Nature, 2013.
6. Elbanhawi, M. and Simic, M., Sampling-based robot motion planning: A review. *IEEE Access*, 2, 56–77, 2014.
7. Emmanuel, I., Fuzzy logic-based control for autonomous vehicle: A survey. *Int. J. Educ. Manage. Eng.*, 7, 2, 41, 2017.
8. Fauzi, F.A., Mulyana, E., Mardiati, R., Setiawan, A.E., Fuzzy logic control for avoiding static obstacle in autonomous vehicle robot, in: *2021 7th International Conference on Wireless and Telematics (ICWT)*, IEEE, pp. 1–5, 2021.
9. Fayjie, A.R., Hossain, S., Oualid, D., Lee, D.-J., Driver-less car: Autonomous driving using deep reinforcement learning in urban environment, in: *2018 15th International Conference on Ubiquitous Robots (UR)*, IEEE, pp. 896–901, 2018.
10. Guo, C., Sentouh, C., Haué, J.-B., Popieul, J.-C., Driver–vehicle cooperation: a hierarchical cooperative control architecture for automated driving systems. *Cognit. Technol. Work*, 21, 657–670, 2019.
11. Ibrahim, H.A., Azar, A.T., Ibrahim, Z.F., Ammar, H.H., A hybrid deep learning based autonomous vehicle navigation and obstacles avoidance, in: *Proceedings of the International Conference on Artificial Intelligence and Computer Vision (AICV2020)*, Springer, pp. 296–307, 2020.
12. Kocić, J., Jovičić, N., Drndarević, V., Sensors and sensor fusion in autonomous vehicles, in: *2018 26th Telecommunications Forum (TELFOR)*, IEEE, pp. 420–425, 2018.
13. Kuutti, S., Bowden, R., Jin, Y., Barber, P., Fallah, S., A survey of deep learning applications to autonomous vehicle control. *IEEE Trans. Intell. Transp. Syst.*, 22, 2, 712–733, 2020.

14. Lai, X.-C., Ge, S.S., Al Mamun, A., Hierarchical incremental path planning and situation-dependent optimized dynamic motion planning considering accelerations. *IEEE Trans. Syst. Man Cybernetics Part B (Cybernetics)*, 37, 6, 1541–1554, 2007.
15. Memon, K.R., Memon, S., Memon, B., Memon, A.R., Muhammad Zaigham Abbas Shah, S., Real time implementation of path planning algorithm with obstacle avoidance for autonomous vehicle, in: *2016 3rd International Conference on Computing for Sustainable Global Development (INDIACom)*, IEEE, pp. 2048–2053, 2016.
16. Montemerlo, M., Becker, J., Bhat, S., Dahlkamp, H., Dolgov, D., Ettinger, S., Haehnel, D., Hilden, T., Hoffmann, G., Huhnke, B., *et al.*, Junior: The stanford entry in the urban challenge. *J. Field Rob.*, 25, 9, 569–597, 2008.
17. Naranjo, J.E., Sotelo, M.A., Gonzalez, C., Garcia, R., De Pedro, T., Using fuzzy logic in automated vehicle control. *IEEE Intell. Syst.*, 22, 1, 36–45, 2007.
18. Nascimento, A.M., Vismari, L.F., Santos Tancredi Molina, C.B., Cugnasca, P.S., Camargo, J.B., de Almeida, J.R., Inam, R., Fersman, E., Marquezini, M.V., Hata, A.Y., A systematic literature review about the impact of artificial intelligence on autonomous vehicle safety. *IEEE Trans. Intell. Transp. Syst.*, 21, 12, 4928–4946, 2019.
19. Ort, T., Paull, L., Rus, D., Autonomous vehicle navigation in rural environments without detailed prior maps, in: *2018 IEEE International Conference on Robotics and Automation (ICRA)*, IEEE, pp. 2040–2047, 2018.
20. Paden, B. and M., A survey of motion planning and control techniques for self-driving urban vehicles. *IEEE Trans. Intell. Veh.*, 1, 1, 33–55, 2016.
21. Park, J.M., Kim, D.W., Yoon, Y.S., Kim, H.J., Yi, K.S., Obstacle avoidance of autonomous vehicles based on model predictive control. *Proc. Inst. Mech. Eng., Part D: J. Automob. Eng.*, 223, 12, 1499–1516, 2009.
22. Pradhan, B., Nandi, A., Hui, N.B., Roy, D.S., Rodrigues, J.JPC., A novel hybrid neural network-based multi robot path planning with motion coordination. *IEEE Trans. Veh. Technol.*, 69, 2, 1319–1327, 2019.
23. Ramirez, J.M., Gomez-Gil, P., Lopez Larios, F., A robot-vision system for autonomous vehicle navigation with fuzzy-logic control using lab-view, in: *Electronics, Robotics and Automotive Mechanics Conference (CERMA 2007)*, IEEE, pp. 295–302, 2007.
24. Saffiotti, A., Fuzzy logic in autonomous navigation, in: *Fuzzy Logic Techniques for Autonomous Vehicle Navigation*, pp. 3–24, Springer, 2001.
25. Subramanian, V., Burks, T.F., Dixon, W.E., Sensor fusion using fuzzy logic enhanced kalman filter for autonomous vehicle guidance in citrus groves. *Trans. ASABE*, 52, 5, 1411–1422, 2009.
26. Tsourveloudis, N.C., Valavanis, K.P., Hebert, T., Autonomous vehicle navigation utilizing electrostatic potential fields and fuzzy logic. *IEEE Trans. Robotics Automation*, 17, 4, 490–497, 2001.
27. Urmson, C., Anhalt, J., Bagnell, D., Baker, C., Bittner, R., Clark, M.N., Dolan, J., Duggins, D., Galatali, T., Geyer, C., *et al.*, Autonomous driving in urban

environments: Boss and the urban challenge. *J. Field Rob.*, 25, 8, 425–466, 2008.
28. Weerakoon, K., Sathyamoorthy, A.J., Liang, J., Guan, T., Patel, U., Manocha, D., *Graspe: Graph based multi-modal fusion for robot navigation in unstructured outdoor environments*, 2022, arXiv preprint arXiv:2209.05722.
29. Zong, W., Zhang, C., Wang, Z., Zhu, J., Chen, Q., Architecture design and implementation of an autonomous vehicle. *IEEE Access*, 6, 21956–21970, 2018.

16

AI-Driven Electric Vehicle Integration for Sustainable Transportation

Loveneet Mishra, Usha Chauhan and Manasi Pattnaik*

DEECE, Galgotias University, Greater Noida, Uttar Pradesh, India

Abstract

Electric automobiles are at the leading edge of sustainable transportation. This effort to render transportation more ecologically friendly, since they reduce, if, not abolish reliance on conventional fuels and emit zero harmful emissions they, do however provide unique obstacles. One of the most significant challenges involves integrating AI along with additional technologies into the sector of transportation as electric vehicles replace more traditional gas powered vehicles. Their role in the transportation system expands and the demand on the power grid grows. This transition necessitates more electricity which poses a number of challenges. There is increased demand on the grid, which can put a strain on grid stability as well as the capacity to supply power reliably necessitating flexible solutions that will allow local grid administrators to successfully control and maximize the distribution of energy. Furthermore, as the number of electric vehicles grows charging infrastructure becomes increasingly important as charging facilities become increasingly critical for promoting electric vehicle use. Competition on power supply service sites increases, potentially affecting accessibility quality of service in public spaces to address these difficulties. An understanding of the dynamics inside the ecosystem of electrical vehicles is required. Together with interventions that use AI smart grid technology this study seeks to delve into these dynamics by exploring why it is crucial to regulate charging loads to ensure an high quality shift toward electric cars, exploring the intricacies of incorporating vehicles and leveraging AI to enhance energy efficiency and manage infrastructure. This study aims to provide insights and recommendations for fostering a robust and effective electric vehicle charging network.

***Keywords*:** Electric vehicles, power system, automobiles, artificial intelligence, transportation

**Corresponding author*: mpattnaik46@gmail.com

Arindam Mondal and Souvik Ganguli (eds.) Controller Design for Industrial Applications, (331–350)

16.1 Introduction

Revitalizing Electric Vehicles is a Contemporary Perspective on Sustainable Transportation and the Role of Artificial Intelligence. While electric vehicles (EVs) were initially introduced decades ago, recent technological advancements and growing environmental concerns related to petroleum-based transportation have sparked a renewed interest, leading to their current surge in popularity. Propelled by electric motors and fueled by onboard batteries, EVs stand in contrast to traditional vehicles, which rely on internal combustion engines and nonrenewable fossil fuels. Notably, the electricity powering EVs can be sourced from diverse origins, incorporating fossil fuels, renewable energy sources approximating solar and wind power, or a hybrid blend thereof. This versatility offers the potential for significantly lower emissions compared to conventional automobiles, particularly when powered by renewable energy sources (L. Diaz Olvera, 2010).

The swift proliferation of EVs heralds an auspicious juncture for the advancement of sustainable transportation, yet it also introduces new challenges, particularly in optimizing the charging infrastructure to meet the growing demand efficiently. Traditional charging methods often lack flexibility and adaptability to dynamic energy supply and demand patterns. Nevertheless, contemporary artificial intelligence (AI) technologies present auspicious solutions to tackle these challenges through the facilitation of intelligent and adaptable EV charging systems. This overview explores the application of AI in optimizing EV charging, discussing key technologies, benefits, challenges, and future directions (A. Nikitas).

Current State of EV Charging Framework: The implementation of EV charging setup has surged in recent years, propelled by governmental incentives, technological progress, and burgeoning consumer exigency. However, challenges persist in managing charging demand variability and grid integration and ensuring equitable access to charging facilities. Traditional charging stations typically offer fixed pricing and charging schedules, leading to inefficiencies and grid congestion during peak demand periods (Hafner).

Role of AI in EV Charging Optimization: AI technologies, including machine learning, optimization algorithms, and predictive analytics, play a vital role in optimizing EV charging infrastructure. Machine learning algorithms can find historical charging data, user behavioral patterns, and environmental parameters to predict future charging demand accurately.

Benefits of AI-enabled EV Charging:

The following are some advantages of integrating AI technology for optimizing EV charging:

i. Dynamic Charging Scheduling: Based on shifting energy costs, grid circumstances, and user preferences, AI algorithms can optimize charging plans in real time. Dynamic scheduling saves billing expenses, lessens grid congestion, and optimizes the usage of renewable energy sources (Adiang *et al.*, 2017).
ii. Demand Balancing: Demand balancing ai-enabled systems can lessen grid imbalances and prevent overloading local distribution networks by spreading the charging demand over various time periods and places as a result there is less need for expensive infrastructure investments and increased grid stability.
iii. Predictive Maintenance: Through the data from a charging station, AI could predict which equipment is prone to failure or breakdown, which gives an opportunity for proactive maintenance, reduces downtime, increases dependability, and upgrades the user experience. AI may personalize charging experiences according to the user, based on his/her previous drive behavior, limitations, and preferences (e.g., preferred charge times and locations, and possible payment methods), thereby increasing users' pleasure and loyalty.

Barriers and Concerns: Although the expected benefits of using AI-based EV chargers, there are many issues and problems that need to be sorted out prior to the implementation. Among them include:

i. Data Privacy and Security AI: Systems work by large databases that have personal information on EV owners, together with sensitive information regarding charging habits. To facilitate trust and conformity with laws, keeping these data confidential and safe is essential.
ii. Interoperability and Standardization: The broadening of the EV charging infrastructure realm with its service providers focuses on the necessity of interoperability regarding communication protocols and standardization for data formats, which ensure harmonious collaboration, compatibility, and full integration across varying charging infrastructures and networks.

iii. Scalability & Deployment: Setting up AI-powered EV charging systems involves significant investments in infrastructure, software development, and employee training. Spreading these technologies across various locations and customer groups can lead to logistical and operational hurdles.
iv. Regulatory Framework: The regulatory landscape governing EV charging infrastructure is complex and evolving. Policymakers must establish clear guidelines and incentives to promote the adoption of AI technologies while addressing concerns related to market competition, consumer protection, and environmental sustainability.

EVs were first introduced many decades ago, but due to technological advancements and the environmental effect of petroleum-based transportation, their resurrection and real popularity have only recently begun. Electric cars are propelled by electric motors and can be power-driven via electricity stored in onboard batteries (L. Diaz Olvera, 2010). Traditional automobiles, on the other hand, are propelled by internal combustion engines and rely on nonrenewable fossil fuels. EVs diverge from conventional counterparts in their ability to draw power from a spectrum of origins, encompassing fossil fuels, renewable dynamism reservoirs for instance solar and wind power, or a composite amalgamation thereof. EVs have lower emissions and other emissions than automobiles if they are powered by renewable energy.

The transportation industry consumes the most oil and so contributes significantly to air pollution. Growing concerns about environmental consequences and energy scarcity have fueled efforts to revitalize the transportation industry and expedite the use of electric mobility. While the adoption of EVs will improve the environment and help alleviate the energy crisis, it will also put a strain on the power infrastructure as demand shifts from gasoline to electricity.

The renewed interest in EVs in the past decade has been directly related to the merger of technological advancement and increased comprehension of the eco-friendly effect of traditional petroleum-built transportation. While the notion of electric automobiles has existed for decades, they have only recently accomplished significant traction and broad acceptance.

Conventional cars use gas-powered motors, operated by batteries that are powered by solar energy, or regular petroleum products. This contributes to environmental sustainability, by way of conventional petroleum products, or solar-powered energy, separately, or a combination of the two, while traditional cars do not. In addition, custom cars have the ability to harness energy from wind power, another environmentally conscious power source (Ehebrecht *et al.*, 2018).

Transportation technology, particularly those fueled by energy from renewable sources, such as electric automobiles (EV), has significant ecological and efficiency benefits. These developments in EV technologies would use AI to optimize energy utilization and create better vehicle performance (Edinger *et al.*, 2018). These cars emit much less CO_2, lowering the entire carbon footprint of a conventionally powered vehicle. This joint usage of modern technology validates EVs' future as eco-friendly and inexpensive transportation alternatives, as well as their overall presentation as part of an upsurge of sustainable transportation advances.

Transportation is being revolutionized by the rebirth of electric vehicles. EVs are driven by environmental consciousness and technological advancements. EV offer adaptability and the integration of AI, making them essential in the sector. Despite their long-standing presence. Their widespread appeal and momentum can be attributed to the combination of environmental concerns and technological advancements.

Studies on energy conservation, artificially intelligent technology, AI, and environmentally friendly transportation stress their interconnectedness and the opportunity for joint solutions. Recent research has focused on overcoming these challenges by enhancing battery technology, creating smart charging tactics such as integrating vehicle-to-grid (V2G) technologies. Furthermore AI technologies such as machine learning and forecasting are increasingly being used to optimize electric vehicle charging schedules control energy usage and improve traffic flow efficiencies.

This research uses a comprehensive examination of various sources, such as respected journals, conference papers, industry reports, and government publications, to clarify the present understanding of energy sustainability, AI, and sustainable transportation. It pinpoints key patterns and developments in EV adoption and usage while evaluating crucial obstacles and opportunities. Furthermore, actual cases and data are used to give practical perspectives on sustainable transportation solutions. By taking this interdisciplinary approach, the study seeks to provide a complete comprehension of the intricate connections among energy sustainability, AI, and sustainable transportation, especially in relation to EV integration and usage.

16.2 Overview of Electric Vehicle Charging

EVs' increasing integration has sparked a flurry of research activities from both business and academia. These initiatives can be divided into two categories, each of which addresses one of the two major effects described above. The initial category pertains to the ramifications associated with the power grid, notably the consequential substantial load impact capable of jeopardizing the stability and reliability of the control grid. The subsequent category concerns consequences oriented toward EV users, specifically the diminution in service quality induced by the competition among EVs for charging resources, including charging stations and power supplies (A. Nikitas) (L. Diaz Olvera, 2010).

A. EV Charging Oriented to the Power Grid
This category of research delves into strategies aimed at mitigating the potential grid ramifications stemming from widespread EV charging. Should the interaction between EVs and the power grid remain unregulated or lacking in coordination, an incongruity between charging requisites and power provisioning ensues (W. Adler, 2019). This phenomenon is exemplified by underscoring a positive correlation between the impact of charging loads and the level of EV penetration.

B. EV Charging with a Focus on User-Centered Design
User satisfaction and the shift toward sustainable mobility are significantly impacted by the adoption of EVs which face challenges such as range anxiety charging infrastructure and user experience necessitating the use of ai. Ai-driven systems are transforming important areas by addressing common deterrents like range anxiety by analyzing driving patterns weather conditions and traffic congestion. ai can provide tailored recommendations enabling drivers to undertake extended journeys without fear of battery depletion thereby reducing anxiety and promoting sustainable transportation (Damaj *et al.*, Apr. 2021).

AI-powered charging stations improve user convenience by adjusting rates based on grid demand and personalized preferences. They streamline charging timetables, reduce waiting times and optimize renewable energy utilization. AI technologies provide a seamless interface enabling hands-free interaction with onboard systems. AI powered autonomous driving enhances safety and convenience and collaborative ecosystems for EV fleets. AI driven predictive maintenance systems prevent breakdowns and optimize vehicle uptime (Li *et al.*, 2020).

AI and EVs are transforming industries, necessitating collaboration and innovation. By adopting a user-oriented approach and leveraging AI's power, we can overcome barriers to EV adoption, accelerate sustainable transportation transition, and create a cleaner, greener future for generations to come (Jia *et al.*, Oct. 2022).

The research group aims to improve EV user satisfaction by regulating competition for charging resources, considering factors like cost, duration, rate, and proximity to stations. This has led to a significant scholarly interest in charging rate control as a subject of numerous research endeavors. The majority of these fall into one of two categories. The first category focuses on electricity grid-related concerns, with the overall purpose of ensuring the grid's performance (Hamid, 2023). The second category is concerned with EV user difficulties, with the overall purpose of benefiting and satisfying EV users. Furthermore, these works include a variety of optimization goals, including load flattening, cost cutting, and fairness, among others. These recent developments were discussed in this section. The debate is organized first according to the category, then according to the optimization.

16.3 Rate Control Oriented at the Power Grid

i. Equalization of Direct Loads: Current initiatives aimed at flattening the load on the power grid typically elevate either the cumulative load or variance in amassed load.
ii. Aggregated Load Minimization (ALM) is a term that refers to the process of reducing the author's attempt to reduce aggregated load by utilizing a convex program to solve the problem. The charge time for each EV is continually adjustable within a particular range in the formulation, and this setting is comparable to the one used in this paper. They use a decentralized algorithm called gradient projection to solve the formulated problem (W. Adler, 2019). Although the approach can reduce the charging station's calculation responsibilities, it comes with a lot of expense in terms of transferring intermediate results between parties and requires each party's computation assistance.
iii. Load Variance Minimization is a technique for reducing load variation. A more straightforward goal for load flattening is to reduce the aggregated load's volatility directly (Zhao and Malikopoulos, Jan. 2022). The authors of this paper compare

global and local techniques to load variance minimization (W. Adler, 2019). The former is described as a global aggregator that coordinates EV loads, while the latter is described as community places that solely manage the loads among their own residences.

iv. Power Loss Minimization is a term used to describe the process of reducing the amount of increased overloads and power losses that can result from uncontrolled and unpredictable PEV charging, especially in the distribution system (Zhu *et al.*, Jan. 2019).

v. Frequency Control: The frequency of the electrical grid deviates from its normal value due to a mismatch between supply and demand, and the purpose of frequency control is to correct this mismatch. Financial incentives are commonly used in existing works to encourage EV customers to engage in frequency regulation.

A. User-Oriented Rate Control (EV User-Oriented Rate Control)
Cost-cutting is a term used to describe the practice of charging as little as possible. It may not be in a customer's best interests to follow power-grid control and meet the necessity of the power grid or indeed the aggregator. Indeed, mindlessly obeying the grid's or aggregator's requirements may not result in the best outcome for an EV customer.

Fairness in Pricing Existing papers that are most relevant to this paper's discussion include. These two papers define EV customer satisfaction as fairness in power distribution and propose distributed methods for determining charging prices that achieve proportionate fairness among vehicles (Hafner).

Certainly, here is a summary of some recent research on user-oriented rate control of EVs. EV charging stations and pricing strategies, dynamic pricing plans to encourage off-peak hours and renewable energy using behavioral analysis and user profiling using machine learning for personalized suggestions, smart grid integration and demand response strategies for grid stability and rate modification, predictive analytics and optimization algorithms for commercial applications like delivery and ride-hailing.

Advanced technologies like machine learning smart grid integration and human-computer interaction are being integrated into user-oriented rate control of EVs to improve charging experiences and promote sustainable transportation.

16.4 Model of the System

The charging station (CS) has a controller that adjusts the charging proportions of EVs within its premises, as depicted in Figure 16.1. Employing a discrete-time model, we partition the command horizon into a succession of equidistant time intervals. In practice, these intervals may span several minutes, such as 5 minutes. The controller governs the charging velocities of altogether EVs by the side of the charging point during each time interval (Hamid, 2023). For the design and analysis of the accompanying algorithm, we operate within a single interval. The entirety of the control process is disaggregated into discrete time intervals, each executing the prescribed algorithm (Eghtessad, 2014).

I. In addition to relying on the power grid, charging stations are progressively tapping into localized renewable energy sources, for instance, solar besides wind power, to fulfill the imperatives of sustainable development. As a result, we take into account both the electricity grid and co-located renewable generation. A transformer provides power to a charging station from the electrical grid.

II. The threshold of the feeding new broom capacity is denoted by the symbol L, representing the utmost power load within the transformer's capability to manage (Hamid, 2023). Please be advised that the design besides scrutiny herein can be seamlessly modified to accommodate a transformer serving

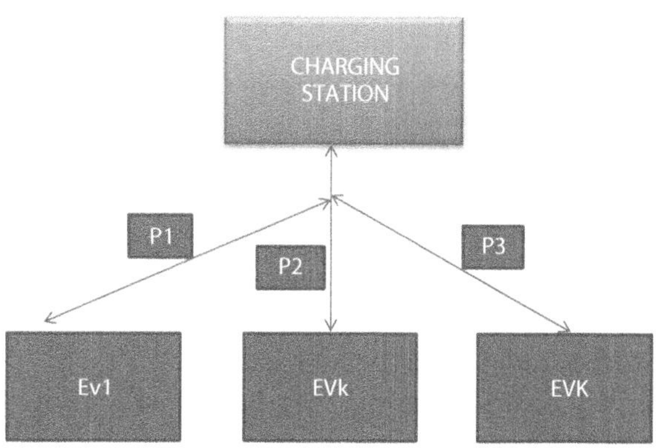

Figure 16.1 A charging rate control scheme. S: charging station (CS) power supply; p-k: EVk charging rate.

multiple charging stations. "R" designates renewable energy generation, sourced from diverse outlets such as solar and wind energy. The charging place initially taps into renewable energy reservoirs before resorting to grid power to compensate for any deficits.

III. Model of a substantial load. The power demand of the feeding transformer can be segmented into two classifications: the flexible EV load originating from the charging station and the inflexible base load from other users. The degree of elasticity is contingent upon the charging station's capacity to regulate the load.

A. Demand Response Charging

Customers are incentivized to adjust their electricity consumption in accordance with the status of battery packs, such as by curtailing their energy usage in rejoinder to signals indicating high price peaks or directives from electric utility providers, facilitated by demand response initiatives (Onat and Kucukvar, Nov. 2022). Consequently, demand rejoinder mechanisms can aid the clout grid in shifting from a supply-driven to a demand-driven paradigm.

Demand response has been underscored as a pivotal element of (Goletz and Ehebrecht, 2018) the prospective smart grid, notably in assimilating decentralized generation (such as co-located renewables) into the electricity grid and meeting postponed demands (such as EV charging) from clientele.

B. Load-Scheduling Charging

Owing to the protracted and recurrent charging cycles inherent in EV charging, a promising control strategy known as park-and-charge has emerged as of late. Under this approach, parking lots are equipped with charging infrastructure, thereby offering dual functionality in parking and charging services (Tengilimoglu *et al.*, Jan. 2023). During the duration of the vehicle's stationary status, the electric car undergoes charging. Patrons can expeditiously resume their engagements subsequent to parking their vehicles in this mode, alleviating concerns pertaining to the charging process. Facilities such as office towers, shopping complexes, and airports stand to reap substantial advantages from the implementation of this park-and-charge methodology (Starkey, March 2016).

Numerous field trials have recently been undertaken to assess the achievability of the park-and-charge methodology. For instance, in conjunction with The Bosch Group, several European entities are engaged in the V-Charge initiative, which endeavors to pioneer an automated valet parking plus charging system to support autonomous local transit. GM, OnStar, and Timber Rock have collaborated to execute an innovative experiment at GM's E-Motor Plant, with the neutral of synchronizing the charging requisites of stationary electric automobiles with nearby renewable energy sources (Badami, 2005).

C. Demand-Balancing Charging
The provision of public charging stations is aimed at diminishing charging duration, thereby encouraging the adoption of EVs. Conversely, competition for charging infrastructure diminishes service standards and poses a threat to this objective. EVs might, for instance, converge on a charging station offering advantageous conditions such as lower energy tariffs and closer proximity. Their collective preference could strain the charging facility, prompting certain users to depart in pursuit of an available charging slot. The inefficiency resulting from this search contradicts their original aims of cost savings or expedited charging intervals (Kumar and Alok, Apr. 2020, Republic of Tanzania).

I. Demand balancing on a global scale. The authors describe here for allocation of EVs across many stations in order to improve service quality. They analyze performance by simulating a collection of charging stations with a multi-queue system.
II. EV Routing (Electric Vehicle Routing). The shortest route in road networks is the focus of traditional vehicle routing problems. Because of the unique properties of EVs, such as their restricted cruising range, recharging times, and capacity to regenerate energy, new routing algorithms are needed to select the most cost-effective routes rather than merely the shortest ones. As a result, the majority of routing algorithms incorporate the above qualities into current shortest-path techniques.

16.5 Future of AI-Enabled EV Charging

i. Edge Computing and IoT Fusion: The amalgamation of edge computing and Internet of Things (IoT) technologies facilitates instantaneous data processing and decision-making at the periphery of the network, diminishing latency and amplifying scalability (The Transport Licensing (Motor Cycles and Tricycles) Regulations, 2010).

ii. Vehicle-to-Grid (V2G) Integration: V2G technology permits EVs to interact with the grid bidirectionally, enabling them to store and discharge energy based on grid conditions and demand signals. AI algorithms can optimize V2G operations to maximize grid stability and revenue generation for EV owners.
iii. Collaborative charging networks enable EV owners to share charging resources, monetize excess capacity, and access additional options during peak demand periods.
iv. The rise of autonomous and connected vehicles presents new opportunities for AI-enabled EV charging, including predictive routing, dynamic energy management, and seamless integration with smart city infrastructure.

16.5.1 Edge Computing and IoT Integration

Edge computing and IoT integration are revolutionizing data processing in distributed systems. Edge computing processes data near its origin, reducing reliance on centralized cloud infrastructure. IoT involves interconnected devices and sensors gathering and conveying data across the internet, transforming the way data is processed and analyzed in distributed systems (Hakak *et al.*, Feb. 2023).

Bringing computing closer to where data are created with internet-connected devices offers advantages like instant processing, reduced waiting times, improved adaptability to more users, and increased privacy and safety, especially for smart cities, automated factories, healthcare, and transportation. Connecting internet devices directly to networks at the edges is ready to totally change transportation by enhancing traffic management, monitoring vehicles, optimizing fleets, and developing intelligent transportation systems (Bishop and Amos, 2015). Computing near the edge in transportation allows instant decision-making near sensors and equipment in vehicles, roads, and infrastructure by placing technology close by. This permits handling data locally and reacting immediately to what happens.

IoT gadgets link up car sensors, video cameras, and (Kuutti *et al.*, Feb. 2021) controllers for traffic signals with the web. This gives valuable information about how our transportation network is doing. IoT and edge computing let us process data and analyze it smartly. Machine learning algorithms can sift through all the real-time data and spot patterns. This means that we can make decisions before problems occur, predicting issues in our transportation systems. It also allows frequent checks on vehicles for mechanical hitches or tired drivers.

One key plus of combining edge computing and IoT in transport is that it cuts down data delays and network jams. This happens by handling data locally at the edge instead of shipping it off to central cloud servers to be studied. This lowers communication lag times and makes sure less bandwidth is used. This is really key for safety-first applications, as even the smallest delay can lead to big problems. Also, edge computing allows for divided computing setups. This means data processing jobs are shared around many edge devices, which eases the load on individual devices and pumps up scalability.

To wrap up, combining edge computing and IoT is changing transport. It is prompting faster choices, smarter data crunching, and faster responses. These tools help those in transport better control traffic, keep an eye on vehicles, boost safety, and allow for fresh, new ideas. Still, there are issues like link challenges, safety, and keeping data private. These need solutions to make the most of edge computing and IoT in transport. But, it is clear that the way we think about travel is guided by these fresh ways of thinking.

16.5.2 Vehicle-to-Grid (V2G) Integration

Vehicle-to-grid (V2G) brings a fresh idea to electric cars and energy rules. It lets electric cars use power from the grid and give extra back if needed. This two-way flow of power changes electric cars into moving energy stores. This offers many benefits to car owners and the broader power network.

V2G works by using clever charging stations and a system so electric cars can talk to the grid in real time. This gives many options across different areas. It is useful in managing the power grid, adding renewable energy, and demand response programs (Ercan *et al.*, Nov. 2022).

V2G integration has many benefits. It helps make the energy grid stronger and more dependable. How? It uses the power from EV batteries during busy times or when the grid is full. This helps utilities avoid blackouts. It also means they do not have to spend as much on updating infrastructure. The grid becomes stronger. Plus, EVs with V2G can give useful services to the grid. These include frequency control, voltage support, and peak shaving. This helps balance the energy supply and demand and keeps the grid stable.

Moreover, a V2G connection lets solar and wind energy sync with the grid better. Easy to grasp, EVs can store extra clean energy when there is a lot of it. They give it back when clean energy is low or when we need more power. Simply put, this can iron out the ups and downs of renewable power. It can also reduce wastage and boost the amount of clean energy we use overall. If you own an EV, V2G integration can put more money in your pocket. This cuts down the cost of owning your vehicle. How? Simply plug your car into demand response programs or energy markets. You give power back to the grid and get paid or get rewards in return. This way, not only do you save money but you also promote clean transportation and renewable energy methods (Huang *et al.*, Dec. 2022).

With V2G, EV batteries last longer and work better. It cuts down the times you have to change them. How? V2G makes the best of charging and discharging based on what the grid needs and what the user wants. So, EV batteries using V2G do not get worn out fast. They remain reliable and high-performing over the long term.

Although V2G brings good things, it comes with its issues. We need to think about how different systems talk to each other, the security of data, and rules, and regulations. We need a strong setup for a two-way energy flow. Plus, the use of V2G may mean changes to business models and rules. It might need motivations for investing and taking part from various parties.

To sum up, V2G integration is an innovative strategy for managing electricity that utilizes EVs to strengthen grid reliability, blend renewable power, and generate new income for owners of EVs. Even though there are many issues to be addressed, the possible benefits of V2G integration are huge and might change our energy generation, distribution, and use in the future. Nowadays, since technology is improving at a faster rate and the adoption rates have also risen, V2G integration is on the verge of becoming a major player in building sustainable, robust energy ecosystems (Liu *et al.*, 2020).

16.5.3 Collaborative Charging Networks

A pioneering technique for addressing the problems that arise from EV charging infrastructure is represented by collaborative charging networks which include limited availability of charging stations, long waiting times, and grid congestion during peak demands. This new way of thinking leverages the power of resource pooling and peer cooperation to optimize charging resources in order to enhance overall efficiency and availability for EV users as detailed in L. Diaz Olvera (2010).

Collaborative charging networks enable ev owners to share their electricity supply in return for rewards utilizing peer-to-peer (p2p) or community-based charging points. This coordination of schedules and resources reduces pressure on the utility grid.

Ev owners benefit from collaborative charging networks which offer key advantages such as increased charging stations monetization of excess capacity and cost-effective infrastructure expansion without new hardware investments. Ev charging networks optimize schedules to minimize unnecessary infrastructure and energy wastage promoting sustainability and resource optimization by prioritizing renewable energy storage and minimizing ecological impact.

Ev charging networks are a crucial means for users to share their power reservation systems and knowledge promoting eco-friendly transport. These networks not only enhance accessibility and sustainability but also provide economic benefits for both owners and providers fostering a sense of community and knowledge sharing.

Collaborative charging networks are promising for optimizing EV charging infrastructure, enhancing transportation convenience, sustainability and economic viability. These types of networks involve peer-to-peer collaboration and resource sharing, transforming how EVs are charged and hastening the transition towards cleaner, energy-efficient and fairer transport systems. However, challenges such as interoperability, cybersecurity and regulatory barriers need to be addressed.

16.5.4 Autonomous and Connected Vehicles

Autonomous vehicles also known as self-driving cars are autonomous vehicles that can navigate and function without human intervention. Leveraging an array of sensors, cameras, radar, lidar, and sophisticated AI algorithms, they discern their environment, formulate decisions, and execute driving maneuvers. These vehicles proffer a plethora of advantages, encompassing heightened safety, diminished traffic congestion, augmented mobility for individuals with disabilities, and heightened productivity during commuting endeavors (Hamid, 2023).

One of the principal merits of autonomous vehicles lies in their capacity to markedly diminish traffic accidents and fatalities. Through the eradication of human error, a predominant catalyst of accidents, autonomous vehicles possess the capability to safeguard thousands of lives annually and avert numerous injuries. Furthermore, autonomous vehicles can enhance traffic fluidity, alleviate congestion, and enhance fuel efficiency by curbing abrupt braking, acceleration, and lane alterations.

On the other hand, connected vehicles employ wireless communication technologies to share information with other cars, infrastructure, and the environment around them instantly. This allows for a wide range of applications and services which include vehicle-to-vehicle (V2V) communication, vehicle-to-infrastructure (V2I) communication, and vehicle-to-cloud (V2C) communication.

These automobiles can interconnect with surrounding vehicles about their speed, position, and trajectory, hence allowing cooperative collision avoidance and traffic management. Also, V2I allows vehicles to interact with traffic signals, road signs, and infrastructure sensors to get real-time traffic updates, navigation help, and hazard warnings. V2C is there to let cars use cloud-based services such as real-time traffic data, weather forecasts, or remote software updates.

The synergy between autonomous vehicle technology and connectedness results in many positive effects. When self-driving cars are combined autonomously with the connectivity of their counterparts, they can access up-to-date environmental data so as to make better decisions more safely while maneuvering through complex driving situations.

What is more, new mobility services and business models such as ride-hailing, car-sharing, and MaaS have been made possible by the advent of autonomous and connected cars. These solutions utilize the opportunities that are presented by self-driven cars which provide on-demand transportation services leading to a reduced need for privately owned vehicles while maximizing the current transport networks.

Notwithstanding their numerous advantages, autonomous and connected vehicles also come with difficulties and considerations to deal with. Stakeholders including infrastructure providers are grappling with widespread cyber risks highlighting the need for collaboration and cooperation to mitigate these potential societal impacts.

The stated safety and values of self-driving cars are transforming the transportation system offering improved efficiency and convenience.

16.6 Conclusion

In this final part, we have focused on the practical aspects of EVs and charging load control needed to make them a viable part of sustainable transport. As we have seen, the era of the electric car is with us at last. Cars are no longer dependent on oil; instead, EVs have arrived, leaving us with no sacrifices to our environmental conscience. Still, the surge in demand has placed a massive strain on existing electricity infrastructure. To relieve it, the strategic deployment of public charging stations among a wider distribution of private charging points had to be strategized. A significant point requiring attention involves the management of power resources to individual EVs at an indicting outlet, which regulates the flow of charge within each dock. Such an approach helps to optimize efficiency and enhance the reliability of EV charging systems, providing seamless operations with no undue stress on power grids. This paper will also discuss various future research directions that are promising in this field. As more people embrace EVs, there is an urgent need for creative solutions that would make them scalable and durable.

References

Nikitas, A., Thomopoulos, N., Milakis, D., The environmental and resource dimensions of automated transport: A Nexus for enabling vehicle automation to support sustainable urban mobility. *Annu. Rev. Environ. Resour.*, 46, 1, 167–192, Oct. 2021.

Hafner, B., Bajpai, V., Ott, J., Schmitt, G.A., A survey on cooperative architectures and maneuvers for connected and automated vehicles. *IEEE Commun. Surv. Tuts.*, 24, 1, 380–403, Dec. 2022.

Adiang, C.M., Monkam, D., Njeugna, E., Gokhale, S., *Projecting Impacts of Two-Wheelers on Urban Air Quality of Douala, Cameroon*, Elsevier, pp. 49–63, Transportation Research Part D 52, 2017.

Ehebrecht, D., Heinrichs, D., Lenz, B., Motorcycle-taxis in subSaharan Africa: Current knowledge, implications for the debate on "informal" transport and research needs. *J. Transp. Geogr.*, 69, 242–256, 2018.

Edinger, H., Mc Donald, K., Schaefer, S., *Africa Automotive Insights - An East African consumer perspective*, Deloitte Africa Automotive, Brochure, 2018, [Online]. https://www2.deloitte.com/content/dam/Deloitte/ug/Documents/consumer-business/ea_2018-A-Consumer-Perspective-Auto.pdf.

Damaj, I.W., Serhal, D.K., Hamandi, L.A., Zantout, R.N., Mouftah, H.T., Connected and autonomous electric vehicles: Quality of experience survey and taxonomy. *Veh. Commun.*, 28, 1–12, Apr. 2021.

Li, K., *et al.*, Cloud control system for intelligent and connected vehicles and its application. *Automot. Eng. China*, 42, 1595–1605, 2020.

Diaz Olvera, L., Plat, D., Pochet, D., Sahabana, M., Motorized Two-Wheelers in Sub-Saharan African Cities: Public and Private Use, in: *Paper presented at the 12th World Conference on Transport Research*, WCTR society, 2010, Lisbon, Portugal, 11-15 July 2010.

Jia, L., *et al.*, On autonomous transportation systems. *Smart Resilient Transp.*, 4, 2, 66–77, Oct. 2022.

Zhao, L. and Malikopoulos, A.A., Enhanced mobility with connectivity and automation: A review of shared autonomous vehicle systems. *IEEE Intell. Transp. Syst. Mag.*, 14, 1, 87–102, Jan. 2022.

Zhu, L., Yu, F.R., Wang, Y., Ning, B., Tang, T., Big data analytics in intelligent transportation systems: A survey. *IEEE Trans. Intell. Transp. Syst.*, 20, 1, 383–398, Jan. 2019.

Eghtessad, M., *Optimale Antriebsstrangkonfigurationen für Elektrofahrzeuge. Ph.D., 2014, dissertation*, Technische Universität Braunschweig, Schriftenreihe des Instituts für Fahrzeugtechnik, TU Braunschweig 35, 2014. ISBN: 978-3-8440-2782-2.

Badami, M.G., Transport and Urban Air Pollution in India, in: *Environmental Management, B*, 36, vol. 2005, pp. 195–204, Springer US, Brooks, 2005.

Goletz, M. and Ehebrecht, D., How Can GPS/GNSS Tracking Data Be Used to Improve Our Understanding of Informal Transport? A Discussion based on a Feasibility Study from Dar es Salaam. *J. Transp. Geogr., 88*, 2018. (in press).

Adler, M.W., Peer, S., Sinozic, T., Autonomous connected electric shared vehicles (ACES) and public finance: An explorative analysis. *Transp. Res. Interdiscipl. Perspect.*, 2, 1–13, Sep. 2019.

Onat, N.C. and Kucukvar, M., A systematic review on sustainability assessment of electric vehicles: Knowledge gaps and future perspectives. *Environ. Impact Assess. Rev.*, 97, 36–49, Nov. 2022.

Tengilimoglu, O., Carsten, O., Wadud, Z., Implications of automated vehicles for physical road environment: A comprehensive review. *Transp. Res. E Logist. Transp. Rev.*, 169, Jan. 2023.

Starkey, P., The Benefits and Challenges of Increasing Motorcycle Use for Rural Access, in: *Paper presented at the International Conference on Transport and Road Research*, Mombasa, Kenya, 16-18 March 2016.

Kumar, R.R. and Alok, K., Adoption of electric vehicle: A literature review and prospects for sustainability. *J. Clean. Prod.*, 253, Apr. 2020.

Republic of Tanzania, *The Transport Licensing (Motor Cycles and Tricycles) Regulations*, 2010.

Hakak, S., *et al.*, Autonomous vehicles in 5G and beyond: A survey. *Veh. Commun.*, 39, 100551, Feb. 2023.

Kuutti, S., Bowden, R., Jin, Y., Barber, P., Fallah, S., A survey of deep learning applications to autonomous vehicle control. *IEEE Trans. Intell. Transp. Syst.*, 22, 2, 712–733, Feb. 2021.

Bishop, T. and Amos, P., Opportunities to Improve Road Safety Through 'Boda-Boda' Associations in Tanzania, in: *Final Report*, p. 16, AFCAP/ASCAP Project Management Unit, Cardno Emerging Market (UK) Ltd, Thame, UK, 2015.

Ercan, T., Onat, N.C., Keya, N., Tatari, O., Eluru, N., Kucukvar, M., Autonomous electric vehicles can reduce carbon emissions and air pollution in cities. *Transp. Res. D Transp. Environ.*, 112, Nov. 2022.

Hamid, U., *Autonomous Connected Electric and Shared Vehicles: Disrupting the Automotive and Mobility Sectors*, SAE International, Warrendale, PA, USA, 2023, [online] Available: https://www.sae.org/publications/books/content/r-517/.

Huang, X., Cen, X., Cai, M., Zhou, R., A framework to analyze function domains of autonomous transportation systems based on text analysis. *Mathematics*, 11, 1, 158, Dec. 2022.

Liu, Y., Yu, Y., Gao, N., Wu, F., A grid as smart as the internet. *Engineering*, 6, 7, 778–788, 2020.

17

Wireless EV Charging System Design

Koushik Majumder[1], Maitrayee Chakrabarty[1], Rakesh Das[2*] and Raju Basak[3]

[1]*Department of Electrical Engineering, JIS College of Engineering, Kalyani, India*
[2]*Department of Electrical Engineering, Calcutta Institute of Technology, Uluberia, Howrah, West Bengal, India*
[3]*Department of Electrical Engineering, Techno India University, West Bengal, India*

Abstract
The technology of electric vehicles (EVs) is changing very fast in order to fabricate the automobile industry to a large extent. The growth of electric automobiles (EAs) is taking place very fast to replace energy from fossil fuels. The biggest challenge in front of EA is its battery capacity and charging infrastructure. Battery charging architecture is a crucial fact for EA consumers. Dynamic wireless charging (DWC) is designated as motion charging. Three distinct power transfer systems—capacitive wireless power transfer (CWPT), inductive wireless power transfer (IWPT), and resonant-inductive wireless power transfer (RIWPT)—are analyzed in this research for a dynamic wireless EA charging system. This study investigates different parameters, and the resonant inductive power transfer charging system is the most well-known and advanced version of the DWC system. The result shows a great dependency on wireless charging systems, with large efficiency. This investigation proposes a dynamic contactless charging station for EAs that utilizes a renewable system on the concept of electromagnetic induction. The system's outline and application require attentive consideration of several limiting factors, as well as energy storage system size, charging pad efficiency, and power translate topology. In addition, the overall cost and practical approaches must be ensured with intelligent decision mechanisms to improve the performance.

***Keywords*:** Wireless, EV charging system, renewable energy, dynamic system

**Corresponding author*: rakesh.das.btech@gmail.com

Arindam Mondal and Souvik Ganguli (eds.) Controller Design for Industrial Applications, (351–374)

17.1 Introduction

Plug-in electric automobiles (PEAs) for stationary applications and high-power uses such as electric vehicles (EAs) have both recommended the use of contactless charging systems (CCSs). When compared to plugin charging methods, CCS offers additional benefits with regard to ease of use, dependability, and simplicity [1]. To improve the two areas of battery depot capacity and range, the dynamic mode of operation of the CCS for ECs has been investigated. In an effort to improve the two areas of range and sufficient volume of battery depots, the dynamic mode of operation of the CCS for ECs has been researched [2]. One issue or constraint of CCS is that it can only be used in stationary settings, like parking lots, carports, or traffic signals, or while the car is immobile. Furthermore, there are several obstacles that stationary CCS must overcome, including problems with electromagnetic compatibility (EMC), restricted power transfer, large buildings, a shorter range, and increased efficiency [3]. This technique allows battery storage devices to be charged while the car is moving. This increases the vehicle's range for transportation and reduces the requirement for expensive battery storage [4]. Nonetheless, a dynamic CCS has to contend with two primary challenges: a wide air gap and coil misalignment [5] prior to its general acceptance. The power transmission efficiency is determined by the arrangement of the coil and air gap between the source and receiver [6]. The normal air-gap distance for small passenger automobiles is between 150 and 300 mm, while it may be greater for larger cars [7]. In the dynamic mode, the vehicle is automatically driven, making it simple to align the transmitter coil at the ideal driving position. Furthermore, in order to lower parasitic losses and boost system efficiency, various compensation procedures—such as combinations of series and parallel—are employed on both the receiving and sending sides. The basic functioning of CCS for EAs, including power transfer techniques, is examined in this research study. Furthermore, a range of contactless transformer architectures are described to increase the effectiveness of power transfer [8]. Additionally, this study describes recent advancements in the commercial and academic domains for WEACS's static and dynamic modes. The Contactless power transfer (CPT) technique can resolve any and all EA charging issues [9]. Parking the car at the charging station makes charging an electrical vehicle simple thanks to CPT technology. Lately, a lot of electrical cars employ the inductively coupled power transfer (ICPT) technology for CPT. Compensating capacitors and a linked inductor with a significant air gap between them make up the ICPT.

The whole flux generated by one inductor winding becomes connected to the other when the windings share a magnetic core. Tightly linked networks are employed to transfer electricity effectively as a result. Alternatively, the flux connecting the secondary coil will be very low if the coils are not separated by a significant air gap or share a magnetic core, which will lower the power transmission via the coils [10]. Nevertheless, the low efficiency of the CPT system can be addressed by maintaining these coils at a resonant frequency. To keep the connected inductors at resonance, a capacitor is provided on both sides of the inductor. We refer to this as compensation [11]. Similarly, Kumar and Padmanaban [12] investigate the impact that EVs can have on the necessary levels of grid capacity, productivity, and efficiency. They also examine the effects of EVs on the environment and the economy. An overview of EV charging techniques is given by Habib *et al.* [13], along with an analysis of how they affect power distribution systems. Furthermore, the writers conduct an examination of coordinated and uncoordinated charging techniques, delayed loading, and thoughtful charge planning. Finally, they investigate the financial benefits of charging strategy–based vehicle-to-grid (V2G) technology.

17.2 Literature Survey

Various reviews of the literature have been conducted based on the body of knowledge on EVs. A CCS for high-power applications, such as EAs and PEAs in stationary applications, has been proposed by Kalwar *et al.* One issue or constraint of CCS [1] is that it can only be used in situations where the car is stationary, like parking lots, car parks, or traffic signals. Furthermore, stationary WCS has a number of difficulties, including limited power transfer, bulkier constructions, lower range, and higher efficiency, as well as EMC concerns. There are different significant advantages that have been highlighted in this paper. The total investigation is based on the theorem of static power transfer methods for wireless power transfer. The "TESLA FUNDA", which is relatable to the great scientist NIKOLA TESLA, was also investigated.

On the other hand, the total investigation could not hold the information regarding the dynamic power coupling system, and this theorem only held the information about the difference between the two types of charging systems. More recently, Das *et al.* [14] provided an assessment of the potential effects of linked EVs and autonomous driving on grid integration and EV charging.

Imagine a future where autonomous electric cars (EAs) require a method for self-driving charging that does not require human assistance. To ensure a viable business model and rapid adoption in electrified transportation systems, a completely automatable, quick, safe, economical, and dependable charging infrastructure is necessary. Wireless charging technologies make it possible to understand these characteristics. CPT, developed by Mahesh *et al.* [2], is an advanced technology that provides benefits such as safety, convenience, flexibility, and the potential for full automation. Resonant inductive wireless charging, which is highly efficient and requires little maintenance, needs to be given more consideration compared to alternative wireless power transfer technologies in WPT methods. Recent developments in Big Data analytics are reviewed by Li *et al.* [15] in order to enable data-driven battery health estimates. More precisely, they talk about their benefits and drawbacks and categorize them according to cost-effectiveness and viability. Reviewing the current state of resonant inductive wireless power transfer charging technologies in this literature emphasizes both its current and potential futures in the wireless EV market. The study begins by providing a quick overview of wireless charging techniques, emphasizing their benefits and drawbacks. The paper makes this easier using a comparative examination of the several inductive pads, rails, and compensation technologies that have been produced thus far. The properties of both static and dynamic charging techniques are also demonstrated. The relevance of power electronics as well as different kinds of converters utilized in different applications is covered.

There are different significant advantages that have been highlighted in this paper. The investigation is more efficient for the research of wireless power transfer and inductive power transfer methods. The perfect description for inductive power transfer and a good investigating model is incorporated here.

Different disadvantages have been found in this paper. The investigation is based only on an inductive power transfer system and more description is required about capacitive and resonance inductive power transfer systems.

The shortage of petroleum supplies has led to the widespread adoption of EVs as a replacement for the traditional transportation system. Over the past 20 years, a lot of research has been done to develop effective EV charging techniques. According to Laskaris *et al.*, early study of a few wireless charging techniques, ICPT is a practical and promising technology for charging EVs wirelessly. The ICPT system's comparable circuit analysis and characteristics are included in this study, which also highlights the advancements in research concerning charging coil designs, misalignment tolerance, power level augmentation, and leakage inductance compensation topologies. The adoption of EV charging has been made possible by the advancement of these elements. There is now a brief description of the ICPT system's control and design process. The technology of EVs is very different from internal combustion engines. There are various features in how EVs are manufactured as compared with ICE. The objective is to protect the engine, on the other hand, protecting the batteries in manufacturing an EV. Designers and engineers are completely modifying the design of EVs, as well as creating new production and assembly methods to build it on a large scale. They are now designing an EV from various considerations taking into mind the concept of aerodynamics, weight, and other energy efficiencies. EV charging stations' connected loads are often very intermittent because of the variation in the charging level of batteries. Controllers like P, PI, and PID do not recommend a quick transient response to the converter [16]. In these cases, EV charging stations face instability regions due to rapid changes in loads. So, investigation of nonlinear controllers for EV battery charging stations is very essential for applications. The challenge is to design a driving system with high performance, reduced speed and increased torque, a large speed range, and excellent overall efficiency [17]. Based on the data taken from the literature, conclusions have been drawn, and some suggestions for the future are given.

So, the total literature survey shows the results that no type of research was held in the past which is based on the comparable study for three types of contactless power transfer methods. That is why there was a comparable study for three types of power transfer systems. Another thing that we have followed from the literature survey is that no single research has been conducted previously that compares solar system–generated power and a moderated bus-bar system. This is the primary investigation that compares three types of contactless power transfer systems with the help of the same power and the same bus-bar system. Liu *et al.* explain the creation and application of a multifunctional TMR sensor matrix for the CCS of EVs [18]. Qin *et al.* describe a 6.6-kW high-frequency wireless power transfer system that charges EVs at megahertz utilizing a multilayer nonuniform self-resonant coil [19]. On the other hand, Zhang *et al.* state that for an EV DWC system and analyze the effectiveness of the coil design [20]. Also, Miller *et al.* state the fundamentals of primary-side power flow control of contactless power transfer for EV charging systems [21]. Additionally, Khan *et al.* also describe wireless power charging on EVs [22].

17.3 Problem Statement

The DWC system of EVs is the foundation of this idea. Due to the issue with electrical vehicles that take too long to recharge their batteries in a static state, dynamic charging is the ideal solution.

Figure 17.1 presents the overall process of the proposed paper for power flow for wireless power transfer:

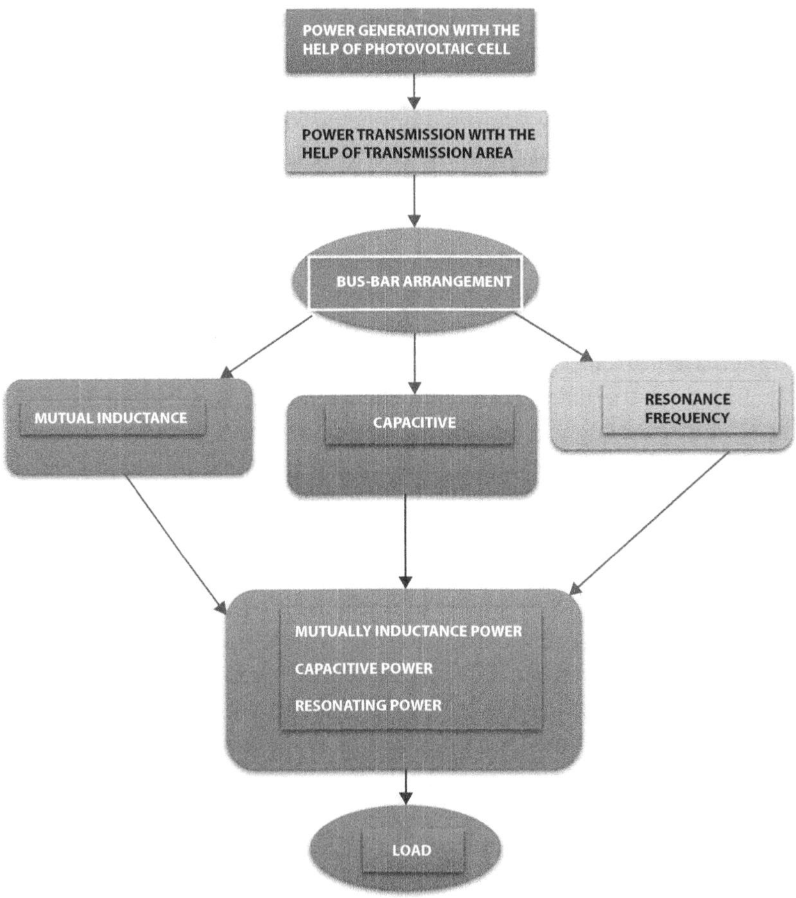

Figure 17.1 Flowchart of the proposed method.

17.4 Methodology

There are numerous resonating frequencies in the system because different reactive components contribute to resonance. Resonant frequencies differ from other frequencies in a few ways. Therefore, to comprehend the CPT system, a thorough study of resonant circuits at different frequencies is necessary. After performing a frequency analysis on a series-series (SS) compensator, one of the other compensators was selected for contactless power transfer. PV arrays can be used in conjunction with CPT to give an environmentally responsible way to charge EA. A solar energy source can be converted into electricity using a solar cell in a quiet, clean, and pollution-free manner. Therefore, the PV array can serve as the CPT system's main energy source. Considering all of the aforementioned elements, Figure 17.2 shows the proposed system.

In the current day, transport networks are essential. It also contributes roughly 37% of the carbon emissions. Because carbon emissions cause global warming to reach dangerous levels, they are essential. In addition to endangering human life, global warming causes significant alterations in the climate.

Figure 17.3 displays the global warming pie chart study from many sources.

Researchers, policymakers, and government officials are working to minimize carbon emissions in an effort to mitigate the effects of global warming. The Paris Agreement for Zero Carbon Emissions includes this endeavor. The primary source of carbon emissions is the combustion of fossil fuels. Fossil fuels are used to provide fuels for transportation systems, like diesel and gasoline. Figure 17.1 shows global carbon emission statistics broken down by sector. The global electric automobile market share is studied from various sources. Figure 17.4 shows the graphical increased market for electric automobiles.

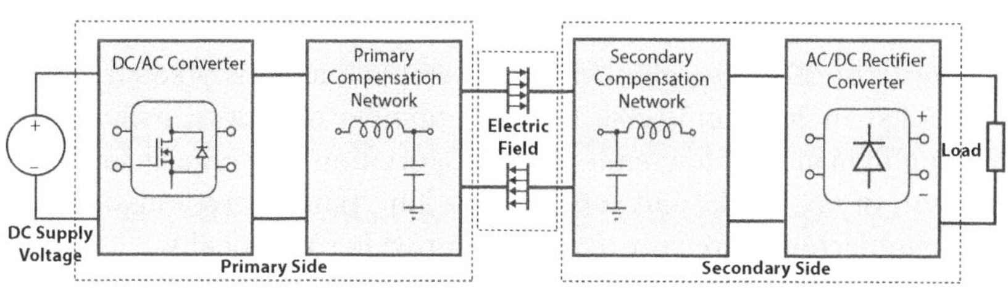

Figure 17.2 Wireless power transfer system (CPT).

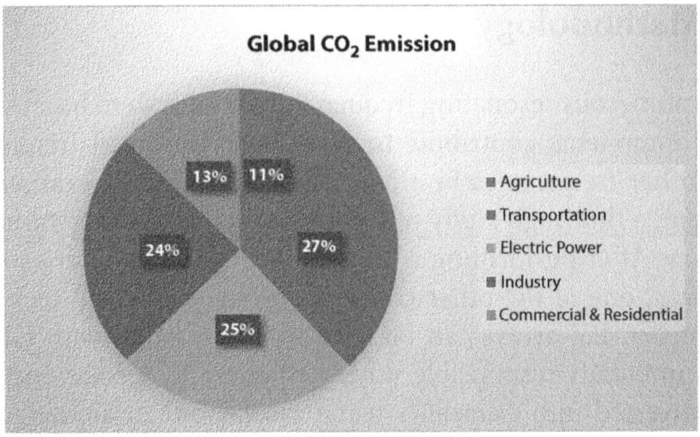

Figure 17.3 Pie chart for the global carbon emission.

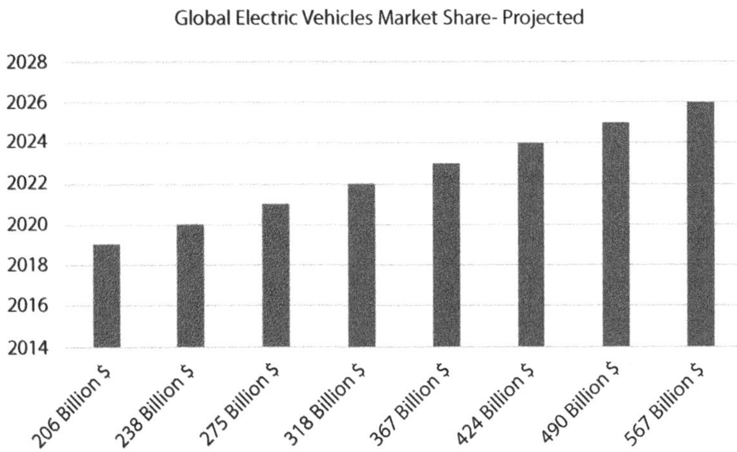

Figure 17.4 Increased market for electric automobiles.

The ideal substitute for conventional transportation systems is EVs. Due to their nearly minimal emissions, EVs are gaining popularity throughout the world. By 2030, the number of EVs on the market is predicted to rise from 8,151 in 2022 to 39,208. Due to a number of factors, including the growing demand for low-emission transportation and official support in the form of tax breaks and subsidies for long-range, zero-emission vehicles, manufacturers are now required to provide EVs globally.

17.5 Principle of WPT

Figure 17.1 shows the Static-WCS diagram for EVs. By using AC/DC and DC/AC converters, high-frequency (HF) AC is generated from AC mains from the grid to increase power transfer from the transmission coil to the reception coil. To improve system efficiency overall, compensation topology based on series and parallel combinations is applied. The oscillating magnetic flux fields are converted into high-frequency AC by the receiving coil, which is typically mounted underneath the vehicle. The continuous DC supply produced by the HF AC is then used by the onboard batteries. To ensure consistent performance and avoid any risks to health and safety, the battery management system (BMS), communications, and power control are also integrated. The standard power outlets provide around 3-kW power, which would imply a 10-h load on average for charging a maximum of 30-kWh energy in a battery [23]. It would save you time to charge the battery if you could immediately charge an electrically powered car while it was driving down the highway. However, since all of the vehicles served by BES should utilize identical batteries, the growing number of EV models, the variety of battery types, and the absence of standardization make the usage of BESs an impractical procedure [24]. At both the transmitter and the receiver, magnetic planar ferrite plates are used to enhance the dispersion of magnetic flux and minimize any detrimental leakage fluxes.

17.6 Advantages

In contrast, the operational costs of an EV are much lower. Instead of using fossil fuels like gasoline or diesel to charge their batteries, EVs use energy. Due to their higher energy density [25], these batteries were utilized in the 1990s; however, they have a significant memory effect and a short life span, and cadmium is an extremely costly and hazardous metal. Compared to buying diesel or petrol for your travel needs, an electric car is more economical due to its higher efficiency and lower cost. When EVs run on renewable energy, they are less harmful to the environment. Charge costs can be further reduced if renewable energy sources, such as solar panels, are installed at the residence. Since they do not have as many moving parts as internal compared to combustion engines, electric cars require a lot less maintenance. Consequently, the annual cost of operating an EV is minimal. Notwithstanding all of the benefits, EV owners are having severe difficulties charging their cars due to a lack of infrastructure. Notwithstanding all of the benefits, EV owners are having severe difficulties charging their cars due to a lack of infrastructure. Just 25% of the anticipated total number of public charging stations will really be in place by 2032 at the rate of current growth. The acute lack of lithium and other precious metals, which are required for EV chargers, is one of the reasons for the delayed growth of charging stations. Thanks to wireless technology, charging an EV is now simple. Static-WCS and DWCS are the two varieties of wireless charging. We will now have a better understanding of each of them.

17.7 WPT Method

Since the invention of wireless charging systems for EVs, three methods have been utilized to build WEVCS: the contemporary IWPT method, the CWPT method, and the RIPT method.

17.7.1 Case Study 1 (IWPT Method)

In 1914, Nikola Tesla altered traditional IWPT to enable wireless energy transfer. The classic IWPT is shown in Figure 17.5 based on multiple topologies for EV charging. IWPT has been tested and applied in a variety of milliwatt to kilowatt applications to transfer contactless power from the source to the receiver. General Motors introduced the Chevrolet S10 EV in 1996. It was powered by the magnet-charge IWPT (J1773) system, which offered fast charging at level 3 (50 kW) and level 2 (6.6 kW) for level 2 applications. The primary coil of the magnet charge, also known as an inductive couple or charging paddle, was inserted into the vehicle's charging port and provided power to the secondary coil, allowing the EV

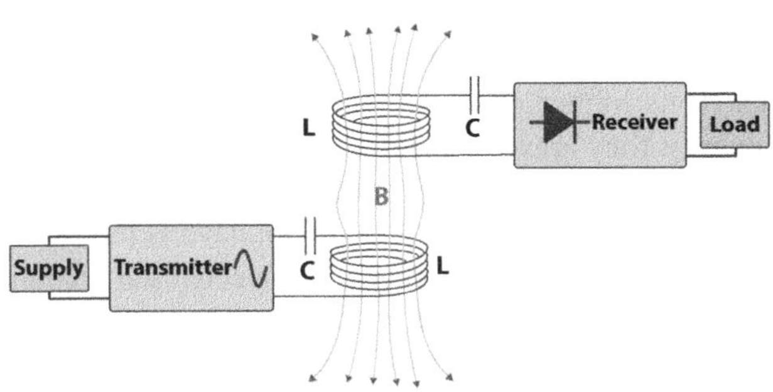

Figure 17.5 Basic block diagram for IWPT system [26].

to be charged. The University of Georgia demonstrated a 6.6-kW level 2 EV charger that could charge batteries with voltages between 200 and 400 V at an operating frequency of 77 kHz. This universal IWPT had a 10-kVA coaxial winding transformer that offered several advantages, such as an easily adjusted power range and a variable inductive coupling architecture.

17.7.1.1 Result Analysis for Case Study 1

Figure 17.6 has been prepared in MATLAB environment version 16.a of IWPT system.

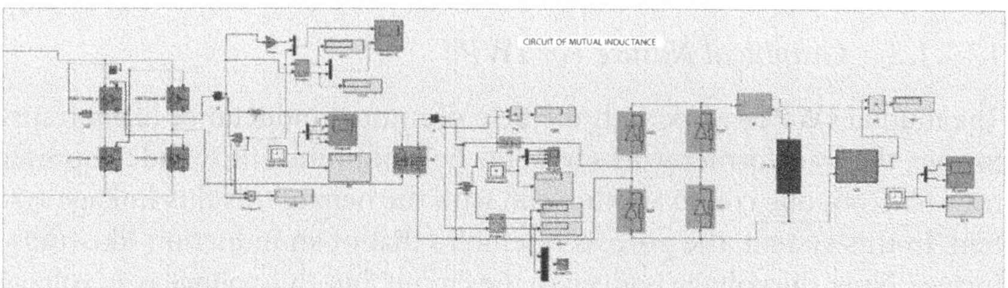

Figure 17.6 MATLAB model for IPT.

In this section, Table 17.1 focuses on the coil calculation of the inductive power transfer system. For simplicity, considering that the time (T) is 0.002 seconds where the primary side of coil power is 34.2 W, the primary side of coil reactive power is -16.1 VAR, the primary side of the coil gain is 50, the secondary side of coil voltage is 115.5-V volt, the secondary side of coil active power is 34.15 W, and the secondary side of coil reactive power is -164.9 VAR.

Table 17.1 Coil calculation for inductive power transfer.

Primary side of coil power in watts	Primary side of coil reactive power in VAR	Primary side of coil gain	Secondary side of coil voltage in volts	Secondary side of coil active power in watts	Secondary side of coil reactive power in VAR
34.2 W	-16.1 VAR	50	115.5 V	34.15 W	-164.9 VAR

Table 17.2 Final observation of the inductive power transfer.

Voltage (primary side) in volts	Current (primary side) in amp	Power (primary side) in watts	Secondary side voltage in volts	Secondary side current in amp	Secondary side power in watts
249.7 V	0.40 A	101.8 W	49.6 V	0.62 A	30.78 W

Table 17.2 focuses on the overall observation of the inductive power transfer system, considering that the time (T) is 0.002 seconds. Data have been given in Table 17.2.

17.7.1.2 Graphical Nature For IWPT

The mutual IWPTs work with the help of mutual inductance like a transformer, the transformer works with the help of mutual flux, and the power transfer from one coil to another coil with the help of this flux linkage system. In this system, the process resembles that of an induction-like transformer. Now, the voltage is given in the circuit, but this voltage is ac voltage because the transformer and the mutual induction system worked with ac. That is why the input graph is fluctuated in the ac condition. The input voltage is given around 249-test voltage, and the voltage fluctuation can observed in the graphical nature. The second graph is the current graph; this graph also fluctuated with respect to the voltage graph and the most important thing is that there was a gain parameter also added in the circuit

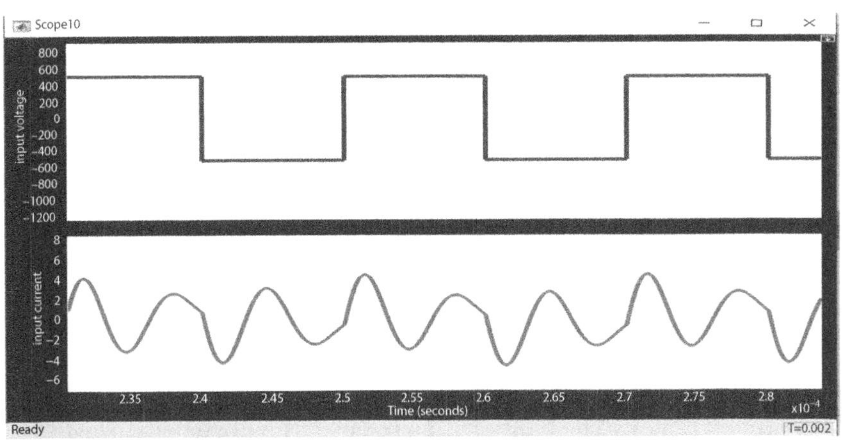

Figure 17.7 Input voltage and current graph for inductive power transfer.

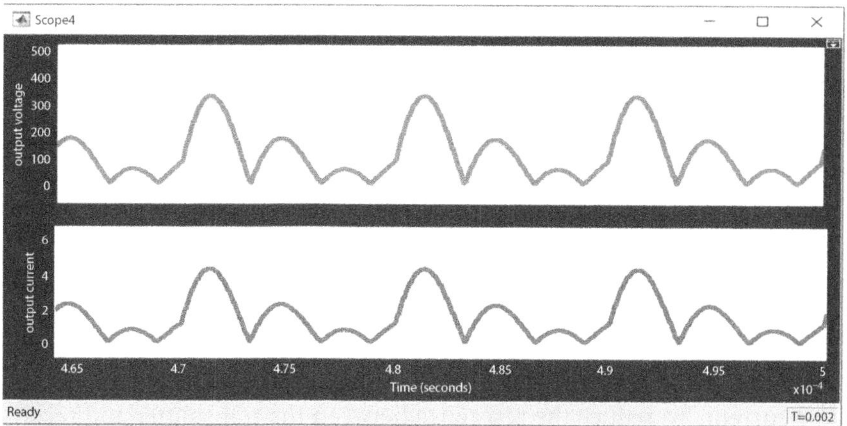

Figure 17.8 Output voltage and current graph for inductive power transfer.

which actually controlled the whole circuit and the feedback for the control system in the total circuit. Figure 17.7 shows the things.

Now, the output voltage and current graph are like dc with fluctuation. The EVs worked with the help of the dc battery. That is why the circuit was modified by the ac-to-dc system after the power transfer system. You can see that the output voltage and current are naturally the dc characteristics but fluctuated because we did not use here a rectification circuit after the mutual power transfer system. Figure 17.8 shows the things.

The efficiency calculation is based on the observation data from the proposed circuit. Table 17.3 shows the efficiency calculation consider the different load resistance values.

Table 17.3 Efficiency calculation.

Load resistance in ohm	Formula = (output/input) *100	Efficiency (%)
100	(25.99/41.43) *100	62.73%
90	(27.8/37.86) *100	73.42%
80	(30.78/34.28) *100	89.78%
76	(32.41/32.85) *100	98.66%

17.7.2 Case Study 2 (CWPT)

A few low-cost, low-complexity applications of CWPT technology are spinning machinery, portable devices, and mobile phone chargers. This technology makes use of the intricate coupler capacitors' mechanical and geometric structures. Figure 17.9 displays an example schematic diagram of the CPWT based on a series resonant circuit. To transport power from the source to the receiver in the CWPT, coupling capacitors are utilized in place of coils or magnets. Power factor adjustment circuitry provides the principal AC voltage to an H-bridge converter. High-frequency AC produced by the H-bridge is routed through coupling capacitors at the receiving end. Unlike the IWPT, the CWPT operates at both high voltage and low current. To reduce the resistance between the transmitter and receiver's sides at the configuration, more inductors are linked in series with the coupling capacitors. Furthermore, this arrangement makes it easier for the electronics to switch on and off gently.

17.7.2.1 Result Analysis for Case Study 2

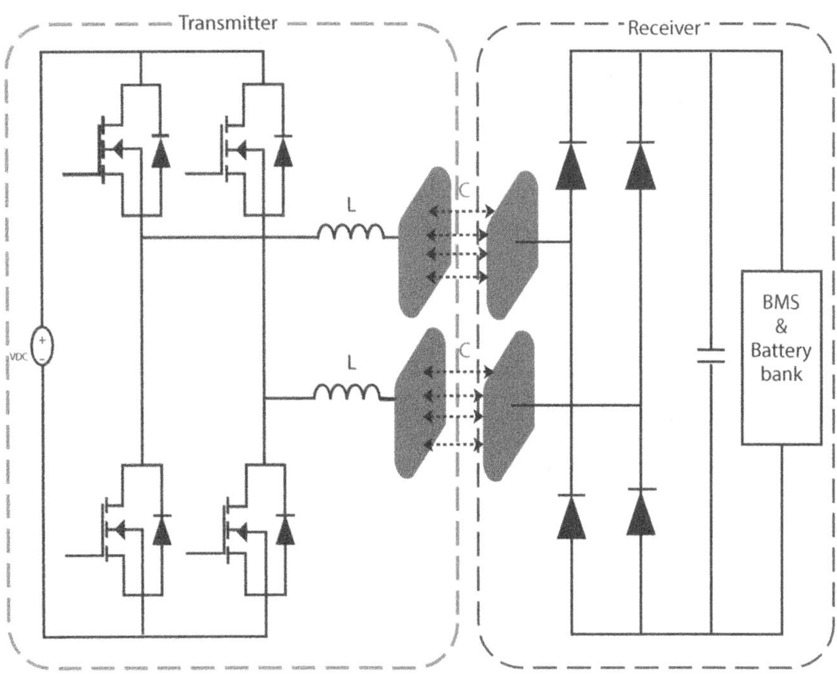

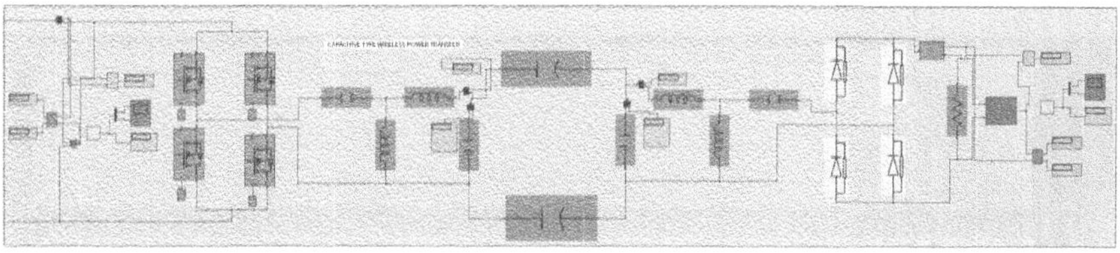

Figure 17.9 MATLAB model for CWPT method.

Figure 17.9 has been prepared in MATLAB environment version 16.a for capacitive power transfer wireless contact system.

Table 17.4 focuses on the overall observation of the inductive power transfer system, considering that the time (T) is 0.002 seconds. Data have been given in Table 17.4. The primary side voltage is 249.7 volt and the primary side current is 2.6×10^{04} A. The primary side power is 6.5×10^{06} W. The secondary side voltage is 0.04826 V. The secondary side current is 9.6×10^{-06} A. The secondary side power is 4.6×10^{-07} W.

Table 17.4 Overall observation of the inductive power transfer system.

Voltage (primary side) in volts	Current (primary side) in amp	Power (primary side) in watts	Secondary side voltage in volts	Secondary side current in amp	Secondary side power in watts
249.7 V	$2.6*10^{04}$ A	$6.5*10^{06}$ W	0.04826 V	$9.6*10^{-06}$ A	$4.6*10^{07}$ W

17.7.2.2 Graphical Nature for CWPT Method

The capacitive-type WPT system uses two capacitive plates incorporated with the road and one is incorporated with the car. Capacitive-type WPT system worked with the assistance of dc. That is why we have used an ac-to-dc converter system because we have to use the same power from the bus-bar which is used for the mutual inductance-type power transfer system, we have used the same power from the same bus-bar that is why ac-to-dc system incorporated here. However, we have not used the dc gain for this circuit that is why there are two types of graphs you have observed. One is a voltage graph and one is a current graph. The capacitor voltage is determined by the capacitor current, which is why you observe a leading nature of voltage in this circuit. This same voltage has been used in both this and the previous circuit. The voltage primarily increases and then decreases which is why the circuit is not a practical use circuit. Figure 17.10 shows the things.

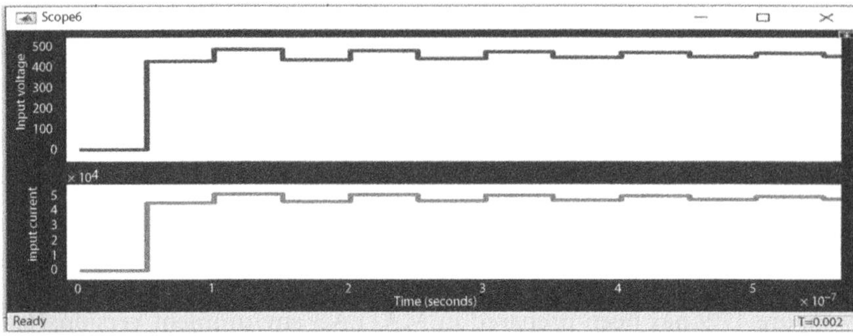

Figure 17.10 Input voltage and current graph for capacitive type power transfer.

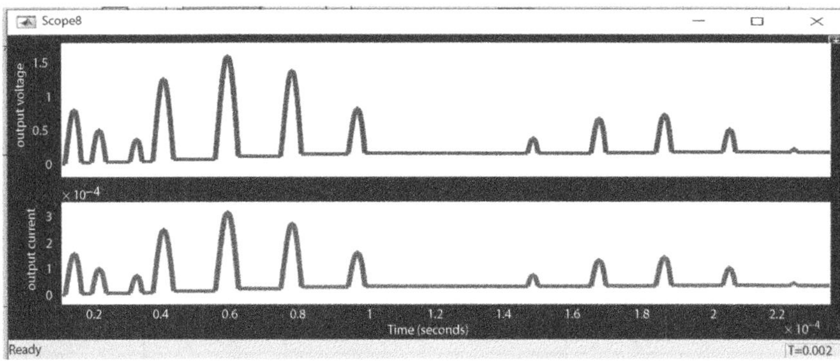

Figure 17.11 Output voltage and current graph for capacitive power transfer.

The output graph mirrors the input nature because no rectifier or converter is used. As a result, the output graph remains identical to the input graph. The entire circuit operates under DC conditions, which is why the output graph resembles a DC curve. Both the output voltage and output current graphs also resemble DC characteristics. Due to the decreasing condition, the same voltage as in mutual inductance is used, but the characteristics differ. This is why it is not practical to use this approach in real conditions. This circuit is only used for industry and testing purposes. Figure 17.11 shows the things.

The efficiency calculation is based on the observation data from the proposed circuit.

$$\text{Efficiency} = [\{(I_r L)^2 \times R_L\}/\{(I_r L)^2 \times R_L\} + \{(IL_2)^2 \times R_2\} + \{(IC_2)^2 \times RC_2\}$$
$$+\{(IC_1)^2 \times RC_1\} + \{(IL_1)^2 \times R_L\}]$$
$$= \{(0.0000455817)*(5000)\}/\{(0.22)+(0.0046)+(0.0882)+(0.0035)+(0.0059)\}$$
$$= \{(0.22/0.32)*100\}$$
$$= 68.75\,\%$$

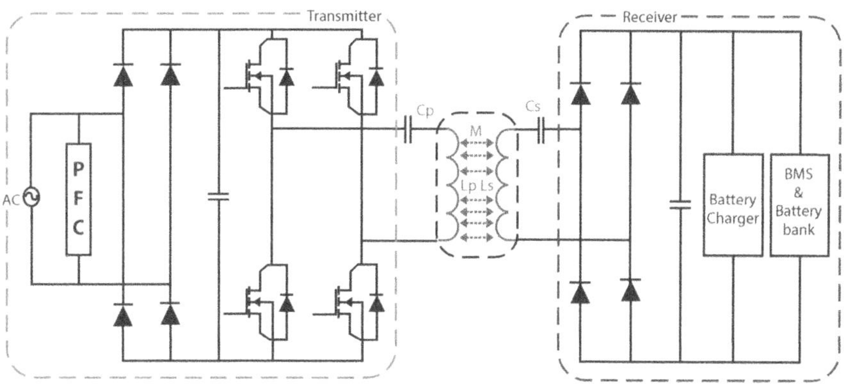

Figure 17.12 Resonance inductive type power transfer.

17.7.3 Case Study 3 (Resonance Inductive Power Transfer)

To construct a WPT coil, the RIPT is among the most intelligent iterations of the conventional IPT. A schematic diagram for EVs is displayed in Figure 17.12. Similar to other WPTs, the transmitter or primary winding receives the main AC voltage after it has been converted to an HF AC source. Different magnetic fields are used to deliver electricity to the secondary coil or receiver. Additive power circuits and filter circuitry transform the received electricity into DC for the EVs' battery bank. Compared to the conventional IPT, more compensation networks are added to the primary and secondary winding in series/parallel typologies in order to decrease extra losses as well as to provide the resonant case, as shown in Figure 17.12.

17.7.3.1 Result Analysis for Case Study 3

Figure 17.13 has been prepared in MATLAB environment version 16.a for resonance inductive power transfer wireless contact system.

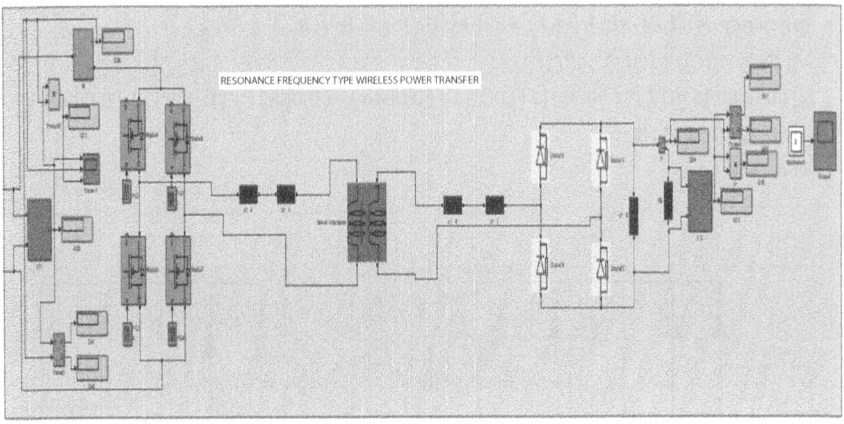

Figure 17.13 MATLAB model for resonance inductive power transfer system.

Table 17.5 Inductive power transfer system details.

Voltage (primary side) in volts	Current (primary side) in amp	Power (primary side) in watts	Secondary side voltage in volts	Secondary side current in amp	Secondary side power in watts
416.3 V	$4.3*10^{04}$ A	38,072 W	4.391 V	2.92 A	132.85 W

Table 17.5 focuses on the overall observation of the inductive power transfer system, considering that the time (T) is 0.002 seconds. Data have been given in Table 17.4. The primary side voltage is 416.3 V. The primary side current is $4.3*10^{04}$ A. The primary side power is 38,072 W. The secondary side voltage is 4.391 V. The secondary side current is 2.92 A. The secondary side power is 132.85 W.

17.7.3.2 Graphical Nature for RIWPT Method

The resonance frequency–type power transfer is more or less the same as mutual inductive–type power transfer; here, we use a separate small transmitting area that transmits the ac power, and the ac-to-dc conversion is incorporated with the circuit. Now, the input graph characteristic is unstable primarily and then it is going to the stable condition; it happens for the inductor which is included in this circuit. Figure 17.14 has been display the input voltage and current graph for resonance inductive power transfer process.

The output current and voltage characteristics are the same as the mutual inductance–type power transfer system, which is why the voltage and current conditions exhibit a dc nature here. Now, the output voltage and current graph are like dc without fluctuation. The dc condition shows that the EVs worked with the help of the dc battery. That is why the circuit was modified by the ac-to-dc system after the power transfer system. You can see natural dc characteristics but fluctuated because we did not use here a rectification circuit after the mutual power transfer system. Figure 17.15 shown the output voltage and current graph for resonance inductive power transfer process.

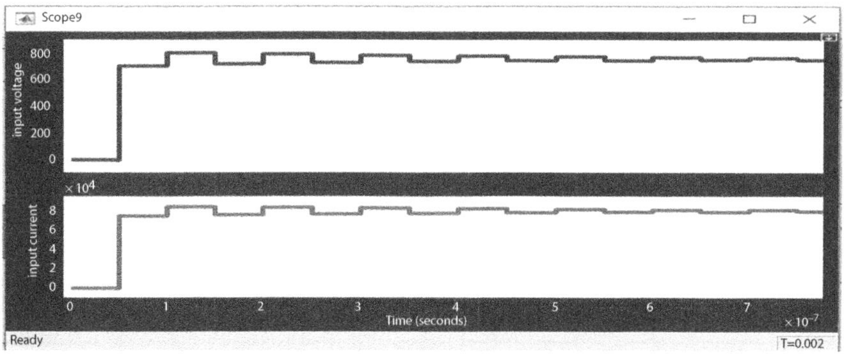

Figure 17.14 Input voltage and current graph for resonance inductive power transfer.

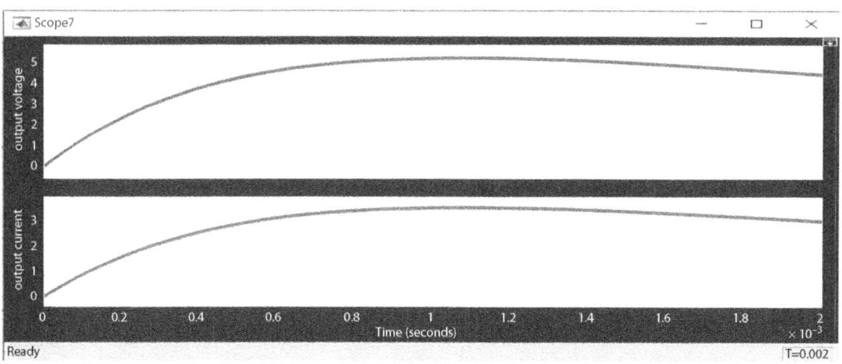

Figure 17.15 Output voltage and current graph for resonance inductive power transfer process.

Table 17.6 has been given the efficiency calculation for resonance inductive power transfer.

Table 17.6 Efficiency calculation.

Load resistance in ohm	Formula = (output/input) *100	Efficiency (%)
1	(12.24/33.72) *100	36.29%
1.5	(12.85/33.72) *100	38.10%
2	(12.49/33.72) *100	37.04%
2.5	(11.79/33.72) *100	34.96%
3	(11/33.72) *100	32.62%

17.8 Performance Analysis for Three Case Studies

Table 17.7 has been given the performance analysis for three case studies.

Table 17.7 Related analysis.

Type	Maximum input voltage	Frequency (HZ/ KHZ)	Efficiency (%)	Conclusion on usefulness
Mutual inductance	350 V	0.5 kHz	60% to 98%	➢ Useful for big EVs as well as small EVs ➢ Construction is easy ➢ High efficiency ➢ Low cost ➢ As the output voltage reaches 50 V, it is good for all types of EVs ➢ EXAMPLE → E-bikes to truck bus
Resonance frequency	300 to 320 V	5 kHz	30% to 40%	➢ Useful for small EVs ➢ Construction is easy ➢ Low efficiency ➢ Low cost ➢ As output voltage is minimal that is why it is useful for small EVs ➢ Example → E-bike, E-cycle
Capacitive	500 to 520 V	0.5 kHz	68.75%	➢ Useful for only big EVs ➢ Construction is complex ➢ Medium efficiency ➢ HIGH COST ➢ As the output voltage is not good that is why this type of power transfer system is not much useful

17.9 Proposed Method for Power Source by Renewable Energy Source

We have used the input power for charging the EVs from the existing generation system, but what if we use the local generation system with the help of a photovoltaic array. That is our proposal, in the project work, we have used the photovoltaic array and the local generating system and we have transferred the power by a transmitting system as well as the bus-bar system.

The benefit of using a photovoltaic array is that the system is eco-friendly and comparatively easy to use for everyone in this country.

In the total system, our modification and the small new idea is the local generation system. The local generation is that system that generates the power with the help of the photovoltaic array and the small transmitting area is incorporated here. The local generating system is like today's traditional coil power generation but with solar. The proposed local generation is converting the local petrol pump into the local generating station and a small transmitting area is included for the purpose of transmitting this power from somewhere to anywhere. Figure 17.16 shows the total system.

17.9.1 MATLAB Model for Proposed System

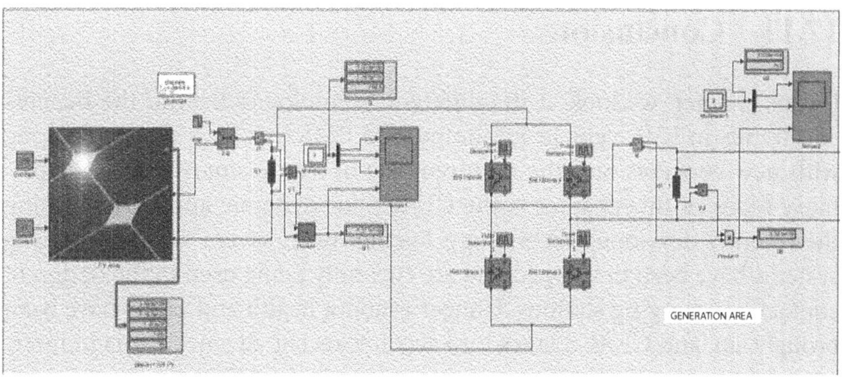

Figure 17.16 MATLAB model for local generation.

The bus-bar system is included here for the testing three of types of WPTs with the same power. This system is also included in the practical model of the system. The bus-bar system helps the total power system from the faulty condition and the bus-bar system helps the power isolation if needed in future proposes. Figure 17.17 shows the transmitting area.

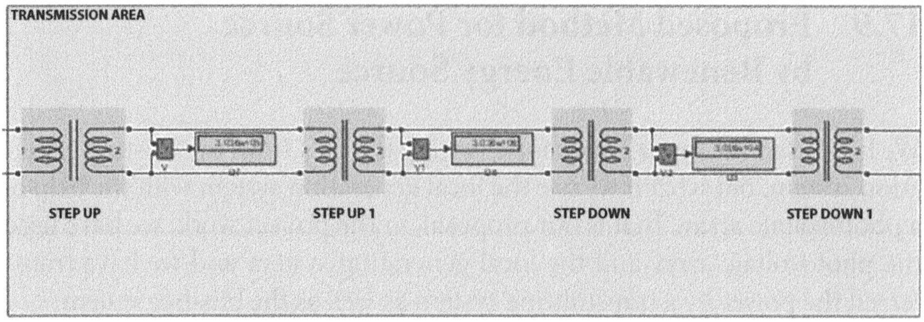

Figure 17.17 Transmitting area.

17.10 Future Scope for Contactless Power Transfer Method

Wireless power transfer is everywhere, where it would be needed. Smoothing performance in the EV sector is crucial for the future in India and other countries. The vehicle-to-grid and grid-to-vehicle (V to G & G to V) systems are incorporated in India and other countries. The smart EV sector is to be built in India and other countries.

17.11 Conclusion

In this chapter, we look at the different technologies behind the benefits and trends in EVs, vehicle autonomy, and BMS that have emerged, along with new concepts and the requirements of future smart cities. The primary focus of this chapter is the CEACS for dynamic applications using the recently developed technology. Furthermore, several local generating systems have been developed and are currently being used in the design of contactless charging stations. Concerns about health and safety have been brought up, and CEACS takes into account recent advancements in international standards. The most recent dynamic CEACS have been examined and presented, incorporating ongoing research and development from both public and commercial entities. In the end, MATLAB is used to research and simulate emerging future technologies.

References

1. Kalwar, K.A., Aamir, M., Mekhilef, S., Inductively coupled power transfer (ICPT) for electric vehicle charging–A review. *Renew. Sustain. Energy Rev.*, *47*, 462–475, 2015.
2. Ombach, G., Design considerations for wireless charging system for electric and plug-in hybrid vehicles, in: *IET Hybrid and Electric Vehicles Conference 2013 (HEVC 2013)*, IET, pp. 1–4, 2013, November.
3. Leskarac, D., Panchal, C., Stegen, S., Lu, J., PEV charging technologies and V2G on distributed systems and utility interfaces. *Vehicle-to-Grid Linking Electr. Veh. to Smart Grid*, *79*, 157–209, 2015.
4. Villa, J.L., Sallán, J., Llombart, A., Sanz, J.F., Design of a high frequency inductively coupled power transfer system for electric vehicle battery charge. *Appl. Energy*, *86*, 3, 355–363, 2009.
5. Moon, S., Kim, B.C., Cho, S.Y., Ahn, C.H., Moon, G.W., Analysis and design of a wireless power transfer system with an intermediate coil for high efficiency. *IEEE Trans. Ind. Electron.*, *61*, 11, 5861–5870, 2014.
6. Sugiyama, R., *Maximum Achievable Efficiency for Multiple-Receiver Inductive Power Transfer Systems*, Nara Institute of Science and Technology, 8916-5 Takayama-cho, Ikoma, Nara 630-0192, JAPAN, 2018.
7. Covic, G.A. and Boys, J.T., Modern trends in inductive power transfer for transportation applications. *IEEE J. Emerging Sel. Top. Power Electron.*, *1*, 1, 28–41, 2013.
8. Lukic, S. and Pantic, Z., Cutting the cord: Static and dynamic inductive wireless charging of electric vehicles. *IEEE Electrif. Mag.*, *1*, 1, 57–64, 2013.
9. Eghtesadi, M., Inductive power transfer to an electric vehicle-analytical model, in: *40th IEEE Conference on Vehicular Technology*, IEEE, pp. 100–104, 1990, May.
10. Fuller, M., Wireless charging in California: Range, recharge, and vehicle electrification. *Transp. Res. Part C Emerging Technol.*, *67*, 343–356, 2016.
11. Van Der Pijl, F., Bauer, P., Castilla, M., Control method for wireless inductive energy transfer systems with relatively large air gap. *IEEE Trans. Ind. Electron.*, *60*, 1, 382–390, 2011.
12. Kumar, R. and Padmanaban, S., Electric vehicles for India: overview and challenges. *IEEE India Inf.*, 14, 139, 2019, 2019.
13. Habib, S., Kamran, M., Rashid, U., Impact analysis of vehicle-to-grid technology and charging strategies of electric vehicles on distribution networks–a review. *J. Power Sources*, *277*, 205–214, 2015.
14. Das, H.S., Rahman, M.M., Li, S., Tan, C.W., Electric vehicles standards, charging infrastructure, and impact on grid integration: A technological review. *Renew. Sustain. Energy Rev.*, *120*, 109618, 2020.

15. Li, Y., Liu, K., Foley, A.M., Zülke, A., Berecibar, M., Nanini-Maury, E., Hoster, H.E., Data-driven health estimation and lifetime prediction of lithium-ion batteries: A review. *Renew. Sustain. Energy Rev.*, *113*, 109254, 2019.
16. Sivaraman, P., Logeswaran, T., Raj, J.S.S., Boopathimanikandan, S., Design and Analysis of Sliding Mode Control for Battery Charging Applications, in: *IOP Conference Series: Materials Science and Engineering*, vol. 995, IOP Publishing, pp. 012002, 2020, November.
17. Lian, J., Zhou, Y., Ma, T., Wang, W., Design for motor controller in hybrid electric vehicle based on vector frequency conversion technology. *Math. Probl. Eng.*, *2010*, 2010. https://scholar.google.com/scholar?hl=en&as_sdt=-0%2C5&q=Lian%2C+J.%2C+Zhou%2C+Y.%2C+Ma%2C+T.%2C+Wang%2C+W.%2C+Design+for+motor+controller+in+hybrid+electric+vehicle+based+on+vector+frequency+conversion+technology.+Math.+Probl.+Eng.%2C+2010%2C+2010.&btnG=
18. Liu, X., Liu, C., Han, W., Pong, P.W., Design and implementation of a multi-purpose TMR sensor matrix for wireless electric vehicle charging. *IEEE Sens. J.*, *19*, 5, 1683–1692, 2018.
19. Qin, R., Li, J., Costinett, D., A 6.6-kW high-frequency wireless power transfer system for electric vehicle charging using multilayer nonuniform self-resonant coil at MHz. *IEEE Trans. Power Electron.*, *37*, 4, 4842–4856, 2021.
20. Zhang, X., Yuan, Z., Yang, Q., Li, Y., Zhu, J., Li, Y., Coil design and efficiency analysis for dynamic wireless charging system for electric vehicles. *IEEE Trans. Magn.*, *52*, 7, 1–4, 2016.
21. Miller, J.M., Onar, O.C., Chinthavali, M., Primary-side power flow control of wireless power transfer for electric vehicle charging. *IEEE J. Emerging Sel. Top. Power Electron.*, *3*, 1, 147–162, 2014.
22. Khan-ngern, W. and Zenkner, H., Wireless power charging on electric vehicles, in: *2014 International Electrical Engineering Congress (iEECON)*, IEEE, pp. 1–4, 2014, March.
23. Ireland, S.E., Hybrid Electric and Battery Electric Vehicles. *AEA Energy Environment.*, 2007. https://scholar.google.com/scholar?hl=en&as_sdt=0%2C5&q=Ireland%2C+S.E.%2C+Hybrid+Electric+and+Battery+Electric+Vehicles.+AEA+Energy+Environment.%2C+2007.&btnG=
24. Jing, W., Yan, Y., Kim, I., Sarvi, M., Electric vehicles: A review of network modelling and future research needs. *Adv. Mech. Eng.*, *8*, 1, 1687814015627981, 2016.
25. Haschka, F.R., II and Schlieck, D., II, High power nickel-cadmium cells with fiber electrodes (FNC), in: *Proceedings of the 32nd International Power Sources Symposium*, 1986.
26. https://www.watelectronics.com/wireless-power-transfer/assessed by 15/02/2024

Index

Also of Interest

Check out these other related titles from Scrivener Publishing

Special Electrical Machinery, Edited by Dr. Jigneshkumar P. Desai, ISBN: 9781394193912. This book is a comprehensive guide to specialized motors, covering everything from switched reluctance motors to axial flux ironless PM motors, giving in-depth information on their operating principles, applications, and controls.

DESIGN, MODELING AND RELIABILITY IN ROTATING MACHINERY, Edited by Robert X. Perez, ISBN: 9781119631682. This broad collection of current rotating machinery topics, written by industry experts, is a must-have for rotating equipment engineers, maintenance personnel, students, and anyone else wanting to stay abreast with current rotating machinery concepts and technology.

MAINTENANCE, RELIABILITY AND TROUBLESHOOTING IN ROTATING AND PROCESS MACHINERY, Edited by Robert X. Perez, ISBN: 9781119631644. For Rotating equipment professionals, maintenance personnel, project engineers, process engineers, mechanical engineering students, and anyone working around critical rotating equipment.

Troubleshooting Rotating Machinery: Including Centrifugal Pumps and Compressors, Reciprocating Pumps and Compressors, Fans, Steam Turbines, Electric Motors, and More, by Robert X. Perez and Andrew P. Conkey, ISBN 9781119294139. A must-read for anyone whose job depends on operating or maintaining safe, reliable, and efficient machinery in a process facility, this volume shows the newcomer and veteran alike how you can start troubleshooting rotating equipment problems immediately by implementing this book's common-sense methodology. *NOW AVAILABLE!*

Operator's Guide to Process Compressors, by Robert X. Perez, ISBN 9781119580614. The perfect primer for anyone responsible for operating or maintaining process gas compressors. *NOW AVAILABLE!*

Operator's Guide to General Purpose Steam Turbines: An Overview of Operating Principles, Construction, Best Practices, and Troubleshooting, by Robert X. Perez and David W. Lawhon, ISBN: 9781119294214. A must-read book for anyone operating or maintaining general purpose steam turbines in a process setting, this is a practical reference that includes a collection of field-tested procedures aimed at ensuring safe general purpose steam turbines start-ups, shutdowns, and over-speed tests. *NOW AVAILABLE!*

Dictionary of Industrial Terminology 2nd Edition, by Michael D. Holloway and Emma Holloway, ISBN 9781119363446. Greatly expanded and revised, this second edition is the most comprehensive and in-depth dictionary of industrial terms available, a must-have for any engineer, scientist, technician, or operator working in industry. *NOW AVAILABLE!*

Basic Electrical and Instrumentation Engineering, by P. Sivaraman, C. Sharmeela, A. Thaiyal Nayagi, and R. Mahendran, ISBN 9781119764465. Offering the most up-to-date coverage on the subject available, this new volume provides detailed concepts of AC circuits and power systems, transformers, DC machines, AC machines, measurement and instrumentation, and many other practical concepts for the engineer or engineering student. *NOW AVAILABLE!*

ELECTRIC VEHICLE DESIGN: Design, Simulation and Applications, Edited by Krishan Arora, Suman Lata Tripathi, and Himashu Sharma, ISBN: 9781394204373. This book will serve as a definitive guide to conceptual and practical knowledge about the design of hybrid electrical vehicles (HEV), battery electrical vehicles (BEV), fuel cell electrical vehicles (FCEV), plug-in hybrid electrical vehicles (PHEV), and efficient EV charging techniques with advanced tools and methodologies for students, engineers, and academics alike.

INDUSTRIAL CONTROL SYSTEMS, Edited by Vipin Chandra Pal, Suman Lata Tripathi, and Souvik Ganguli, ISBN: 9781119829256. This volume serves as a comprehensive guide in the journey of industrial control systems with a multidisciplinary approach to the key engineering problems in the 21st century.

MODERN AUTOMOTIVE ELECTRICAL SYSTEMS: Theory and Applications, Edited by Pedram Asef, Sanjeevikumar Padmanaban, and Andrew Lapthorn, ISBN: 1119801047. Presenting the concepts and advances of modern automotive electrical systems, this volume, written and edited by a global team of experts, also goes into the practical applications for the engineer, student, and other industry professionals.

NANODEVICES FOR INTEGRATED CIRCUIT DESIGN, Edited by Suman Lata Tripathi, Abhishek Kumar, K. Srinivasa Rao, and Prasantha R. Mudimela, ISBN: 9781394185788. Written and edited by a team of experts in the field, this important new volume broadly covers the design of nano-devices and their integrated applications in digital and analog integrated circuits (IC) design.

DESIGN AND DEVELOPMENT OF EFFICIENT ENERGY SYSTEMS, edited by Suman Lata Tripathi, Dushyant Kumar Singh, Sanjeevikumar Padmanaban, and P. Raja, ISBN 9781119761631. Covering the concepts and fundamentals of efficient energy systems, this volume, written and edited by a global team of experts, also goes into the practical applications that can be utilized across multiple industries, for both the engineer and the student.

Electrical and Electronic Devices, Circuits, and Materials: Technical Challenges and Solutions, edited by Suman Lata Tripathi, Parvej Ahmad Alvi, and Umashankar Subramaniam, ISBN 9781119750369. Covering every aspect of the design and improvement needed for solid-state electronic devices and circuit and their reliability issues, this new volume also includes overall system design for all kinds of analog and digital applications and developments in power systems.

Printed and bound by CPI Group (UK) Ltd, Croydon, CR0 4YY

17/09/2026

14964262-0003